Traffic Safety Data

Other related titles:

You may also like

- PBTR0070 | Meng Lu | Evaluation of Intelligent Road Transport Systems: Methods and results | 2016
- PBTR023 | Ling Huang and Jianping Wu | Behavioural Modelling and Simulation of Bicycle Traffic | 2021

We also publish a wide range of books on the following topics:
Computing and Networks
Control, Robotics and Sensors
Electrical Regulations
Electromagnetics and Radar
Energy Engineering
Healthcare Technologies
History and Management of Technology
IET Codes and Guidance
Materials, Circuits and Devices
Model Forms
Nanomaterials and Nanotechnologies
Optics, Photonics and Lasers
Production, Design and Manufacturing
Security
Telecommunications
Transportation

All books are available in print via https://shop.theiet.org or as eBooks via our Digital Library https://digital-library.theiet.org.

IET TRANSPORTATION SERIES 280

Traffic Safety Data

Sources, analysis and applications

Edited by
Aliaksei Laureshyn, Niels Agerholm and Matúš Šucha

The Institution of Engineering and Technology

About the IET

This book is published by the Institution of Engineering and Technology (The IET).

We inspire, inform and influence the global engineering community to engineer a better world. As a diverse home across engineering and technology, we share knowledge that helps make better sense of the world, to accelerate innovation and solve the global challenges that matter.

The IET is a not-for-profit organisation. The surplus we make from our books is used to support activities and products for the engineering community and promote the positive role of science, engineering and technology in the world. This includes education resources and outreach, scholarships and awards, events and courses, publications, professional development and mentoring, and advocacy to governments.

To discover more about the IET please visit https://www.theiet.org/.

About IET books

The IET publishes books across many engineering and technology disciplines. Our authors and editors offer fresh perspectives from universities and industry. Within our subject areas, we have several book series steered by editorial boards made up of leading subject experts.

We peer review each book at the proposal stage to ensure the quality and relevance of our publications.

Get involved

If you are interested in becoming an author, editor, series advisor, or peer reviewer please visit https://www.theiet.org/publishing/publishing-with-iet-books/ or contact author_support@theiet.org.

Discovering our electronic content

All of our books are available online via the IET's Digital Library. Our Digital Library is the home of technical documents, eBooks, conference publications, real-life case studies and journal articles. To find out more, please visit https://digital-library.theiet.org.

In collaboration with the United Nations and the International Publishers Association, the IET is a Signatory member of the SDG Publishers Compact. The Compact aims to accelerate progress to achieve the Sustainable Development Goals (SDGs) by 2030. Signatories aspire to develop sustainable practices and act as champions of the SDGs during the Decade of Action (2020–2030), publishing books and journals that will help inform, develop, and inspire action in that direction.

In line with our sustainable goals, our UK printing partner has FSC accreditation, which is reducing our environmental impact on the planet. We use a print-on-demand model to further reduce our carbon footprint.

Published by The Institution of Engineering and Technology, London, United Kingdom

The Institution of Engineering and Technology (the "**Publisher**") is registered as a Charity in England & Wales (no. 211014) and Scotland (no. SC038698).

First published 2024

The Institution of Engineering and Technology
Futures Place,
Kings Way, Stevenage,
Herts, SG1 2UA, United Kingdom

www.theiet.org

British Library Cataloguing in Publication Data
A catalogue record for this product is available from the British Library

ISBN 978-1-83953-045-6 (hardback)
ISBN 978-1-83953-046-3 (PDF)

Typeset in India by MPS Limited

Cover image credit: SilviaJansen/E+ via Getty Images

Contents

About the editors xvii

1 Introduction **1**
Niels Agerholm, Matúš Šucha and Aliaksei Laureshyn
Reference 3

2 Accident data and records **5**
Carmelo D'Agostino and Piotr Olszewski
2.1 Definition of road accidents 5
2.1.1 Introduction 5
2.1.2 Formal definitions of accident events and accident frequency 5
2.1.3 Statistical definition of accident frequency 7
2.2 Accident data collection 8
2.2.1 Introduction 8
2.2.2 Accident reporting process 8
2.2.3 Self-reported accidents 11
2.2.4 Accident reports and in-depth accident investigations 12
2.3 Accident databases 14
2.3.1 Introduction 14
2.3.2 International databases 15
2.3.3 The international experience 17
2.3.4 Problem of data harmonization 18
2.4 How to interpret accident data 19
2.4.1 Introduction 19
2.4.2 Identification of hazardous locations 22
2.4.3 What underreporting is and why it is a problem 23
2.4.4 Concluding remarks 23
References 24

3 Accident modelling: an overview **29**
Carmelo D'Agostino and Bhagwant Persaud
3.1 Introduction 29
3.2 Why accident modelling? 30
3.2.1 Basic definitions 30
3.2.2 Regression to the mean effect and relationship with exposure 30

3.3 Regression technique, model form and goodness-of-fit evaluation 31
3.3.1 Introduction 31
3.3.2 The generalized linear modelling (GLM) approach 32
3.3.3 Goodness-of-fit measures 34
3.3.4 Empirical Bayes estimation and the role of the NB dispersion parameter 36
3.3.5 Key points on accident modelling 38
3.4 Available knowledge 39
3.4.1 Introduction 39
3.4.2 The Highway Safety Manual 39
3.4.3 Scientific literature: old and new challenges 40
References 42

4 In-depth accident investigations 47
John-Fredrik Grönvall
4.1 Background 47
4.2 Methodologies 48
4.2.1 Different needs of methods and techniques 48
4.2.2 Background investigation 48
4.2.3 On-scene investigation 49
4.2.4 Retrospective investigation 52
4.2.5 Special Crash Investigations (SCIs, severe accidents with special attendance) 52
4.3 Selection/sampling criteria 52
4.4 Interviews of involved persons 52
4.5 Variables in the database 54
4.5.1 Crash details 54
4.5.2 Vehicle information 54
4.5.3 Human factors 54
4.5.4 Environmental factors 55
4.5.5 Post-crash outcomes 55
4.5.6 Legal and regulatory variables 56
4.5.7 Other variables 56
4.6 Case analysis 56
4.7 Representativity of collected data 57
4.7.1 Comprehensive data collection 57
4.7.2 Random sampling 57
4.7.3 Demographic representation 57
4.7.4 Geographic and environmental factors 58
4.7.5 Standardized data collection methods 58
References 58

5 Healthcare data 59
Niels Bos, Aliaksei Laureshyn and Carl Johnsson
5.1 Limitations of police accident records 59

5.2 Healthcare data 62
5.2.1 Potentials of healthcare data for road accident research 63
5.2.2 Types of hospital records, availability 64
5.2.3 Injury classifications 66
5.2.4 Linking police and hospital records 71
5.2.5 Hinders for access and use of hospital data 76
5.3 Case example: the Netherlands 76
5.3.1 History 76
5.3.2 The linked database 77
5.4 Case example: Sweden 83
5.4.1 Historical note 83
5.4.2 Strada database 84
5.5 Conclusion 88
References 89

6 The use of self-reported data in road safety research 97
Wouter Van den Berghe and Uta Meesmann
6.1 Introduction 97
6.1.1 Using self-reported data in research 97
6.1.2 Sources and references 97
6.1.3 Overview of main data collection methods for self-reported data 98
6.2 Advantages and limitations of using questionnaire surveys in road safety research 100
6.2.1 Advantages of using questionnaire surveys 100
6.2.2 Additional advantages of using online panels for questionnaire surveys 101
6.2.3 Limitations of questionnaire surveys 102
6.2.4 Specific limitations of using online panels 104
6.2.5 Suitability of self-reported data collection for road safety research 105
6.3 Measuring road user behaviour through self-reporting 105
6.3.1 Methods for measuring road user behaviour 105
6.3.2 Example comparing self-reported and observed behaviour 107
6.3.3 International acceptance of data on self-reported behaviour 109
6.3.4 Future directions for self-reported data on behaviour in traffic 109
6.4 Good practice in self-reported data collection 110
6.4.1 Survey design 110
6.4.2 Data analysis 113
6.4.3 Ethical considerations when questioning people in surveys 114
6.5 Some examples of studies using self-report data 114
6.5.1 The ESRA initiative 114
6.5.2 Public support for policy measures 116

6.5.3 Traffic Safety Culture Index 117
6.5.4 Measuring drowsy driving 118
6.6 Conclusion 118
References 119

7 Automated driving and safety data 125
Andrew Tarko and Vamsi Bandaru
7.1 Components of driving task 126
7.1.1 Perceiving the surrounding environment 127
7.1.2 Communicating with nearby road users and/or infrastructure 128
7.1.3 Predicting near-future positions of nearby road users (risk-free situation) 128
7.1.4 Planning own motion to eliminate the risk of collision 129
7.1.5 Execution of the planned trajectory 129
7.2 AV collected data 130
7.2.1 Own perception sensors data 130
7.2.2 Data received from other vehicles and roadside units 131
7.2.3 Prediction and planning results 131
7.2.4 Motion and vehicle control data 131
7.3 Safety-relevant incidents and their causes 131
7.3.1 Causes of conflicts and crashes 132
7.3.2 AV data for crash analysis 133
7.3.3 AV data for traffic conflicts analysis 134
7.4 Data sharing and users 136
7.4.1 Government 137
7.4.2 Insurance companies 138
7.4.3 Vehicle owners 139
7.4.4 Individual AV owners 139
7.4.5 AV fleet owners 140
7.4.6 AV manufacturers 140
7.5 Conclusion 141
References 142

8 Naturalistic (driving/cycling/walking) studies 147
Niels Agerholm
8.1 Theoretical background 147
8.2 What are naturalistic driving studies? 147
8.3 Methods 148
8.4 Data for NDS 149
8.4.1 GNSS data in NDS 150
8.4.2 Video recording in NDS 150
8.5 Examples of projects based on GNSS data 151
8.6 Examples of NDS based on video recording 152

8.7 Type of analyses 152
8.8 Pitfalls regarding data analyses 153
References 154

9 Surrogate measures of safety: site-based observations 157
Aliaksei Laureshyn and Nicolas Saunier
9.1 Why surrogate measures? 157
9.1.1 Limitations of accident data 157
9.1.2 Historical note 158
9.2 Theoretical background 159
9.2.1 Hierarchy of events 159
9.2.2 Concept of severity 161
9.2.3 Traffic conflicts and exposure 162
9.2.4 Traffic conflicts and accidents 164
9.2.5 Reliability and validity 166
9.3 Indicators, measures and techniques 166
9.3.1 TTC family 167
9.3.2 PET family 170
9.3.3 Measures related to evasive actions 171
9.3.4 Outcome-based indicators 172
9.3.5 Combined indexes 173
9.3.6 TCTs 173
9.4 Data collection tools 174
9.4.1 Human observations 174
9.4.2 Video recording 175
9.4.3 Video analysis and other emerging technologies 176
9.5 Future directions 177
References 180

10 Behavioural observations 189
Matúš Šucha and Ralf Risser
10.1 Theoretical background 189
10.2 Methods/types of observations 192
10.2.1 Obtrusive vs. unobtrusive observation 192
10.2.2 Stationary observation 193
10.2.3 Observation with technical help or solely by technical equipment 194
10.2.4 Behaviour observation on the traffic simulator 195
10.3 Description of procedures and data 195
10.3.1 In-car observation with a focus on the Vienna driving test 195
10.3.2 Traffic conflict registrations on site 199
10.3.3 On-site observation of behaviour and interaction of road users 201
10.3.4 FOT/natural driving studies (NDS) 204

10.4 Some results of studies with the help of observation methods 206
10.4.1 In-car observations: some applications and results 206
10.4.2 Traffic conflict registrations: some applications and results 207
10.4.3 Some results derived from on-site observation studies 208
10.4.4 FOT/NDS: some applications and results 209
10.5 How to understand traffic behaviour with help of the model of Michon 211
10.5.1 Ride along and road-user following 212
10.5.2 Behaviour observation on-site and traffic conflict registrations 213
10.5.3 FOT 213
10.5.4 NDS 214
10.5.5 Conclusion – Understanding what road users do and why they do it 214
References 215

11 Road Safety Inspection and Road Assessment Programme 221
András Várhelyi
11.1 Road Safety Inspection 221
11.1.1 Why RSI? 222
11.1.2 Benefits and costs of RSI 222
11.1.3 When should RSI be carried out? 223
11.1.4 Partners involved in the RSI process 224
11.1.5 Who can perform RSI? 224
11.1.6 Organisational issues 224
11.1.7 Performing an RSI 225
11.1.8 Good RSI practice guidelines 227
11.1.9 Typical safety deficiencies identified by RSIs 228
11.2 Road Assessment Programme 231
11.2.1 Star Rating 232
11.2.2 Risk mapping 233
11.2.3 Performance tracking 234
11.3 Concluding remarks 235
References 235

12 Safety performance indicators 237
Wouter Van den Berghe, Victoria Gitelman and Shalom Hakkert
12.1 Introduction 237
12.1.1 What are safety performance indicators? 237
12.1.2 The role of SPIs in road safety management 238
12.2 Methodological rules for using SPIs 239
12.2.1 The basic model 239
12.2.2 Criteria and conditions for using SPIs 241
12.2.3 SPIs' subdivisions 242

12.2.4 Principles for data collection and analysis for estimating SPIs 245
12.3 Current practices on the use of SPIs 249
12.3.1 International recommendations 249
12.3.2 Good practice in using SPIs 251
12.3.3 Examples of SPI evaluations 253
12.3.4 SPIs as predictors of road safety outcomes 253
12.4 Discussion: challenges and future needs 256
References 258

13 Insurance data **263**
Magdalena Lindman and Anders Kullgren
13.1 Introduction 263
13.2 Analyzing trends of crashes and injured traffic participants to supplement official statistics 263
13.2.1 Differences between insurance statistics and official road traffic statistics 263
13.2.2 Example of insurance data reports of trends for crashes and injured traffic participants 264
13.2.3 The advantage of a common long-term injury classification, a Swedish method 265
13.3 Conducting retrospective studies on the effectiveness of safety interventions 265
13.4 Providing consumer information on losses incurred by used cars 266
13.4.1 Example IIHS, HLDI: insurance losses by make and model and driver death rates by make and model 266
13.4.2 Example: Folksam's How safe is your car? 267
13.5 Enrichment of insurance data: creation of crash databases and crash datasets 269
13.5.1 UDB – the German Insurance Association crash database 269
13.5.2 VCTAD – Volvo Cars traffic Accident Database 270
13.5.3 PAV – the People Around the Vehicle crash database by If 270
13.5.4 Insurance datasets with information from crash recorders 271
References 274

14 Traffic safety footprint and safe organisations **277**
Matilda Magnusson and Carlos Viktorsson
14.1 Traffic safety footprint 277
14.2 Organisational traffic safety data 278
14.3 Challenges and opportunities related to organisations and traffic safety 279
14.4 Future and potential 281
References 282

15 Background data **285**
Jenny Eriksson and Maria Varedian
15.1 Total vehicle mileage – definition and theory 288
15.2 Design of a study for the estimation of total vehicle mileage 288
15.3 Speed – approach and analysis 291
References 292

16 A brief introduction to research ethics **295**
Ulf Görman
16.1 Good research conduct 295
16.2 Ethics-related control 295
16.3 A risk-based approach 296
Reference 299

17 Use of personal data and artificial intelligence–ethical and legal issues **301**
Ulf Görman
17.1 Introduction 301
17.2 Proviso 301
17.3 Historical aspects of human rights and data protection 302
17.4 Personal integrity – why it matters 303
17.5 The EU General Data Protection Regulation (GDPR) 304
17.5.1 What are personal data? 304
17.5.2 What does processing of personal data involve? 305
17.6 General rules for the management of personal data 305
17.7 Lawfulness 308
17.7.1 Special categories of data 308
17.7.2 Personal data relating to criminal convictions and offences 309
17.8 Fairness and transparency 309
17.8.1 The rights of the data subject 310
17.9 Consent 311
17.9.1 Withdrawal of consent 311
17.9.2 Information and consent to participation in activities vs. data processing 312
17.9.3 Safeguards and derogations related to personal data in the public interest or scientific purposes 312
17.10 GDPR on profiling 313
17.11 Anonymous information 313
17.12 International projects 314
17.13 Open access to research data 315
17.14 What is new with AI? 316
17.14.1 The EU AI Act 316
17.14.2 Ethics guidance for trustworthy AI 320

17.14.3 Foundations of Trustworthy AI 321
17.14.4 Requirements for trustworthy AI 322
References 326

18 Conclusion **329**
Matúš Šucha, Niels Agerholm and Aliaksei Laureshyn

Index **335**

About the editors

Aliaksei Laureshyn is an associate professor in traffic safety at Lund University, Sweden. His core competence is within theory and practical application of proactive (non-accident based) methods in safety analysis. Other research interests cover emerging technologies for data collection in traffic, safety of vulnerable road users, and policy and practice within traffic safety management, both in mature contexts and in low- and middle-income countries. He is a passionate educator about the Safe System approach to traffic safety, giving lectures and courses on this subject in different arenas and in different countries. Since 2021, he has acted as the editor-in-chief of the *Traffic Safety Research* journal (2004–3082).

Niels Agerholm is the executive manager for Road Sector & Authority at The Danish Road Directorate. His departments work with traffic planning. Road administration, road design, road freight transport and legal support. He holds a Master's Degree and a Doctoral Degree from Aalborg University in Denmark. He has worked with traffic safety research for 14 years, but since 2020, he has worked for the Danish Road Directorate.

Matúš Šucha is a Czech traffic psychologist and professor of traffic psychology at Palacky University in Olomouc. His professional focus is on the field of traffic psychology, specifically on issues of mobility, sustainability and safety. He is a forensic expert and a Europsy mentor in the field of traffic psychology. He is the author of many publications on the human factor in transport. He is an active member of many prestigious international organizations, among others, vice-president of ICTCT – International Co-operation on theories and concepts in traffic safety.

Chapter 1
Introduction

Niels Agerholm[1], Matúš Šucha[2] and Aliaksei Laureshyn[3]

Traffic accidents pose a significant global challenge, causing nearly 1.2 million fatalities annually (WHO 2023). This alarming statistic places traffic accidents among the leading causes of death worldwide. In most countries, the police are responsible for compiling official accident statistics. They collect data at the scene of the accident, including details about the vehicles involved, the circumstances leading to the accident, and the resulting casualties. However, these data often do not encompass all information about traffic safety. For instance, minor accidents or those not reported to the police are usually not included in these statistics. This leads to the existence of 'dark figures' in traffic safety data, referring to unreported or underreported incidents.

The data collected can be presented in various ways, such as fatalities per motor vehicles or per population. This flexibility in data presentation can sometimes be misused for political purposes, manipulating the perceived road safety status. For instance, presenting data as fatalities per motor vehicles might show a decreasing trend due to the increasing number of vehicles, even if the absolute number of fatalities remains unchanged. It is therefore essential to use standardized indicators for international comparisons, such as the rate of fatalities per 100,000 population or per 10,000 motor vehicles.

Relying solely on police-reported traffic accidents does not provide a complete picture of traffic safety due to, among others, underreporting. Not all accidents are reported to the police. Minor accidents, especially those without injuries, often go unreported. Also, police reports may lack detailed information on the causes and circumstances of accidents. Additionally, traffic safety also involves conflicts, traffic violations, and other incidents that do not result in accidents but indicate potential safety issues. To get a comprehensive understanding of traffic safety, it is hence essential to combine police reports with other data sources. This multi-faceted approach can help identify patterns and areas for traffic safety improvement more effectively.

[1]Danish Road Directorate, Denmark
[2]Department of Psychology, Polacky University Olomouc, Czech Republic
[3]Department of Technology and Society, Faculty of Engineering, LTH, Lund University, Sweden

As indicated, no data sets are sufficiently comprehensive to clarify all aspects of traffic safety, but the datasets and methods mentioned below together give a holistic view of traffic safety.

After this introductory chapter, we invite you to dive further into the world of traffic safety data and explore the complexities of data and how they can underpin each other with various advantages and the opposite.

Carmelo D'Agostino (Sweden) and Piotr Olszewski (Poland) present the police accident records, and we discover the intricacies of police-reported accident data, the backbone of official accident statistics worldwide. Learn about the precision and limitations of these data, and the dark figures that lurk beneath the surface.

Carmelo D'Agostino (Sweden) and Bhagwant Persaud (Canada) turn the perspectives towards statistical methods for accident data analysis, uncover the random nature of traffic accident occurrence and delve into the principles of designing and performing safety evaluation studies. This chapter illuminates the state-of-the-art tool of regression modelling.

From the wide perspective of a full traffic safety data set, John-Fredrik Grönvall (Sweden) turns into more details concerning in-depth investigations and takes us on a journey into the heart of in-depth investigations, a crucial aspect of understanding traffic accidents.

An even wider data set in healthcare records is presented by Niels Bos (the Netherlands), Aliaksei Laureshyn and Carl Johnsson (Sweden) and explores the untapped potential of healthcare records, a treasure trove of under-reported accidents. Learn about the challenges and benefits of cross-matching police and hospital registers.

Wouter Van den Berghe and Uta Meesmann (Belgium) then present the perspectives of self-reporting of behaviour and attitudes and step into the realm of self-reporting, a unique window into the behaviour and attitudes of road users. This chapter draws on experiences from the ESRA Europe-wide questionnaire.

Andrew Tarko and Vamsi Bandaru (USA) take us into an emerging data field – data from automated vehicles and venture into the future of autonomous driving, a rapidly evolving field with safety at its core. Understand the central role of safety in the regulation and acceptance of autonomous vehicles.

Niels Agerholm (Denmark) introduces us to naturalistic driving/cycling/walking studies and presents how to work with the data of naturalistic studies.

Aliaksei Laureshyn (Sweden) and Nicolas Saunier (Canada) then bring us closer to the location of the traffic accidents by presenting the site-based traffic conflict observations and witnessing the dynamics of traffic conflicts, near-miss incidents that serve as surrogate measures of safety. Learn about the recent advancements in video analysis techniques that make data collection more accurate and cost-efficient.

Then, Matus Sucha (the Czech Republic) and Ralf Risser (Austria) introduce behavioural observations by observing the systematic recording of specific domains of behaviour in typical environments, a method that provides replicable descriptions of behaviour.

Andras Varhelyi (Sweden) presents the road safety inspection and EURORAP, where we can learn about the systematic on-site review conducted by road safety experts to identify hazardous conditions, faults, and deficiencies that may lead to serious accidents across the main road network of Europe.

Shalom Hakkert, Victoria Gitelmann (Israel) and Wouter Van den Berghe (Belgium) then present the perspectives of safety performance indicators.

Magdalena Lindman and Anders Kullgren (Sweden) unravel the content of insurance data in traffic safety and how they can be used in the traffic safety work.

Then, M. Magnusson, Carlos Viktorsson, and Tania Willstrand from Sweden introduce us to the road safety footprint and how we can take it into consideration across the many datatypes presented.

Also, the social aspects of traffic safety are presented by Hanna Wennberg, who presents the big pattern on how any traffic accidents or injured affect the broader perspectives of the society.

Further, embark on a journey with Jenny Ericsson and Maria Varedian (Sweden) as they delve into the role of background data in traffic safety. Discover how exposure data, traffic flow, and speed variations contribute to risk assessment and safety measures. Yet, also navigate the challenges of data quality, privacy, and management.

Ulf Görman (Sweden) introduces us then as the last topic to the legal issues including GDPR related to data treatment and how to navigate the sensitive area of road accidents involving legal liability and personal health data. They help understand the ethical and personal data protection issues involved in safety data collection.

Finally, Aliaksei Laureshyn, Niels Agerholm, and Matus Sucha present the conclusion and outlook and reflect on the range of accident data types and surrogate safety measures presented, their advantages and disadvantages. Look forward to the upcoming tendencies in the field of traffic safety and their potential effects.

A final note is related to the terminology used in this book, particularly to the usage of 'accidents' vs. 'crashes'. It is a heated (overheated?) debate, with reasonable arguments found on both sides. As editors, we chose not to enforce any of the terms and allowed the authors to decide on the wording that resonates best with their knowledge, experience, and, in a way, beliefs.

Good appetite for reading.
Aliaksei, Matus and Niels
Autumn 2025

Reference

WHO (2023), '*Global status report on road safety*' (World Health Organization), https://iris.who.int/bitstream/handle/10665/375016/9789240086517-eng.pdf.

Chapter 2
Accident data and records

Carmelo D'Agostino[1] and Piotr Olszewski[2]

2.1 Definition of road accidents

2.1.1 Introduction

Accident data analysis is the common basis for any decision-making process on traffic safety strategies and management. The quality of the data, in terms of completeness (accuracy and precision of information about accidents) and coverage (severity of accidents recorded, but also the level of accident reporting or under-reporting), is always a concern because it can lead to wrong decisions. An accident is defined as a set of events that result in injury or fatality due to the collision involving one motorized vehicle or a motor vehicle and another motorized vehicle, a cyclist, a pedestrian or an object. The terms 'crash', 'collision' and 'accident' are often used interchangeably.

The adoption of Vision Zero and the Safe System approach in the entire EU is often leading the European countries to focus on severe accidents, admitting that accidents can happen, but severe injuries or fatalities are not an option[*]. This concept, even if it is of immediate interpretation, requires, among other things, common definitions of accident severity and an effort of each EU Member State to reduce or eliminate underreporting for the fatal and severe accidents.

2.1.2 Formal definitions of accident events and accident frequency

The term 'injury road accident' is commonly defined in Europe [1] as an incident on a public roadway involving at least one moving vehicle and at least one casualty (an individual injured or killed). Consequently, incidents that result in property damage only (PDO accidents) are not categorized as 'accidents' and are often not recorded by law enforcement in various countries. Nonetheless, there is ambiguity

[1]Department of Technology and Society, Faculty of Engineering, LTH, Lund University, Sweden
[2]Faculty of Civil Engineering, Warsaw University of Technology, Poland

[*]EU Road Safety Policy Framework 2021–2030 – Next steps towards 'Vision Zero', https://transport.ec.europa.eu/system/files/2021-10/SWD2190283.pdf

surrounding the definition of an 'injury' – local standards and practices may differ regarding whether minor scratches and bruises are considered injuries [41].

Defining what precisely constitutes a fatal accident also presents challenges. Luoma and Sivak [2] examined fatal road accident databases from some 20 countries and their accessibility. While the 30-day fatality criterion is commonly employed, other standards for including fatalities in databases vary from one country to another. For instance, some countries incorporate accidents occurring on private roads accessible to the public, while others do not. Additionally, fatalities resulting from suicide or medical conditions may or may not be counted. The inclusion of incidents involving motor vehicles colliding with trains, trams and animal-drawn carriages is also inconsistent across different nations.

The Vienna Convention adopted in 1968 defined a road fatality as any individual killed instantly or dying within 30 days due to a road traffic incident. Ever since, there have been ongoing efforts to standardize the definitions and criteria used to classify accidents and the severity of injuries. This effort has become increasingly urgent with the establishment of international road accident databases that aggregate data from multiple countries. A WHO report [3] highlights significant advancements: by the year 2011, 92 countries (out of 182) were using the 30-day fatality definition. However, substantial disparities still exist in how the numbers of road accident injuries are calculated and categorized.

High-quality data on road accidents, fatalities and injuries is essential for effective safety evaluations and the development of successful safety enhancement programs. Nevertheless, the quality of accident data presents various challenges, which have been documented and scrutinized in numerous studies [3–5]. From both administrative and legal perspectives, the definitions related to different road accident severity levels encompass other crucial operational aspects tied to reporting procedures, as well as data storage and management.

In its policy documents [6,7], the European Commission (EC) asserts that the total number of individuals seriously injured in road traffic incidents is likely to be significantly underestimated. This discrepancy arises because road accident databases often lack relevant and comparable data on serious injuries. The primary issues are the absence of uniform definitions and the prevalent underreporting and misreporting of accidents.

In road safety statistics, injury severity is sometimes classified based on medical criteria and, at other times, according to the long-term consequences of the injury. Frequently, it is measured by the duration of hospitalization. The EC has recommended adopting a standardized definition of severe injury based on level 3 or more on the Maximum Abbreviated Injury Scale (MAIS3+) medical injury classification. However, this approach is only implemented in certain countries, and there are considerable practical challenges to establishing it as a universal standard. The Abbreviated Injury Scale (AIS), a specialized trauma injury classification system, ranges from 1 (minor injuries) to 6 (fatal injuries) and can be derived from the International Classification of Diseases. Since a single individual may sustain multiple injuries, MAIS represents the highest AIS score among all injuries diagnosed for that person (this will be further elaborated later in the chapter).

2.1.3 Statistical definition of accident frequency

The statistical distribution of accident frequency is not part of this chapter, but its basic concept is needed to fully understand the phenomena and the research effort required to achieve better accident prediction. The nature of road accidents is that they are random events, and the distribution of their counts at a certain location per unit time (e.g. per year) can be represented by the Poisson probability distribution [43]. The probability of observing n accidents during one year $P(A = n)$ can be expressed by the following formula:

$$P(A = n) = \frac{\lambda^n e^\lambda}{n!}$$

where λ is the average number of accidents per year, n is the non-negative integer number (0, 1, 2, ...), and A is the actual accident count in a year (or observed accident frequency).

A special property of the Poisson distribution is that its variance, Var(A), is equal to the mean, λ. Therefore, standard deviation (s) is equal to the square root of the mean, $s = \sqrt{\lambda}$. This property makes it possible to quickly evaluate the confidence in estimates based on accident counts. The expected value of accident frequency based on a series of Y years of accident counts is calculated as an arithmetic mean:

$$E(A) = \frac{\sum A}{Y} = \lambda$$

where Y is the number of years of observation.

The standard error of this mean (s_E), given by the general statistics formula as $s/\sqrt{Y}$, in the case of the Poisson distribution, reduces to

$$s_E = \frac{s}{\sqrt{Y}} = \sqrt{\lambda/Y}$$

Let us assume, for example, that during a 3-year period 180 accidents were recorded at site 1 and 15 at site 2. This gives the mean counts per year: $\lambda_1 = 60$ and $\lambda_2 = 5$. The standard errors of the mean are as follows: $s_{E1} = (60/3)^{0.5} = 4.47$ and $s_{E2} = (5/3)^{0.5} = 1.29$. We can then say that the expected number of accidents per year will be $E(A1) = 60 \pm 4.47$ at site 1 and $E(A2) = 5 \pm 1.29$ at site 2. In the first case, this margin of error (precision) represents 7.5% of the mean and in the second case 25.8%.

The fact that the relative margin of error increases as the average accident frequency declines has important implications. First, as the general road safety situation improves, there are fewer accidents but also less precision in terms of the expected number of accidents. This problem is especially evident in countries that are leaders in improving road safety. In this case, underreporting and misreporting play an even more important role. Second, the precision of accident estimates decreases when the number of accidents is divided by type or severity, i.e. if we consider only fatal accidents or pedestrian accidents rather than all accidents. This second aspect is, unfortunately, inevitable in traffic safety analysis that aims to identify accident contributing factors since the entire process has to be based on specific accident types and severities.

When analysing accident count statistics from several sites, the count variability is often greater than required by the Poisson distribution; that is, the variance of accident counts is greater than the mean. This phenomenon is known as 'over-dispersion'. In such cases, it is better to model accident numbers with the negative binomial distribution. This distribution is more general than Poisson and has two parameters. The variance is related to the mean in the following way:

$$\text{Var}(A) = \text{Mean}(A) + \varphi\,\text{Mean}(A)^2$$

where Mean(A) is the average number of accidents per site and φ is the over-dispersion parameter[†].

The value of parameter φ relates to data dispersion in the following way. When φ is small, variance is close to the mean, so the distribution is close to Poisson. As φ gets increasingly larger, the data becomes more and more dispersed. The value of φ can be estimated from the above equation using accident data from a reporting period of at least 3 years.

2.2 Accident data collection

2.2.1 Introduction

Since accident frequency is used as a performance measure of the safety level of the road network [8,9], the main issue for road safety engineers is to identify the long-term mean accident frequency for that specific part of the road by using historical data. In this context, the quality of data plays a fundamental role. The definitions of precision and accuracy are, in a broad way, provided by ISO 5725 [10]. In the case of road accidents, accuracy identifies how close the estimated accident frequency is to the 'true value' that long-term mean needed as a base for road safety interventions, while precision represents the dispersion, or variability in time due to the random nature of accident events. There are techniques to account for bias due to precision and accuracy issues. At the base of all road safety studies, there are accident counts or other correlated safety indicators that need to be observed, measured and reported. The reporting phase is not a trivial task and must be organized and formalized – beginning from the accident scene so that all the necessary information items are recorded correctly. To achieve this, the personnel need to be trained and must use the proper forms and procedures.

2.2.2 Accident reporting process

It is worth explaining the flow process, common to many European countries and North America, in which accident reporting and accident database formation are organized in

[†]The definition used here follows that of Elvik [42]. In some textbooks, a related but different parameter, $k = 1/\varphi$, is used, which is also called the "overdispersion parameter".

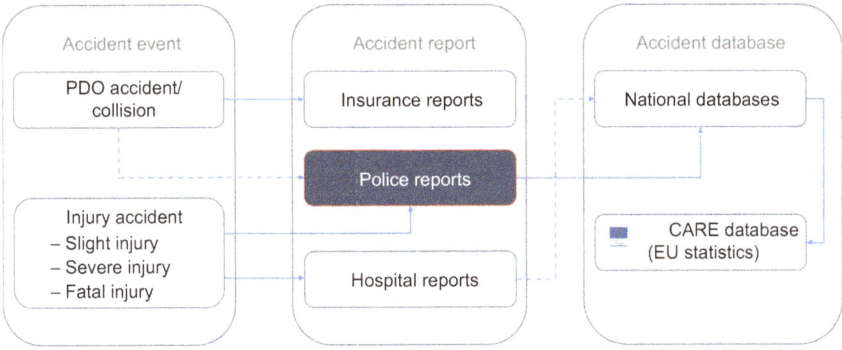

Figure 2.1 Typical schematic workflow from accident occurrence to accident data storage (CARE stands for 'Community Database on Accidents on the Roads in Europe' and collects national statistics from the EU countries)

two distinct phases. Figure 2.1 shows the workflow in a simplified way. This kind of separation between accident reporting and accident data storage can be different (more or less marked) from one country to another, but it is present in the majority of countries worldwide. Figure 2.1 also clarifies that the text below refers mainly to the accident reports and databases while the accident statistics are not treated in this chapter. For the sake of completeness, it is worth clarifying that the structure of the accident databases is very diverse among countries while they are more homogenous in terms of content and information. For example, in Europe, the basic information on accidents is more or less present in any national database. The main differences in the stored information in different parts of the world are due to the database organization [11].

Generally, the information reported includes the accident itself [position of the vehicle(s) and persons involved, injuries, damages, etc.], the accident scene, including the infrastructure conditions, and the personal data of the victims and other participants. Additional data, such as the traffic conditions at the time of the accident, the drivers' conditions, use of alcohol or drugs, can be reported but often are not included in the national statistics and databases.

Following Figure 2.1 workflow, when a road accident happens and the police is called to the scene, in theory, they are obliged to prepare a written road incident report. This report is the basis of the legal prosecution of the guilty party (in case of acts considered to be a criminal offence) and insurance claims. Naturally, the level of detail and accuracy of the incident report depends on accident severity. For fatal and severe injury accidents, the report must be detailed as it may be used in court to prove which party was to blame. For fatal injury accidents, the information items to be included in the report are actually listed in the EU Directive on Road Safety [12,13]. For slight injury accidents, the reports are less detailed and sketchy but, in most cases, are still created – the rate of underreporting varies from country to country. For PDO collisions, the police report is only created in a fraction of cases – where it is needed to support insurance claims.

From a practical point of view, this may create non-negligible differences in the way accident data is collected and how they are processed before joining the national statistics (in terms of quality, precision, accuracy and completeness of information). This also implicitly creates a difference in terms of data quality based on the road functional category. In some countries, in the case of high-speed roads, almost all accidents are reported by police officers who are highly specialized. These roads usually have quite a low accident rate (the ratio of accident frequency to an exposure measure, often traffic count) compared with other roads (this is mainly due to the higher design standard and maintenance condition of the infrastructure) but a higher accident rate for severe accidents. In some urban or suburban areas where severe accidents are rarer, police officers who intervene can have a different background and not be a specialized task force for accident reporting. These two possible scenarios represent the two extreme conditions, namely (1) all the accidents reported by the same and well-trained task force from the police and (2) accidents reported by different police departments in the same area who have different specializations. Several intermediate conditions may take place as well.

However, virtually every country or territory collects data on road accidents from reports recorded by the police. Over the years, the systems of data collection and processing have developed in each country independently; hence, the major differences between these systems. According to a report on the SAU project (Urban Accident Analysis System) [4], the differences in the numbers and types of accidents included in different national databases stem from the differences in the following aspects:

- traffic accident definitions assumed by the national/local administration,
- normative criteria used for including accidents in the database and classifying them by type,
- real practice of the persons in charge of data recording and processing.

The three issues listed above, together with the common definition that road accidents are exclusively the ones involving a 'moving vehicle', contribute to the following problems with data quality and reliability in national road accident databases:

- underreporting of accidents by the police,
- misclassification of injury severity,
- inaccurate or erroneous data,
- incomplete or missing data.

Besides the police, which remains the main actor in accident reporting, there are also other possible sources of data. Insurance companies, for example, report PDO accidents in their databases and hospitals generally know the nature of patients' injuries to plan the right treatments, but these data are difficult to access for General Data Protection Regulation (GDPR) reasons and often incomplete. However, in a specific national context, such as Sweden, the national database merges Police and Hospital data, creating an almost unique example of a comprehensive accident database, which will be described later in the chapter. Another method of accident data collection is the self-reporting. Mainly used for non-injury or slight injury accidents, it deserves a more detailed description due to the growing attention and importance that this method is gaining around the world. It is used above all for accidents (often single-road user accidents) involving cyclists and pedestrians.

2.2.3 Self-reported accidents

The core idea behind self-reporting accidents involves asking individuals about their traffic accident history to obtain insights that do not rely solely on official records maintained by police and/or hospitals [46]. The approaches to collecting this information vary: individuals may be asked to complete paper questionnaires, participate in interviews (either in-person or via telephone) or report their accidents using a mobile app. The methods for gathering self-reported data thus differ significantly – as does the information requested from respondents. In many studies, only the number of accidents in which the participant was involved is relevant to the researcher. In others, respondents may be questioned about potential causal factors of accidents, while some studies focus on the accuracy of the participants' recall of accident details. In short, the objectives of self-reporting vary widely depending on the research question being addressed.

The selection of survey respondents also varies considerably based on the research focus. Some studies target specific groups of drivers (e.g. stroke survivors [14] or older drivers [15]), who voluntarily take part in a scientific research project. Other studies include individuals who have already registered an accident, as the research aims to gain further insights into the known accidents and their contributing factors [16,17]. Some research is designed to assess the level of underreporting, whereby the respondents are selected either from a representative sample of the entire national population (as seen in [18] or [19]) or all residents of an urban area are simply given access to an app through which they can report their accidents.

The advantages and disadvantages of self-reporting are closely intertwined. Since self-reporting provides firsthand information from the individual involved in the accident, it offers a chance to gather insights into factors that are difficult to measure otherwise, such as whether the person involved attributes the accident to feeling tired or stressed. Yet, this is also the method's primary drawback: we only receive one side of the story, and that depends on what the individual chooses to disclose [20]. There is no assurance that the respondents are being truthful – and even if they are, there is no guarantee that their memory of the event fully aligns with what had actually occurred. These aspects are central to the discussion of the limitations of self-reports: memory (as studied by Versteegh [17]) and the willingness to provide truthful answers (as discussed by Kazemzadeh *et al.* [20]).

The advantage of using self-reports to estimate the extent of underreporting of accidents is evident, as it is challenging to assess the number of accidents unreported to the police when only police records are available. All other official records (insurance, hospital, etc.) are inherently incomplete (e.g. [21–23]), as not all accidents are reported. In theory, self-reports should provide the most comprehensive estimate of the total number of accidents. However, the disadvantage of using self-reports for this purpose still relates to memory and truthfulness – can we trust that respondents accurately remember all their accidents, and are there no incentives for them to omit or fabricate details? As with using self-reporting to understand accident causation factors, there is no guarantee of this, but efforts can be made to reduce the uncertainties, such as shortening the recall period to ease memory strain and carefully framing questions to minimize the influence of social desirability in the responses [40]. The choice of respondents also affects the outcome – various biases may arise from respondent selection. One such bias, associated with non-responders (non-response bias), is examined by Tivesten and co-workers [16,24] in the context of accident self-reporting.

2.2.4 Accident reports and in-depth accident investigations

Accident investigations can be categorized into different levels, primarily based on the detail of the investigation and the specific areas or situations they address (Table 2.1). Statistical accident investigations are conducted at the national level to monitor existing issues and trends in accident patterns. At the EU level, national statistics are compiled in the CARE database, considered the most representative

Table 2.1 Categories of road traffic accident investigations

Level	Definition	Examples
Statistical data collection	Collection of anonymous accident data primarily for monitoring trends and identifying priorities	• National statistics • CARE database at the EU level • IRTAD
Intermediate level	Medium-level investigations between statistical and in-depth, suitable for managing black spots. It provides results in a subsequent report.	• Qualified police reports • Insurance reports
In-depth investigations (independent/ non-independent)	Detailed multidisciplinary investigations with a high number of variables (hundreds to over a thousand) aimed at developing safety measures by identifying structural failures, preferably representative of the country in question	• GIDAS (Germany) • Special committee from Trafikverket (Sweden)
Special accident investigations	Serious incidents with significant social impact are usually conducted by government accident agencies. Multidisciplinary investigations using case-specific methodologies aimed at preventing similar serious accidents	• Investigations following the Mont Blanc Tunnel fire in 1999

Source: [28].

for each country. On a global scale, the International Traffic Safety Data and Analysis Group (IRTAD) provides a database that currently comprises data for 35 countries [25].

At the intermediate level, specific locations are analysed in greater detail compared to national statistics, particularly in areas with a higher frequency of accidents (locations commonly referred to as black spots).

In-depth accident investigations represent a more advanced level of detail [46]. These investigations differ fundamentally from analyses based solely on police data; the scientific goal is to comprehend accident causation and the injury process rather than to establish legal accountability. A multidisciplinary team of specialists collects extensive information at the accident site and conducts interviews with involved parties. The numerous variables – typically several hundred or more – allow for a thorough examination of the mechanisms behind injury occurrences and their connections to infrastructure or vehicle design. This methodology aims to pinpoint factors contributing to collisions and injuries, taking into account various parameters such as vehicle design, road conditions, traffic management and human factors. Statistical analysis of in-depth accident data reflects current accident characteristics while examining the evolution of these characteristics over time can provide insights into the effectiveness of previously implemented safety measures.

It is important to recognize that variable definitions are naturally shaped by the investigation's purpose. Furthermore, the scope of investigations can vary significantly, often focusing more on the vehicles involved. It is essential to differentiate between independent approaches (e.g. the GIDAS project in Germany) and non-independent investigations, such as the in-depth studies conducted by the Audi Accident Research Unit of Audi AG, where data collection occurs only if a new Audi vehicle (less than 2 years old) is involved [26,27].

In addition to directly assessing the effectiveness of safety systems, another approach is to utilize present accident data to evaluate future systems through the reconstruction and simulation of accidents during the pre-crash phase (refer to the pre-crash matrix, PCM) [29]. Simulating PCM accidents with generic vehicle safety systems allows for the assessment of a system's effectiveness in specific scenarios.

A significant concern in in-depth investigations is establishing a procedure to protect personal data once informed consent is obtained from accident participants. Data is collected on-site and/or retrospectively through observations, measurements and photographs, necessitating a defined data collection procedure, such as standard methods for photography and alignment. Medical data is gathered through interviews with accident participants and witnesses (following a defined interviewing protocol) and discussions with medical personnel, if relevant, including a review of medical records. A range of technical data is documented regarding vehicle damage, infrastructural factors and other crash-related parameters. To be suitable for in-depth accident investigations, data must be detailed, thus limiting the use of this method to severe accidents – typically only fatal and serious injury incidents are included.

In summary, while in-depth investigations are effective in identifying contributing factors to road accidents, they require trained personnel and a systematic, independent approach. If not applied systematically to severe accidents, in-depth investigations should be preceded by a broader network analysis to identify specific accident types and form the basis for fully understanding accident mechanisms, which can enhance infrastructure and regulatory measures.

2.3 Accident databases

2.3.1 Introduction

In most countries, road accident data is primarily collected and maintained by the police. In some countries, hospitals (Denmark, the Netherlands, Greece, Sweden, Spain, Slovenia) or governmental organizations (the Netherlands, Belgium, Portugal, Hungary) also play a role [45]. In the case of Sweden, accident data is managed by the STRADA (Swedish Traffic Accident Data Acquisition) system, which utilizes Geographic Information Systems (GIS) and incorporates information about accidents from both police reports and hospitals[‡].

The STRADA system in Sweden is considered a benchmark for accident data repositories. It compiles data on road accidents and injuries from reports submitted by the police and medical records from hospitals. STRADA was developed in collaboration with various entities, including the Swedish Police, the Federation of Swedish County Councils, the National Board of Health and Welfare, the Swedish Association of Local Authorities, the Swedish Institute for Transport and Communications Analysis and Statistics Sweden. The Swedish Transport Agency oversees STRADA.

Participation in STRADA is mandatory for both police and hospitals. Nationwide reporting to STRADA by law enforcement has been ongoing since 2003. By 2012, 19 of the 21 counties had all hospitals registered in the system, totalling 68 hospitals in STRADA that year. The data entered by police and hospitals is subsequently matched, yielding more comprehensive details on traffic accidents. In 2013, approximately 30% of all injured individuals recorded by hospitals were also noted in the police databases. Hospital records classify injuries using the AIS, MAIS, Injury Severity Score (ISS), International Classification of Diseases (ICD 10) and Reaction Level Scale. STRADA data are utilized by national, regional and local authorities, as well as by road safety researchers [43].

In the Netherlands, a data linkage project has been developed that integrates information from various sources, including emergency hospital services, ambulance services, fire departments, forensic services, mortality records and insurance companies [30]. Another notable example of a national database is the FARS (Fatality Analysis Reporting System) in the USA, which provides a disaggregated database with detailed information on traffic accidents and open access to raw data files.

[‡]A useful link for finding the sources of information in this respect is: https://road-safety.transport.ec.europa.eu/index_en

Several international accident databases are prominent (only the most widely used are noted here):

- CARE: established by the EC in 1993, CARE aims to identify road safety issues and enhance road safety across the European road transport network. It is based on police accident records from EU member countries.
- IRTAD: a resource managed by the International Transport Forum, an intergovernmental organization of Organization for Economic Cooperation and Development (OECD) countries.
- Eurostat Database: contains statistical data on fatalities from road accidents, aggregated by country; data are provided by EU member states.
- UNECE (United Nations Economic Commission for Europe) Statistical Database: includes information on fatalities and injuries in road traffic accidents, organized by country, user category, accident type, age group and time of accident.
- WHO (World Health Organization) Mortality Database: offers data on traffic fatalities aggregated by country, year, sex and age, sourced from member states' civil registration systems since 1979, with published safety reports providing an overall view of road fatalities. Access to the database is available through the WHO website, which includes tools for analysis and downloadable raw data files.

A comparison of databases and their accessibility is presented in Table 2.2.

2.3.2 International databases

CARE is a disaggregated database, meaning that it contains detailed information on individual accidents reported by countries in the Common Accident Dataset (CADaS) format. CADaS was developed to create a unified framework for the collection of road accident data across Europe. The minimum set of data elements

Table 2.2 Comparison of international road accident databases

Database	**CARE**	**IRTAD**	**Eurostat Database**	**UNECE Statistical Database**	**The WHO Mortality Database**
Coverage	EU countries	35 OECD countries	EU countries	56 UNECE countries	182 Countries
Access	Limited	Open	Open	Open	Open
Type of data	Disaggregated	Aggregated	Aggregated	Aggregated	Aggregated
Information on vulnerable road user accidents	Yes	Yes	No	Yes	No
Exposure data	None	Vehicle-kilometers	None	Population, vehicle fleet	Population

Source: [43].

selected for CADaS was derived from comprehensive research on data sources and systems in 25 European nations, along with the needs and priorities of stakeholders for national-level accident data analysis [31]. After over 4 years of consultations with road safety data specialists, the CADaS data elements were finalized and are now being utilized in the European CARE database. This common dataset underwent review by experts and practitioners in various countries and was adjusted for relevance and practicality. The implementation of this common dataset aims to assist countries in enhancing and standardizing their own road accident databases. The CADaS glossary offers detailed information on the variables that should be submitted to the CARE database.

Accessing the database requires permission from the EC. The CARE database collects the following classes of information for each accident [31,32]:

- accident information
  - date and time of accident
  - weather conditions
  - light conditions
  - accident type
- road information
  - accident location
  - road class (first, second)
  - junction
  - speed limit
  - urban area
- traffic unit information
  - vehicle group
  - vehicle maneuver
- person information
  - road user type: pedestrian, driver, passenger
  - gender
  - date of birth
  - injury severity

The disaggregation of data in CARE allows for more detailed and comprehensive safety analyses. However, the absence of exposure data, variations in injury severity definitions and inconsistencies in accident information reported by different countries limit the potential for benchmarking studies.

The IRTAD gathers and compiles international data on road accidents from 35 OECD countries. IRTAD includes safety and traffic data aggregated by country and year from 1970 to the present [30]. All data are formatted uniformly based on definitions developed and agreed upon by the IRTAD Group. Most data from IRTAD is available in the Road Safety Annual Reports published by IRTAD. Subscribers can access the complete IRTAD database online through the OECD statistics portal. The IRTAD database encompasses the following categories of information, including exposure data:

- Accident data: fatalities, injury accidents, hospitalized victims, injuries classified by
  - road type (motorways, urban roads, rural roads)
  - road user (pedestrians, cyclists, car occupants, powered two-wheeler users, others)
  - age
  - gender
  - seat position in the vehicle
- Exposure data
  - vehicle-kilometers
  - modal split
  - vehicle fleet, categorized by type
  - population
  - driving licence holders
- Other safety data
  - seatbelt-wearing rates
  - helmet-wearing rates

The IRTAD database is aggregated, facilitating the analysis of trends in accidents by road user type, gender and age. However, it does not support analyses of accident causation based on this data.

One example of other initiatives aimed at establishing a common cross-border database to enhance road safety research in Europe is the web-based Road Safety Knowledge System, which was developed within the DaCoTA research project [33]. This system includes not only accident data but also exposure, safety performance and socioeconomic indicators, alongside information regarding road user attitudes and traffic laws. These datasets are employed in road safety analyses to generate Basic Fact Sheets and annual statistical reports. Since 2012, this information has been accessible on the EC Mobility and Transport webpage and is currently being further refined by the DaCoTA research group.

2.3.3 The international experience

When we compare Europe with Australasia and the USA, several common aspects emerge, related mainly to the types of information items collected, while there are differences in accessibility to the data and the entities responsible for their storage and management. Data accessibility may vary from state to state in the USA and Australia due to different policies on data protection and the European GDPR. In Australia, each state has its own accident database and is in charge of data collection and storage. New Zealand has a national database reported in a GIS system with complete maps of accidents in the entire country. It is used for analysing accident-prone locations and general trends in accident data at the national level. In the USA, the FARS stores all the fatal accidents that happened in the entire country. Each state is in charge of accident data collection with its own dataset based on the so-called Model Minimum Uniform Crash Criteria Guideline (MMUCC).

However, the MMUCC has no standard for coded information; it only provides a guide, and each state has different information items in its accident database.

Australia and New Zealand classify injury accidents, as severe and minor. However, there are differences in the way of defining the two categories and the non-injury accidents. In particular, these differences are related to the post-accident care needed (if a person involved in an accident has been hospitalized or required medical treatment on-site or in a hospital – but was not hospitalized). Similarly, in most EU countries and those in the OECD [34,35], the two levels of nonfatal injury are serious injury and slight injury.

In the United Kingdom, a person seriously injured is a person detained in a hospital as an in-patient or suffering any of the following injuries whether detained in hospital or not: fracture, concussion, internal injuries, crushing, burns (excluding friction burns), severe cuts and lacerations, severe general shock requiring medical treatment and injuries causing death 30 or more days after the accident [36].

In the Netherlands, a seriously injured road casualty is a person with injuries assessed at Level 2 or more on the MAIS2+. Because of the greater objectivity of the classifications based on the International Classification of Diseases, the IRTAD proposes to define a seriously injured road casualty as a person with injuries assessed at Level 3 or more on the MAIS, that is, MAIS3+. The NCHRP Project 17-57 in the USA is actively working to adapt the national severity classification system to the one proposed by the International Classification of Diseases.

Several accident severity datasets still use a simpler severity classification. In some of the EU countries, a person is considered seriously injured if she/he has been hospitalized, other than for observation, for more than 24 h. This is also the case with CADaS, which uses a similar definition based on the hours of hospitalization. Currently, length of stay in hospital is also the indicator most often used in countries of the OECD. However, this classification presents a number of drawbacks, including the fact that it does not necessarily reflect the severity of injuries from the medical point of view [18].

2.3.4 Problem of data harmonization

International databases such as CARE are created with the goal of harmonizing accident information between countries to make international comparisons more meaningful. However, in most EU countries, there is a lack of detailed information about the collision type (vehicle manoeuvers), and there are also different subcategories of the junction type. National accident data collection systems across the EU use the CADaS format on a voluntary basis. There are a lot of differences between the national databases, and some values and variables may not be compatible with the CADaS format. Data transformations are very often difficult, which explains why many entries in the CARE database are currently reading: 'not available' or 'other'.

The definition criteria used to classify injury severity vary from country to country and only fatalities can be included in benchmarking studies. As a title of example, these are the CARE definitions of injury severity

- Injury road accident – incident on a public road involving at least one moving vehicle and at least one casualty (person injured or killed).
- Fatally injured – death within 30 days of the road accident; confirmed suicide and natural death are not included.
- Seriously injured – injured (although not killed) in the road accident and hospitalized at least 24 h.
- Slightly injured – injured (although not killed) in the road accident and hospitalized less than 24 h or not hospitalized.
- Not injured – person participating in the accident although not injured.

At present, there is an attempt to harmonize the definitions of traffic accident injury severity by using one of the medical injury scales

- the Abbreviated Injury Scale (AIS)
- the Maximum Abbreviated Injury Scale (MAIS)
- the Injury Severity Score (ISS)
- the New Injury Severity Score (NISS)

IRTAD [30] proposed defining a serious injury as one assessed at Level 3 or more on the MAIS, i.e. MAIS 3+. The AIS is a medical scale describing the severity of injury for each of nine regions of the body as: 1 Minor, 2 Moderate, 3 Serious, 4 Severe, 5 Critical and 6 Unsurvivable. The body regions are: 1 Head, 2 Face, 3 Neck, 4 Thorax, 5 Abdomen, 6 Spine, 7 Upper Extremity, 8 Lower Extremity, 9 External and others.

As briefly mentioned previously in the chapter, the MAIS is the maximum of the AIS scores for all regions of the body while MAIS 3+ – MAIS with a score of 3 or more is to be used for the definition of serious injuries. The EC proposes using MAIS 3+ as a common scale score among EU countries for serious road traffic injuries instead of non-medical definitions based on the length of hospital stay or the need for hospital treatment.

Some researchers advocate using the disability-adjusted life years (DALYs) as an alternative to MAIS 3+ [43]. DALY expresses the number of life years lost due to ill-health, disability or early death. It conveys additional information about the influence of an accident on the future life of the person involved in the accident. However, such an assessment would only be possible at the end of patient's treatment – sometimes many months after the accident.

2.4 How to interpret accident data

2.4.1 Introduction

There is a growing attention to road safety in Europe. New regulations applied on Trans-European Transport Networks push Agencies to introduce new methodological approaches to Road Safety, monitoring the treatment and controlling the level of safety on the road network under their management. There are two commonly used measures of safety based on accident counts:

- ‘Accident frequency’ is defined as the number of accidents occurring at a particular site in a reference time period.
- ‘Accident rate’ is the number of accidents that occurs at a given site during a certain time period in relation to a particular measure of exposure (e.g. per million vehicle kilometers of travel for a roadway segment or per million entering vehicles for an intersection).

However, the accident rate has to be related to a suitable measure of exposure based on the specific accident type and road users involved. For example, relating accidents to vehicle-km can be misleading if pedestrians and/or bicyclists are also present in the traffic environment in question. This is mainly because it is the motor vehicles that hurt unprotected road users, and relating the number of victims to the number of perpetrators is not a good indicator of the risk level. Accident rates may be interpreted as the probability (based on past events) of being involved in an accident per unit of the exposure measure.

Accident counts observed at a site (road segment, intersection, interchange) are commonly used as the fundamental indicator of safety when performing road safety analysis. Because accidents are random events, accident frequencies naturally fluctuate over time at any given site. The randomness of accident occurrence implies that short-term accident frequencies alone are not reliable estimators of the long-term mean accident frequency. If a 3-year period of accidents was used as the sample to estimate the mean yearly accident frequency, it would be difficult to know if this 3-year period represents a high, average or low accident frequency at the site.

This year-to-year variability in accident frequency (Figure 2.2) adversely affects accident estimation based on data collected over short periods. The short-term average accident frequency may differ significantly from the long-term average accident frequency. This effect is magnified at study locations with low accident frequencies, where changes in accident numbers due to random variability represent an even larger fluctuation relative to the expected mean accident frequency.

The accident number fluctuation over time makes it difficult to determine whether changes in the observed accident frequency are due to changes in site conditions or are due to natural random fluctuations. When a period with a comparatively high accident frequency is observed, it is statistically probable that the following period will have a comparatively low accident frequency. This tendency is known as regression-to-the-mean (RTM) and is also evident when a low accident frequency period is followed by a higher accident frequency period (see, e.g. period 2 compared to period 1 in Figure 2.2).

Failure to account for the effects of RTM introduces a potential ‘RTM bias’, also known as ‘selection bias’. Selection bias occurs when sites are selected for treatment based on short-term trends in observed accident frequency. A long period of observation, in the same way, can introduce bias related to uncontrolled factors (changes in the infrastructure, in the traffic rules, etc.) that may negatively affect the estimate of the long-term mean.

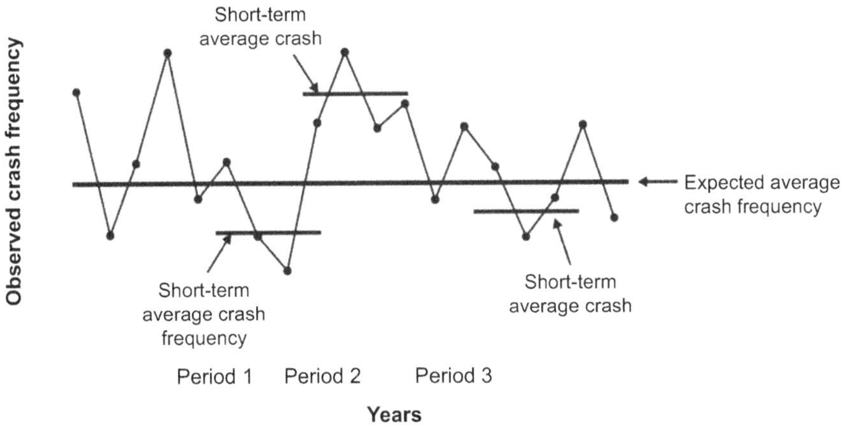

Figure 2.2 Variation in short-term average accident frequency at a particular site (source: [44])

RTM bias can also result in the overestimation of the effectiveness of a treatment (i.e. the change in expected average accident frequency). Without accounting for RTM bias, it is not possible to know if an observed reduction in accidents is due to the treatment or if it would have occurred naturally without the modification.

Another key analytical issue arises when accident rates are used in evaluating safety performance to, e.g. flag locations for safety investigation during the network screening process. AADTs are used directly in the computation of this measure, i.e. accident rate = accident frequency/AADT (or some scalar multiple of this). If accident rates are based on the observed accident counts, then the RTM difficulty discussed above will still apply. In addition, there is a further problem that renders this method of screening dubious. The problem is that the relationship between accident frequency and AADT is typically not linear [8,9].

Specifically, comparing accident rates of two locations with different traffic levels to judge their relative safety may lead to erroneous conclusions. Inferring that when the two rates are equal, they indicate that equivalent levels of hazard may be completely false if different AADT volumes are involved.

In Figure 2.3, two different curves (Curve1 and Curve2) are plotted to give an example. The two curves show a nonlinear relationship between entering AADT and the expected accident frequency for two different groups of intersections. Accident rates (the slopes of straight lines corresponding to each point on the curves) can be compared to judge the relative safety of points A and B, as well as A and C. Such comparisons may lead to erroneous conclusions because points B or C could be considered far safer than point A. Moreover, considering points B and C, the two corresponding rates are equal, indicating equivalent levels of hazard, but this appears to be completely wrong because different AADT levels are involved.

The upshot of all this is that the use of accident rates to compare sites in regard to their safety levels is potentially problematic. When the slope

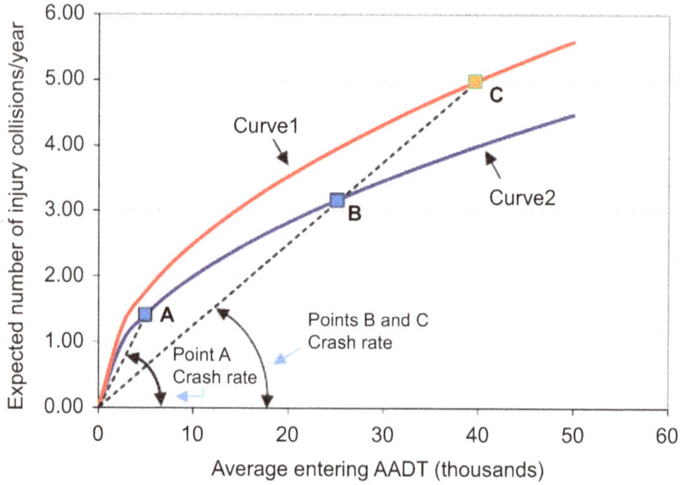

Figure 2.3 Relationship between accident frequency and AADT

corresponding to points on the accidents vs. AADT relationship decreases with increasing traffic volume levels, as is often the case, network screening by accident rates will tend to identify low AADT sites for further investigation. The most valid basis for comparison using accident rates is for the relatively rare cases when the traffic volume levels are the same or when the relationship between accidents and AADT is linear.

Conventional procedures for identifying sites for safety investigation tend to select sites with high accident counts and/or accident rates. However, accident counts could be high or low in a given period solely due to random fluctuations. This can lead to many sites being either incorrectly identified or overlooked and, consequently, to an inefficient allocation of safety improvement resources. In addition, selection on the basis of accident rates tends to wrongly identify hazardous sites with low volumes.

2.4.2 Identification of hazardous locations

The process of pinpointing hazardous locations entails the identification and analysis of black spots, along with the safety assessment of road networks. The distinction between these two tasks can be explained as follows:

- Black spot analysis refers to a technique for identifying high-risk accident sites (such as intersections or short road segments, like hazardous curves), meaning areas with a significant concentration of accidents. This analysis typically forms part of a black spot management program.
- Road network safety analysis is defined as a method for ranking road segments that experience high accident rates. It serves to identify, evaluate and prioritize sections of the road network where a substantial number of accidents occur

relative to traffic flow and road length. This analysis is generally the initial phase in a Road Network Safety Management process.

As previously noted, road accidents are relatively infrequent, random occurrences and their frequency can fluctuate randomly over time. If hazardous locations are identified using short-term (usually 3-year) accident counts, their selection may be influenced by the so-called 'regression-to-the-mean' effect, as discussed earlier in this chapter. As illustrated in Figure 2.2, accident frequency at a given site is elevated during period 2 due to random variation. If this site is classified as a black spot, the accident frequency is likely to decrease during period 3, even without safety interventions, due to natural regression to the long-term mean. Consequently, if black spots are chosen for safety improvements solely based on high accident counts, the effects of the treatment will likely be overestimated, as a natural random decline in accidents will be inaccurately attributed to the intervention. In countries that are leaders in traffic safety, the notion of black spots has largely been discarded. It is now recognized that there are no specific accident-prone locations where most accidents are clustered; rather, the entire network must be assessed, and treatments should be applied at the network level.

2.4.3 What underreporting is and why it is a problem

Misreporting and underreporting often occur because most EU countries rely solely on police reports for their national road traffic injury databases [7]. However, police are not called for all traffic accidents, and they are not equipped to conduct a medical assessment – their evaluation of injuries is merely a rough on-site estimation. This initial evaluation by the police is not consistently cross-checked with subsequent medical reports regarding injury severity. Numerous studies (e.g. [37] or [38]) have demonstrated that underreporting level changes with injury severity and the type of road user involved. Cyclist victims are the least likely to be reported by police (particularly in single-user accidents), followed by pedestrians and motorcyclists. Consequently, it can be inferred that police records tend to be biased against vulnerable road user victims [41]. Due to the inadequacies of police databases, it is essential to merge the records collected by the police with hospital data, a practice already in place in some countries (e.g. the STRADA system in Sweden). This method is the most effective way to address the underreporting of both serious and minor injuries.

2.4.4 Concluding remarks

Road traffic accidents are random events, and therefore, their counts should be analysed with proper statistical tools and always interpreted with caution. Any estimates and predictions based on accident counts are subject to a statistical error, and the relative magnitude of this error increases as the accident counts get smaller. It is a paradox that, as the road transport system becomes safer and the accident numbers decrease, predictions and conclusions become less and less precise as they are based on fewer accident observations. The same problem occurs if a more detailed analysis is attempted, such as looking at a particular accident subcategory (e.g. fatal cyclist accidents at signalized intersections). This would mean looking at accidents fewer in numbers, which inevitably yields less precise results.

The results of road accident data analysis should be interpreted according to the study objectives. These objectives can be grouped into two broad categories:

- Assessment and monitoring of the road safety situation.
- Identification of hazardous locations and their detailed analysis.

In the first category, accident statistics for whole countries, regions or cities are used. Absolute numbers of accidents and their victims are important because they show the magnitude of the problem and indicate where most accidents occur. For the purpose of benchmarking and comparing countries or regions, accident rates are more appropriate than absolute numbers. Fatality rates calculated as the average number of fatalities per million population give an indication of the extent to which road accidents constitute a public health problem. Fatality rates expressed per million km travelled can show the level of safety of different transport modes and the degree of personal risk of the different road users.

In the second category – identification and analysis of hazardous locations – the problem of random accident count variations becomes more apparent. Therefore, it is essential to use statistical tools recommended later in the book, namely the empirical Bayes method, before coming to conclusions on the safety level of individual sites.

General accident statistics alone are not sufficient to assess the safety performance of a road transport system, as they do not tell us anything about the road safety policy or safety measures and treatments introduced. For that purpose, various safety performance indicators were introduced [39], such as proportion of fatal accidents resulting from alcohol use, percentage of vehicles exceeding the speed limit, seat belt wearing rates, vehicle fleet roadworthiness and availability of emergency medical services. Road safety performance indicators are useful tools for policy making, as they allow information to be gathered on the effectiveness of safety interventions in specific areas.

References

[1] IDABC (2004). CARE: Community Road Accident Database. Retrieved from: https://road-safety.transport.ec.europa.eu/statistics-and-analysis/methodology-and-research/care-database_en.

[2] Luoma, J., and Sivak, M. (2007). Characteristics and availability of fatal road-accident databases in 20 countries worldwide. *J. Safety Res.*, *38*(3), 323–327. doi:10.1016/j.jsr.2007.05.001.

[3] WHO (2013). Global status report on road safety 2013: supporting a decade of action. Retrieved from: https://iris.who.int/handle/10665/78256.

[4] Chisvert, López-de-Cózar, and Ballestar. (2007). Quality and representativity of the traffic accident data in urban areas. State of the Art. Retrieved from:

https://ec.europa.eu/transport/road_safety/sites/roadsafety/files/pdf/projects_sources/sau_deliverable_1_state_of_art_final.pdf.

[5] WHO (2010). *Data Systems: A Road Safety Manual for Decision-Makers and Practitioners.* WHO, Geneva, pp. 1–164.

[6] EC (2019). Commission staff working document EU Road Safety Policy Framework 2021–2030 – Next steps towards "Vision Zero" https://transport.ec.europa.eu/system/files/2021-10/SWD2190283.pdf.

[7] EC (2013). Commission staff working document on the implementation of objective 6 of the European Commission's policy orientations on road safety 2011–2020 – First milestone towards an injury strategy. doi:10.1017/CBO9781107415324.004.

[8] Hauer, E. (1997). *Observational Before–After Studies in Road Safety*. Pergamon Press, Elsevier Science, Oxford.

[9] Hauer, E. (2015). *The Art of Regression Modeling in Road Safety*. Springer, Berlin.

[10] ISO 5725-4:2020 Accuracy (trueness and precision) of measurement methods and results. International Organization for Standardization, 2020.

[11] Montella, A., Andreassen, D., Tarko, A. P., *et al.* (2013). Accident databases in Australasia, the European Union, and the United States. *Transp. Res. Rec.: J. Transp. Res. Board*, 2386(1), 128–136. doi:10.3141/2386-15.

[12] ETSC (2018). An overview of road death data collection in the EU. PIN Flash Report 35. Retrieved from: https://etsc.eu/an-overview-of-road-death-data-collection-in-the-eu-pin-flash-35/.

[13] EU (2008). Directive 2008/96/EC of the European Parliament and of the Council of 19 November 2008 on road infrastructure safety management. Retrieved from: https://eur-lex.europa.eu/legal-content/PL/ALL/?uri=CELEX%3A32008L0096.

[14] Finestone, H.M., Guo, M., O'Hara, P., *et al.* (2011). Department of transportation vs self-reported data on motor vehicle collisions and driving convictions for stroke survivors: Do they agree? *Traffic Inj. Prev.* 12, 327–32. doi:10.1080/15389588.2011.581716.

[15] McGwin, G., Owsley, C., and Ball, K. (1998). Identifying crash involvement among older drivers: Agreement between self-report and state records. *Accid. Anal. Prev.* 30, 781–791. doi:10.1016/S0001-4575(98)00031-1.

[16] Tivesten, E., and Wiberg, H. (2013). What can the drivers' own description from combined sources provide in an analysis of driver distraction and low vigilance in accident situations? *Accid. Anal. Prev.* 52, 51–63. doi:10.1016/j.aap.2012.12.016.

[17] Versteegh, S. (2013). The accuracy of driver accounts of vehicle accidents, *J. Chem. Inform. Model.*, 1689–1699. doi:10.1017/CBO9781107415324.004.

[18] Gustafsson, S., and Thulin, H. (2000). Gående och cyklister – Exponering och skaderisker i olika trafikmiljöer för olika åldersgrupper.

[19] Lahrmann, H., Madsen, T. K. O., Olesen, A. V., Madsen, J. C. O., and Hels, T. (2018). The effect of a yellow bicycle jacket on cyclist accidents. *Safety Sci.*, 108, 209–217. https://doi.org/10.1016/j.ssci.2017.08.001.

[20] Kazemzadeh, K., Camporeale, R., D'Agostino, C., Laureshyn, A., and Winslott Hiselius, L. (2020). Same questions, different answers? A hierarchical comparison of cyclists' perceptions of comfort: in-traffic vs. online approach. *Transp. Lett.*, 13, 531–539. doi:10.1080/19427867.2020.1737373.

[21] Wåhlberg, A.E. af, Dorn, L., and Kline, T. (2010). The effect of social desirability on self reported and recorded road traffic accidents. *Transp. Res. Part F: Traffic Psychol. Behav.* 13, 106–114. doi:10.1016/j.trf.2009.11.004.

[22] Roberts, S.E., Vingilis, E., Wilk, P., and Seeley, J. (2008). A comparison of self-reported motor vehicle collision injuries compared with official collision data: an analysis of age and sex trends using the Canadian National Population Health Survey and Transport Canada data. *Accid. Anal. Prev.* 40, 559–566.

[23] Elvik, R., and Mysen, A. (1999). Incomplete accident reporting: meta-analysis of studies made in 13 countries. *Transp. Res. Rec. J. Transp. Res. Board* 1665, 133–140.

[24] Tivesten, E., Jonsson, S., Jakobsson, L., and Norin, H. (2012). Nonresponse analysis and adjustment in a mail survey on car accidents. *Accid. Anal. Prev.* 48, 401–415. doi:10.1016/j.aap.2012.02.017.

[25] IRTAD (2021). Road Safety Annual Report 2021.

[26] Otte, D. (2015). In-Depth-Unfallforschung und deren Möglichkeiten der Nutzung von Daten-Ein historischer und prospektiver Rückblick, in: Rupp, A.P.-I., and Kolke, R.P.-I. (eds.) *Unfallforschung 2015, 1. ADAC Symposium Für Unfallforschung Und Sicherheit Im Straßenverkehr.* Cuvillier Verlag, Göttingen.

[27] Otte, D., Krettek, C., Brunner, H., and Zwipp, H., 2013. Scientific approach and methodology of a new in-depth investigation study in Germany so called GIDAS, in: *Proc. ESV Conf.,* Nagoya, Japan.

[28] Moncleus, J., Löwenadler, L.-G., Maier, R., and Repussard, J.-P. (2007). Independent Indepth Road Accident Investigation in the EU.

[29] Helmer, T. (2015). State of scientific and technical knowledge on pre-crash evaluation, in: Development of a Methodology for the Evaluation of Active Safety Using the Example of Preventive Pedestrian Protection.

[30] IRTAD (2010). Reporting on serious road traffic casualties. *Int. Traffic Saf. Data Anal. Gr.* 107.

[31] IDABC (2004). CARE: Community Road Accident Database [WWW Document]. URL http://ec.europa.eu/idabc/en/document/2281/5926.html (accessed 22.12.22).

[32] Yannis, G., Evgenikos, P., Chaziris, A., *et al.* (2008). D.1.14 CADaS – The Common Accident Data Set (SafetyNet Deliverable).

[33] Yannis, G., Thomas, P., Muhlrad, N., Martensen, H., and Aarts, L., 2014. Development of the European Road Safety Knowledge System.

[34] Road Safety Annual Report (2011). *International Traffic Safety Data and Analysis Group*, International Transport Forum, Paris, 2012.

[35] Reporting on Serious Road Traffic Casualties: Combining and Using Different Data Sources to Improve Understanding of Non-Fatal Road Traffic

Crashes. International Traffic Safety Data and Analysis Group, International Transport Forum, Paris, 2012.

[36] *Road Safety Statistics: Notes and Definitions*. UK Department for Transport, London, 2012. www.dft.gov.uk/statistics/series/road-accidents-and-safety.

[37] Alsop, J., and Langley, J. (2001). Under-reporting of motor vehicle traffic crash victims in New Zealand. *Accid. Anal. Prev.* 33, 353–359. doi:10.1016/S0001-4575(00)00049-X.

[38] Amoros, E., Martin, J.L., and Laumon, B. (2006). Under-reporting of road crash casualties in France. *Accid. Anal. Prev.* 38, 627–635. doi:10.1016/j.aap.2005.11.006.

[39] Hakkert, A. S., Gitelman, V., and Vis. M. A. (2007). "Road Safety Performance Indicators: Theory," Deliverable D3.6 of the EU FP6 Project Safety Net.

[40] Yastremska-Kravchenko, O., Laureshyn, A., D'Agostino, C., and Varhelyi, A. (2022). What constitutes traffic event severity in terms of human danger perception? *Transp. Res. Part F: Traffic Psychol. Behav.*, 90, 22–34. https://doi.org/10.1016/j.trf.2022.08.001.

[41] Olszewski, P., Osińska, B., Szagała, P., Skoczyński, P., and Zielińska, A. (2016). Problems with assessing safety of vulnerable road users based on traffic accident data. *Arch. Civil Eng.*, 62(4/II), 149–168. doi:10.1515/ace-2015-0113.

[42] Elvik, R. (2011). Chapter 27 – Traffic safety, *Handbook of Transportation Engineering*. Vol. 2, Applications and Technologies (2nd edn.), McGraw-Hill, New York.

[43] Olszewski, P., Osińska, B., and Szagała, P. (2018). Road accident statistics and available analysis techniques, in: Polders, E., and Brijs, T. (eds.) *How to Analyse Accident Causation? A Handbook with Focus on Vulnerable Road Users,* Hasselt University.

[44] AASHTO (2010). Highway Safety Manual. American Association of State Highway and Transportation Officials.

[45] ETSC (2006). Road Accident Data in the Enlarged European Union. Learning from Each Other.

[46] Olszewski, P., Osińska, B., Szagała, P., *et al.* (2016). Review of current study methods for VRU safety. Part 1–Main report. Deliverable 2.1 of project InDev, Retrieved from: https://lup.lub.lu.se/search/publication/403c2502-ae84-426b-b909-f8a21c6ed7ec.

Chapter 3
Accident modelling: an overview

Carmelo D'Agostino[1] *and Bhagwant Persaud*[2]

3.1 Introduction

The goal of this chapter is to provide an overview of key concepts related to modelling accident frequency, considering the random nature of traffic accidents. This includes topics such as the expected number of accidents (the 'true' yet unobserved measure of safety), regression-to-the-mean (RTM) and the safety-in-numbers phenomenon, which were briefly introduced in the chapter on accident data.

The main objective here is to explain the foundational principles for designing and conducting safety evaluation studies, with regression modelling being the state-of-the-art tool. However, the RTM effect and the complex relationship between crashes and exposure factors pose challenges in accurately estimating expected accident frequency and, consequently, assessing the impact of safety interventions.

Accurately calculating the long-term mean of accident frequency is essential for identifying the effects of safety countermeasures. This remains one of the most critical aspects of traffic safety analysis. Reliable accident frequency prediction models are fundamental for identifying hazardous locations and evaluating possible solutions to address safety issues.

Therefore, this chapter provides an extensive discussion of the main research efforts since the introduction of modern road safety principles, largely credited to Ezra Hauer in 1997 [1]. It addresses the challenges posed by RTM and the non-linear relationship between accident frequency and exposure. The final section introduces the methodological approach for accident modeling, with a particular focus on tackling the RTM bias. The chapter has been inspired by the Ph.D. thesis 'A stochastic approach to safety management of roadway segment' of Carmelo D'Agostino [2].

[1]Department of Technology and Society, Faculty of Engineering, LTH, Lund University, Sweden
[2]Department of Civil Engineering, Toronto Metropolitan University (Ryerson), Canada

3.2 Why accident modelling?

3.2.1 Basic definitions

A crash is defined as a set of events that result in injury or fatality due to the collision involving one motorized vehicle or a motor vehicle and another motorized vehicle, a bicyclist, a pedestrian, or an object. Single accidents involving any road users are included in the definition (e.g. a bicyclist or a pedestrian fall is included in the definition). The terms ‘crash’, ‘collision’ and ‘accident’ are typically used interchangeably, and it will be in the present chapter.

‘Crash frequency’ is defined as the number of crashes occurring at a particular site in a reference time period.

‘Crash rate’ is the number of crashes that occur at a given site during a certain time period in relation to a particular measure of exposure (e.g. per million vehicle miles of travel for a roadway segment or per million entering vehicles for an intersection). Crash rates may be interpreted as the probability (based on past events) of being involved in an accident per unit of the exposure measure.

Accident count observed at a site (road segment, intersection, interchange) in a given time frame is commonly used as a fundamental indicator of safety when performing road safety analysis methods.

3.2.2 Regression to the mean effect and relationship with exposure

Since crashes are random events, their frequency tends to fluctuate over time at any given location. This randomness means that short-term crash data are not a reliable indicator of long-term crash frequency. For example, using 3 years of crash data as a sample might not accurately reflect whether that period represents a high, average or low crash rate at a specific site.

The variability in crash frequency from year to year complicates estimates based on short-term data. Short-term averages may differ significantly from long-term averages, particularly at locations with low crash frequencies, where fluctuations have a larger impact relative to the expected average.

This fluctuation makes it challenging to determine whether changes in observed crash frequency are due to actual changes in site conditions or just natural variability. When a period with a relatively high crash frequency is followed by a period with a lower frequency, it is likely due to a statistical phenomenon called RTM. This effect is also seen when a low-frequency period is followed by a higher one.

If the RTM effect is not considered, it can lead to ‘RTM bias’ or ‘selection bias’. This occurs when locations are chosen for safety treatments based on short-term trends in crash data. Consequently, RTM bias can lead to an overestimation of the treatment’s effectiveness, as it may appear that the treatment reduced crashes when, in reality, the change could have occurred without the intervention.

Another critical analytical issue arises when accident rates are used to evaluate safety performance, such as flagging locations for investigation during the network screening process. Accident rates are typically calculated using average annual daily traffic (AADT), with the formula: accident rate = accident frequency/AADT (or a scalar multiple of this). However, if accident rates are based solely on observed counts, the RTM problem mentioned earlier still applies.

Moreover, another challenge makes this screening method questionable: the relationship between accident frequency and AADT is not linear. This phenomenon, widely recognized in the literature as the 'Safety in Numbers' effect, is one of the most common sources of error (along with RTM bias) in comparative safety analyses [3].

Specifically, comparing accident rates at different traffic volumes to assess relative safety can lead to incorrect conclusions. Assuming that two equal accident rates reflect equivalent hazard levels is potentially misleading if the AADT values differ.

As introduced in chapter 2, refer to Figure 2.3, where two curves (Curve1 and Curve2) are plotted to illustrate the non-linear relationship between entering AADT and the expected crash frequency, possibly representing two different groups of intersections. It becomes evident that when comparing crash rates (the slopes of each point along the curves) between points A and B, as well as A and C, conclusions about their relative safety could be misleading. Points B and C may appear much safer than point A based on their rates. Additionally, though the crash rates at points B and C are equal, suggesting similar hazard levels, this is inaccurate because the AADT levels at those points differ.

The key takeaway is that using accident rates to compare the safety of different sites is inherently problematic. When the slope of the accident-to-AADT relationship decreases as traffic volume increases, which is often the case, network screening based on accident rates tends to flag low AADT sites for further investigation. However, accident rates are only a valid basis for comparison when traffic volumes are the same or when the relationship between accidents and AADT is linear, which is rare.

Main takeaway of the paragraph: Traditional methods for identifying sites for safety investigation often focus on sites with high accident counts and/or accident rates. However, accident counts can vary significantly due to random fluctuations, which may result in some sites being incorrectly identified or overlooked, leading to inefficient allocation of safety improvement resources. Additionally, selecting sites based on accident rates tends to wrongly flag locations with lower traffic volumes.

Empirical Bayes methods have been proposed as a solution to overcome these issues [1,2,4].

3.3 Regression technique, model form and goodness-of-fit evaluation

3.3.1 Introduction

First, a note on terminology is needed to clarify some terms used in the following section of the chapter. The term 'accident prediction models' (APMs), which is often used interchangeably with 'safety performance functions' (SPFs), typically refers to multivariate models fitted to accident data. These models estimate the statistical relationship between the number of accidents and various factors believed to be related to their occurrence. The term 'predictive' can be somewhat misleading; 'explanatory' might be a more accurate term. Prediction implies forecasting future events, while accident prediction models are based on historical data and thus can only describe and potentially explain past occurrences. [5]. Other models, which will not be treated here, aim to predict accident severities given (i) a non-zero probability of an accident [6] and (ii) exogenous conditions [7–9].

Moreover, the choice of explanatory variables that potentially affect the safety performance of a site should be grounded in theory [8]. For example, a theoretical basis for selecting explanatory variables might involve using a causal model [10]. In practice, however, a theoretical foundation for identifying explanatory variables is rarely stated explicitly [6,11–14]. Instead, the choice of variables often relies primarily on data availability. While it is evident that any analysis is constrained by the available data, the selection of explanatory variables should ideally not be based solely on this factor. Instead, explanatory variables should

- reflect findings from previous studies that indicate a major influence on the number of accidents;
- be measurable in a valid and reliable manner;
- avoid being endogenous, meaning they should not depend on other explanatory variables included in the model or on the dependent variable.

Historically, two statistical methods have been used to develop accident prediction models: conventional linear regression and generalized linear regression [12]. Recently, however, generalized linear regression has become the predominant method for developing collision prediction models. Research studies by Jovanis and Chang [16], Miaou and Lum [17] and Hauer [18] have highlighted the limitations of conventional linear regression for modeling discrete, non-negative and rare events such as traffic accidents. These studies demonstrated that the standard conditions for conventional linear regression – normal distribution of model errors, constant error variance, and a linear relationship between the response and explanatory variables – do not hold when modeling traffic accident occurrences.

3.3.2 The generalized linear modelling (GLM) approach

The GLM approach has the advantage of overcoming the limitations of conventional linear regression in accident frequency modelling. In particular, it also allows a Poisson or negative binomial error structure, distributions that are more pertinent to accident frequency modelling, to be assumed. The generalized linear interactive modelling (GLIM) approach is described in the following [18,20,21].

Suppose Y_i is a random variable that describes the number of crashes at a given location i in a given period of time. Then, Y_i is assumed to possess Poisson distribution and can be expressed as follows:

$$P(Y_i = y_i) = \frac{\lambda_i^{y_i} \cdot e^{-\lambda_i}}{y_i!} \tag{3.1}$$

where $P(Y_i{=}y_i)$ is the probability of occurring y crashes on the roadway section i in a given period of time and λ_i is the expected number of crashes on section i (i.e. $E(y_i)$). In addition, the mean or expected value of Y_i is assumed to be equal to its variance, that is

$$E(Y_i) = \text{Var}(Y_i) = \lambda_i \tag{3.2}$$

where $E(Y_i)$ is the expected number of crashes on section i and $\text{Var}(Y_i)$ is the variance of observed number of crashes. For a given set of explanatory variables (highway geometrics, traffic and other data), λ_i can be estimated using the formulation:

$$\ln(\lambda_i) = \beta X_i \tag{3.3}$$

where X is a vector of explanatory variables and β is a vector of parameters to be estimated.

However, in some cases, this method has limitations when applied to real-world data due to the assumption of equal mean and variance. When the data is overdispersed or underdispersed (i.e. mean is note equal to the variance), the use of this method would overestimate or underestimate the parameters. Many previous studies have found that crash data tends to be overdispersed in many situations with the variance being significantly higher than the mean [13]. In such cases, any inferences made based on Poisson model estimations may lead to wrong conclusions. As a result of this, many researchers recommend using alternative methods in analysing crash data, especially when the data is overdispersed. One such method is to utilize a NB distribution because it does not require the equal mean and variance assumption. In this method, the mean or expected value itself is assumed to be a random variable, which can be described by NB distribution. In this case, λ_i can be now written as follows:

$$\ln(\lambda_i) = \beta X_i + \varepsilon_i \tag{3.4}$$

where ε_i is the unobservable error term with a gamma distribution. The variance of this distribution can be expressed as follows:

$$\text{Var}[Y_i] = \mu_i + k\mu_i^2 \tag{3.5}$$

where μ_i is the expected number of crashes on section i and k is called the NB dispersion parameter. The overdispersion occurs when the value of k is greater than 1 and when its value is zero, NB distribution reduces to Poisson distribution with which the variance is equal to the mean. The corresponding probability distribution under the NB assumption is given by

$$P(Y_i = y_i) = \frac{\Gamma(y_i + 1/k)}{\Gamma(y_i + 1)\Gamma(1/k)} \cdot \frac{(k \cdot \mu_i)^k}{(1 + k \cdot \mu_i)^{(y_i + 1/k)}} \tag{3.6}$$

where $\Gamma(\cdot)$ is the gamma function. The NB model can be estimated using a maximum likelihood method to obtain the model parameters or b values and the dispersion parameter k. This can be carried out through maximization of the likelihood function (L), where N is the total number of sections

$$L(y_i, \mu_i, \alpha) = \sum_{i=1}^{N} \left[y_i \cdot \log(\alpha \cdot \mu_i) - (y_i + 1/\alpha)\log(1 + \alpha \cdot \mu_i) + \log\left(\frac{\Gamma(y_i + 1/\alpha)}{\Gamma(y_i + 1)\Gamma(1/\alpha)} \right) \right] \tag{3.7}$$

The general form of the accident prediction model for road segments adopted is

$$E(Y) = e^{a_0} \cdot L \cdot AADT^{a_1} \cdot e^{\sum_{j=1}^{m} b_j \cdot x_j} \tag{3.8}$$

where $E(Y)$ is the expected accident frequency (accidents/time period); L is the length of the segment under consideration (km); AADT is the average annual daily traffic (veh/day); x_j is the any of m-additional variables and a_0, a_1 and b_j are model coefficients.

This model form was selected because it is generally accepted as the form that better describes the accident phenomenon [22–24,26].

In particular, it logically estimates zero accidents if one of the two exposure variables (AADT or L-in case of road segments) is equal to zero and it allows the non-linear relationship between traffic volume and accident frequency by means of a suitable calibration of the traffic volume coefficient.

In the case of road intersections, the length (L) is not a relevant variable, but the entering flows can be modelled separately for the conflicting traffic directions. In this case, the regression terms for the two flows must be estimated separately in the model.

3.3.3 Goodness-of-fit measures

Several measurements are usually used to assess the goodness-of-fit of the model and the significance of the model parameters [25,28]. These are the t-ratio for the model parameters, the scaled deviance (SD) and the Pearson χ^2 statistics for the model overall. The SD, defined as the likelihood ratio test statistic, is estimated as twice the difference between the log-likelihoods of the studied model and the full or saturated model where the full model has as many parameters as the observations so that the model fits the data perfectly. Therefore, the full model, which possesses the maximum log-likelihood achievable under the given data, provides a baseline for assessing the goodness-of-fit of an intermediate model with specified parameters. Zhang and Ivan [24] showed that for negative binomial error structure the scaled deviance is as follows:

$$\mathrm{SD} = 2\sum_{i-1}^{n} \left[yi \ln\left(\frac{yi}{\widehat{E}(yi)} \right) - (yi + k)\ln\left(\frac{yi + k}{\widehat{E}(yi) + k} \right) \right] \tag{3.9}$$

where SD is the scaled deviance, *yi* is the observed number of accidents of the *i*th segment, $\widehat{E}(yi)$ is the predicted number of accidents of the *i*th segment, *k* is the negative binomial dispersion parameter and *n* is the number of the observations in the sample.

The Pearson χ^2 can be calculated by means of the following formula:

$$\text{Pearson } \chi^2 = \sum_{i=1}^{n} \frac{\left[y_i - \widehat{E}(y_i)\right]^2}{\mathrm{Var}(y_i)} \tag{3.10}$$

where $\mathrm{Var}(y_i)$ is the variance of the *i*th segment.

In order to verify the goodness-of-fit of the model, these two measures must be compared with the value obtained from the χ^2 distribution with the model's degrees of freedom and for a given level of significance. The model can be considered significant, if both measurements are less than the χ^2 critical value.

The Akaike's Information Criterion (AIC) was also used as goodness-of-fit measure [13,15,27]. The AIC value is calculated as follows:

$$\mathrm{AIC} = -2 \cdot \log L + 2 \cdot p \tag{3.11}$$

where $\log L$ is the maximum log-likelihood of the fitted model and *p* is the number of parameters in the model.

The smaller the value of AIC, the better the model data fit. Therefore, AIC can be used also to compare and rank different models calibrated on the same dataset.

Another measure of goodness-of-fit is $R^2{}_k$. This measure was calculated as follows [18,19]:

$$R_k^2 = 1 - \frac{k_0}{k} \tag{3.12}$$

where k_0 is the overdispersion parameter estimated in the negative binomial model with only a constant term and *k* is the NB overdispersion parameter estimated in the full model.

$R^2{}_k$ is simple to calculate, and it yields a value between 0 and 1. Since it is based explicitly on the overdispersion parameter, it is especially applicable in evaluating negative binomial models. The higher the $R^2{}_k$ value, the better the fit.

An adjusted $R^2{}_k$ is a modification of $R^2{}_k$ that adjusts for the number of explanatory terms in a model:

$$R_k^2 adj = 1 - (1 - R_k^2) \cdot \frac{(n-1)}{(n-p-1)} \tag{3.13}$$

where *n* is the total number of observations in the sample and *p* is the total number of the parameters in the model (not counting the constant term).

A tool that can be adopted for the evaluation of the appropriateness of the model form chosen is the cumulative residual analysis [11,29,30]. The residual for each data observation of the sample is equal to the difference between the observed and estimated values of the dependent variable. A standardized residual for the *i*th observation, e.g. road segment (SR_i), is computed from the following equation:

$$SR_i = \frac{(y_i - \hat{y}_i)^{\ 2}}{\sqrt{\hat{y}_i + k \cdot \hat{y}_i}} \tag{3.14}$$

where y_i is the observed number of accidents of the ith segment, $\hat{y}i$ is the estimated number of accidents of the ith segment and k is the dispersion parameter.

The cumulative standardized residual value for the jth element is obtained as follows:

$$CSR_j = \sum_{i=1}^{j} SR_i \tag{3.15}$$

for $j = 1, \ldots, m$ and with m equal to the total number of the observation.

Cumulative standardized residuals analysis is carried out by plotting CSR_i vs. exposure values, computed as the product of AADT and the length for each homogeneous section. The more the points of the CSR curve stay close to the x-axis, avoiding significant increase or decrease of the shape, the closer the estimates are to the observations and the more limited the over/under-estimation phenomenon. Moreover, a CSR curve contained into the two standard deviation interval (e.g. 2s) indicates the appropriateness of the model form chosen [31].

3.3.4 Empirical Bayes estimation and the role of the NB dispersion parameter

Ultimately, a safety performance function provides an estimate of the expected number of accidents for a roadway element with a specific combination of characteristics. Typically, these characteristics include traffic volume and highway geometry. However, most safety performance functions do not account for all factors that contribute to systematic variations in accident counts. As a result, the estimates generated by a safety performance function represent average values for sites with a particular set of traits. Consequently, the expected number of accidents at a specific site will generally differ from the average value for other sites with similar general characteristics.

What is the best estimate of the long-term expected number of accidents for a given roadway element, given the fact that some, but not all the factors affecting accident occurrence, are known? According to the empirical Bayesian method [1], the best estimate of safety is obtained by combining two sources of information:

1. The accident record for a given site.
2. A safety performance function showing how and how much various factors affect accident occurrence.

As previously emphasized, safety performance functions are commonly calibrated using NB regression in which a dispersion parameter that represents extra-Poisson variation is estimated. The negative binomial error structure is preferred to the conventional normal distributed structure assumed in conventional regression modelling because crash data are non-negative counts. The Poisson distribution is a special case of the negative binomial but is only applicable when all entities with identical values of the independent variables can be assumed to have identical means, a near impossibility because of the practical difficulty of accounting for all effects in crash prediction models. Indeed, the size of the calibrated dispersion parameter is a measure of how well a model represents the data and can be thus used to compare competing models for the same data.

Actually, the primary use of the dispersion parameter is not in assessing goodness of fit but in empirical Bayes (EB) estimation [32], in which the posterior mean of an entity with a known crash history is estimated as a weighted average of the crash counts and an SPF prediction. The dispersion parameter is used to calculate the relative weights for each component and is such that the smaller the parameter, the better the model is and the greater the weight assigned to it relative to that assigned to the crash counts. Other types of models do not require the estimation of the dispersion parameters, but despite the effective accident frequency estimation, such analytical stochastic models cannot be easily transferred to different sites [33].

To be more specific, let O be the observed number of accidents and E be the normal, expected number of accidents estimated by a safety performance function. The best estimate EB of the expected number of accidents for a given site is given as follows:

$$\mathrm{EB} = w \cdot E + (1 - w) \cdot O \tag{3.16}$$

The parameter (EB calculation weight) w determines the weight given to the estimated normal number of accidents for similar sites when combining it with the observed number of accidents to estimate the expected number of accidents for a particular site. Usually, w is estimated as follows:

$$w = \frac{1}{1 + \frac{\mathrm{Var}(E)}{E}} \tag{3.17}$$

where Var(E) is the variance of the expected number of accidents E estimated by a safety performance function and is given as follows:

$$\mathrm{Var}(E) = E + k \cdot E^2 \tag{3.18}$$

where k is the value of the dispersion parameter characterizing the safety performance function and is estimated as the ‘shape parameter’ of the negative binomial distribution adopted for calibration in the regression model.

These EB estimates are used in safety management applications in such application tools as the Highway Safety Manual (HSM) [32] and SafetyAnalyst [34], for which the two primary uses are:

1. To estimate the safety of sites for network screening to prioritize them for safety investigation;
2. To estimate the expected safety of a site had it not been treated.

It stands to reason, therefore, that the importance of precise estimation of the dispersion parameter should be established. This need is complicated by findings in a recent flurry of research papers that suggest that the dispersion parameter, contrary to earlier research, is not constant for a given data set but varies from site to site, depending on site characteristics such as segment length. Therefore, the real question is: Does it matter that the dispersion parameter varies and, if so, does it matter how it varies? The question is especially topical since dispersion parameters in the HSM and SafetyAnalyst either are constant or have a very simple form.

The varying form in the HSM and SafetyAnalyst is such that the dispersion parameter for certain classes of road segments is inversely proportional to segment length, as first suggested by Hauer [18] who argued logically that shorter segments have a higher accident frequency variance and consequently should have a higher dispersion parameter than longer segments, and that this variation should influence the long-term estimate of a segment's safety. Since this argument does not apply to intersection models, it is not surprising that dispersion parameters for these SPFs are constant in the HSM and Safety Analyst. However, others [35–38] have suggested that other variables can affect the dispersion parameter, and thus this parameter may also vary for entities such as intersections and rail-highway crossings.

El-Basyouny and Sayed [35] proposed a method that accounts for variation in the dispersion parameter for urban arterials by estimating the variance of the prediction using a similar form to the one used for estimating the mean. In this approach, the dispersion parameter is calculated as the square of the model's prediction divided by the estimated variance. While this method offers considerable flexibility in estimating the dispersion parameter's variability, the chosen form for the variance is somewhat arbitrary, and the rationale for including factors beyond length has not been fully justified. Overall, the variable dispersion parameter improved the model's goodness of fit by accommodating more variability. However, when applied to the identification and ranking of accident-prone locations, the differences from using a constant dispersion parameter were minimal or limited.

3.3.5 Key points on accident modelling

Based on the existing literature, several key points about statistical accident modelling merit emphasis.

First, the GLM technique proves highly effective for calibrating safety performance functions. Unlike conventional linear regression, GLM addresses its limitations in accident frequency modeling by accommodating Poisson or NB error structures, which are more suitable for such analyses [42]. Additionally, GLM allows for the application of advanced calibration techniques.

Second, the model form must account for the nonlinear relationship between accident frequency estimates and traffic volume. It should also ensure that the dependent variable can logically be zero when exposure variables are zero.

Third, the definition and selection of independent variables should encompass road geometric features, consistency measures and road context-related characteristics.

Fourth, contemporary research not only focuses on the selection of independent variables but also on developing guidelines for defining and obtaining homogeneous segments. This includes methodologies for segmenting roadways into sections with uniform safety-related features, enhancing the accuracy and relevance of the safety performance functions [22–24,41,42].

3.4 Available knowledge

3.4.1 Introduction

The development of safety performance functions – often referred to as 'modern' accident prediction models – over the past 20 years marks a significant advancement in road safety research. Today, road safety research is emerging as a well-established scientific discipline [39,40], one that is taught in universities at a master's or graduate level and provides a solid foundation for a rational approach to road safety management.

Development in the field of accident modelling has been so rapid that some models considered state of the art only 10–15 years ago, look somewhat primitive today. Another issue, as pointed out by Lord *et al.* already in 2005 [41] (and Hauer even earlier [18]), is related to moving too far in the direction of mathematical sophistication and the perfect fitting of models. The risk is that research tends to consider the axiom that since accidents are complex phenomena, so models must also be sophisticated to accurately reflect their key characteristics. However, the essence of effective model-building lies in making the right simplifications. A good model is not necessarily one that is highly complex and fits the data perfectly. Instead, it is the simplest model that sufficiently represents the data and captures relationships that are generally valid.

3.4.2 The Highway Safety Manual

The HSM in its first edition (2010) introduced a new common approach to the modelling of crash analysis with the aim of standardizing the methodology.

In Part C, the HSM provides a predictive method for estimating the expected average crash frequency (including by crash severity and collision types) of a network, facility or individual site. The estimate can be made for existing conditions, alternatives to existing conditions (e.g. proposed upgrades or treatments) or proposed new roadways. The predictive method is applied to a given time period, traffic volume and constant geometric design characteristics of the roadway. The predictive method provides a quantitative measure of expected average crash frequency under both existing conditions and conditions that have not yet occurred. This allows proposed roadway conditions to be quantitatively assessed along with other considerations such as community needs, capacity, delay, cost, right-of-way and environmental considerations. The predictive method can be used for evaluating and comparing the expected average crash frequency of situations like

- Existing facilities under past or future traffic volumes,
- Alternative designs for an existing facility under past or future traffic volumes,
- Designs for a new facility under future (forecast) traffic volumes,
- The estimated effectiveness of countermeasures after a period of implementation,
- The estimated effectiveness of proposed countermeasures on an existing facility (prior to implementation).

Each chapter in Part C of HSM provides the detailed steps of the predictive method and the predictive models required to estimate the expected average crash frequency for a specific facility type like

- rural two-lane two-way roads
- rural multilane highways
- urban and suburban arterials

However, the application of the HSM does not always provide adequate results outside North America, and its application should be validated and adapted to the local conditions.

The issue of transferring safety performance functions (SPFs) illustrates how models developed in one country often fail to accurately capture the safety levels of infrastructures in other countries, due to different conditions. The key point is that quantification of the expected reduction of crashes related to different treatments can affect choices and play a fundamental role in the decision-making process. The transferability in HSM is assessed using a factor of calibration to local conditions. In this way using an equation calibrated on 'standard condition', the application of the models in the whole North America is assessed.

Together with SPFs the HSM introduced the 'crash modification factor' (CMF). A CMF is a multiplicative factor used to compute the expected number of crashes after implementing a given countermeasure at a specific site. The CMF is multiplied by the expected crash frequency without treatment. A CMF greater than 1.0 indicates an expected increase in crashes, while a value less than 1.0 indicates an expected reduction in crashes after the implementation of a given countermeasure.

The methodologies of estimation of CMFs have their roots in the methods developed for SPFs since researchers faced the same issues. Indeed, the problem of regression to the mean and the selection bias can be controlled using a sophisticated probabilistic approach already introduced in the 'Observational Before/After Studies in Road Safety – Estimating the Effect of Highway and Traffic Engineering Measures on Road Safety' by Hauer in 1997 and developed by various author in the last three decades.

3.4.3 Scientific literature: old and new challenges

There is a vast body of scientific literature on traffic accident modelling, reflecting decades of research across various disciplines. Early studies primarily employed basic statistical models to identify factors influencing accident frequency and accident rate, while more recent work has embraced probabilistic models, spatial and temporal analysis and advanced machine learning techniques. The proliferation of data from sensors, GPS and smart city infrastructures has fueled this growth, enabling more complex, real-time predictions and accident severity assessments. Today, the literature encompasses diverse approaches, from traditional statistical methods to cutting-edge AI-driven simulations, highlighting the field's rich and ongoing evolution.

If we look at the past 25–30 years, significant advances have been made in the field of traffic accident modelling, building upon the foundational concepts that remain relevant today. The early models, which focused on classical approaches such as Poisson and NB regression models, laid the groundwork for the comprehensive research that has followed. A primary focus has been addressing issues like regression to the mean, enhancing the accuracy of exposure data and understanding the correlations between various explanatory variables.

One of the most notable advances in this domain has been the transition from traditional accident frequency models to more sophisticated methods that account for both accident frequency and severity. Researchers have introduced non-linear models, which are better suited to capturing the complex relationships between traffic conditions, environmental factors and accident outcomes. These models, such as generalized additive models (GAMs) and machine learning techniques, provide a more nuanced understanding of how accidents occur under different conditions.

In addition to these advancements, a major focus has been on the refinement of CMFs used in safety performance functions (SPFs). CMFs provide a quantitative measure of how specific safety interventions or environmental changes impact accident rates, and recent research has sought to improve their accuracy by incorporating more granular data and advanced statistical techniques. This helps policymakers and engineers make better decisions regarding roadway safety improvements.

However, with these advancements, several challenges have emerged. For instance, counterintuitive effects have been observed in some regression models. These occur when model predictions do not align with expected outcomes based on prior knowledge or theory. Such anomalies are often attributed to issues like endogeneity (where independent variables are correlated with the error term) or omitted variable bias, which can lead to misleading conclusions. Furthermore, comparing predictions across different models can yield inconsistent results due to differences in the underlying assumptions and methodologies used in the modeling process.

Addressing these issues has become a key focus for researchers, as resolving them is essential for developing more reliable and interpretable models. New techniques are being explored to correct for these problems, such as the inclusion of instrumental variables and the use of Bayesian approaches to account for uncertainty in model parameters.

In conclusion, while classical accident modeling continues to provide valuable insights, recent advancements in severity modeling, non-linear approaches and improved CMFs are paving the way for more accurate and actionable predictions. Ongoing research aims to address the remaining challenges, ensuring that traffic safety models become ever more robust and useful for practical applications in road safety management.

References

[1] Hauer, E., *Observational Before-After Studies in Road Safety: Estimating the Effect of Highway and Traffic Engineering Measures on Road Safety*, Pergamon Press, Oxford, 1997.

[2] D'Agostino, A., A stochastic approach to safety management of roadway segment. *PhD Thesis*, Ryerson University, Toronto, Canada and University of Catania, Catania, Italy, 2013.

[3] Johnsson, C., Laureshyn, A., D'Agostino, C., and De Ceunynck, T., "The 'safety in density' effect for cyclists and motor vehicles in Scandinavia: An observational study", *IATSS Research*, 45, 2021, 169–175. https://doi.org/10.1016/j.iatssr.2020.08.003.

[4] Hauer, E., *The Art of Regression Modeling in Road Safety*, Springer International Publishing, Cham, 2015.

[5] Elvik, R., State-of-the-Art Approaches to Road Accident Black Spot Management and Safety Analysis of Road Networks. TØI Report 883, Oslo, 2007.

[6] Wang, Y., and Nihan, N., "Estimating the risk of collisions between bicycles and motor vehicle at signalized intersections", *Accident Analysis and Prevention*, 36, 2004, 313–321.

[7] Scarano, A., Riccardi, M. R., Mauriello, F., D'Agostino, C., Pasquino, N., and Montella, A., "Injury severity prediction of cyclist crashes using random forests and random parameters logit models", *Accident Analysis and Prevention*, 192, 2023, 107275. https://doi.org/10.1016/j.aap.2023.107275.

[8] Zhankun, C., Yastremska-Kravchenko, O., Laureshyn, A., Johnsson, C., and D'Agostino, C., "Stochastic method based on copulas for predicting severe road traffic interactions", *Analytic Methods in Accident Research*, 44, 2024, 100347.

[9] Fridstrøm, L., Ifver, J., Ingebrigtsen, S., Kumala, R., and Thomsen, L. K., "Measuring the contribution of randomness, exposure, weather, and daylight to the variation in road accident counts", *Accident Analysis and Prevention*, 27, 1–20, 1995.

[10] Asher, H. B., *Causal Modelling*, Sage Publications, Beverly Hills, CA, 1976.

[11] Hauer, E., "Statistical road safety modelling", *Transportation Research Record: Journal of the Transportation Research Board*, No. 1897, Transportation Research Board of the National Academies, Washington, DC, 2004, pp. 81–87.

[12] Maher M. J., and Summersgill I. A., "Comprehensive methodology for the fitting of predictive accident models", *Accident Analysis and Prevention*, 28 (3), 1996, 281–296.

[13] Shankar, V., Mannering, F., and Barfield, W., "Effect of roadway geometrics and environmental factors on rural freeway accident frequencies", *Accident Analysis and Prevention*, 27, 2005, 371–389.

[14] Cafiso, S., Montella, A., D'Agostino, C., Mauriello, F., and Galante, F., "Crash modification functions for pavement surface condition and geometric design indicators", *Accident Analysis & Prevention*, 149, 2021, 105887. DOI: https://doi.org/10.1016/j.aap.2020.105887.

[15] Sayed, T., and de Leur, P., "Appendix 2F – Evaluation of road improvement program", Road Improvement Program, *Evaluation Report*, 2001, pp. 51–56.

[16] Jovanis, P. P., and Chang, H. L., "Modeling the relationship of accidents to miles traveled", *Transportation Research Record*, 1068, *Transportation Research Board*, National Research Council, Washington, DC, 1986, pp. 42–51.

[17] Miaou, S., and Lum, H., "Modeling vehicle accident and highway geometric design relationships", *Accidents Analysis and Prevention*, 25 (6), 1993, 689–709.

[18] Hauer, E. "Overdispersion in modelling accidents on road sections and in Empirical Bayes estimation", *Accident Analysis & Prevention*, 33 (6), 2001. 799–808.

[19] Hauer, E., Ng, J., and Lovell, J., "Estimation of safety at signalized intersections", *Transportation Research Record*, 1185, Transportation Research Board, National Research Council, Washington, DC, 1988, pp. 48–61.

[20] Kulmala, R., "Safety at Rural Three- and Four-Arm Junctions. Development of Accident Prediction Models", Espoo 1995, Technical Research Centre of Finland, VTT 233, 1995.

[21] Dissanayake, S., and Ratnayake, I., "Statistical modelling of crash frequency on rural freeways and two-lane highways using negative binomial distribution", *Advances in Transportation Studies an Inter-national Journal Section B*, 9, 2006, 106345.

[22] Abdel-Aty, M. A., and Essam Radwan, A., "Modeling traffic accident occurrence and involvement", *Accident Analysis & Prevention*, 32 (5), 2000, 633–642.

[23] Pardillo Mayora, J. M., and Llamas Rubio, R., "Relevant variables for crash rate prediction in Spain's two lane rural roads", *Transportation Research Board 82nd Annual Meeting*, Washington, DC, 2003.

[24] Zhang, C., and Ivan, J. N., "Effects of geometric characteristics on head-on crash incidence on two-lane roads in Connecticut", *Transportation Research Record: Journal of the Transportation Research Board*, No. 1908, TRB, National Research Council, Washington, DC, 2005, pp. 159–164.

[25] Sutterhwaite, S., "A survey of research into relationship between traffic accidents and traffic volumes", Supplementary Report 692. Transport Research Laboratory. Crowthorne Berks, UK, 1981.

[26] McCullagh, P., and Nelder, J. A. *Generalized Linear Models*. Chapman and Hall, New York, 1989.

[27] Miaou, S. P., "Measuring the goodness of fit of accident prediction models", FHWA-RD-96-040, Federal Highway Administration, Washington, DC, 1996.

[28] Miaou, S. P., Lu, A., and Lum, H. S., "Pitfalls of using R2 to evaluate goodness of fit of accident prediction models", *Transportation Research Record: Journal of Transportation Research Board*, No. 1542, National Research Council, Washington, DC, 1995, pp. 6–13.

[29] Lord, D., and Persaud B. N., "Accident prediction models with and without trend: application of the generalized estimating equations procedure", *Transportation Research Record: Journal of the Transportation Research Board*, No. 1717, TRB, National Research Council, Washington, DC, 2000, pp. 102–108.

[30] Cafiso, S., and D'Agostino, C., "Safety performance function for motorways using generalized estimation equations", *Procedia – Social and Behavioral Sciences*, 53 (3), 2012, 901–910.

[31] Hauer, E., and Bamfo, J., "Two tools for finding what function links the dependent variable to the explanatory variables", *Proceedings of ICTCT 97 Conference*, November 5–7 1997, Lund, Sweden.

[32] *Highway Safety Manual (HSM)*. First Edition. FHWA American Association of State Highway and Transportation Officials, USA, 2010.

[33] D'Agostino, C., Cafiso, S., and Kiec, M., "Comparison of Bayesian techniques for the before–after evaluation of the safety effectiveness of short 2+1 road sections", *Accident Analysis and Prevention*, 127, 2019, 163–171. DOI:10.1016/j.aap.2019.02.009.

[34] SafetyAnalyst: Software Tools for Safety Management of Specific Highway Sites, White Paper for Module 1 – Network Screening, December 2002. http://www.safetyanalyst.org/whitepapers/module1.pdf.

[35] El-Basyouny, K., and Sayed T., *Comparison of two negative binomial regression techniques in developing accident prediction models*. Transportation Research Record: Journal of the Transportation Research Board, No. 1950, Washington, DC, 2006, pp. 9–16.

[36] Miranda-Moreno, L. F., Fu, L., Saccomanno, F. F., and A. Labbe, *Alternative risk models for ranking locations for safety improvement*, Transportation Research Record: Journal of the Transportation Research Board, No. 1908, Washington, DC, 2005, pp. 1–8.

[37] Lord, D., and Park, P. Y., "Investigating the effects of the fixed and varying dispersion parameters of Poisson-gamma models on empirical Bayes estimates", *Accident Analysis & Prevention*, 40, 2008, 1441–1457.

[38] Geedipally, S. R., Lord, D., and Park, B.-J., *Analyzing different parameterizations of the varying dispersion parameter as a function of segment length*, Transportation Research Record: Journal of the Transportation Research Board, No. 2061, Washington, DC, 2008, pp. 46–54.

[39] Hauer. E., "The road ahead", *Journal of Transportation Engineering*, 131, 2005, 333–339.

[40] Hauer E., Lovell J., and Persaud B., "New directions for learning about safety effectiveness", Report FHWA/RD-86-015, Federal Highway Administration and National Highway Traffic Safety Administration, January 1986.

[41] Lord, D., Washington, S. P., and Ivan, J. N., "Poisson, Poisson-gamma and zero-inflated regression models of motor vehicle crashes: balancing statistical fit and theory", *Accident Analysis and Prevention*, 37, 35–46, 2005.

[42] Cafiso, S., and D'Agostino, C., "Assessing the stochastic variability of the benefit-cost ratio in roadway safety management", *Accident Analysis & Prevention*, 93, 2016, 189–197. DOI:10.1016/j.aap.2016.04.027.

Chapter 4

In-depth accident investigations

John-Fredrik Grönvall[1]

4.1 Background

The first accident occurred as soon as the first automobiles were introduced to the market. Since then, with increased traffic volumes and intensity, the number of traffic accidents has increased. Also, the number of severe and fatal injuries is increasing.

Therefore, a better understanding of the causes of these accidents is essential. In-depth accident investigations are aimed at obtaining a profound understanding of the pre-, during- and post-crash phases of individual accidents. By reconstructing the accident process and considering detailed information on all aspects of the accident, a better understanding of the factors contributing to the occurrence and/or severity of the accident can be obtained. While expensive, an in-depth accident investigation can lead to unprecedented insights into how and why the accident occurred. Systematic in-depth investigations of severe accidents have a high potential for discovering hidden accident causes and suggesting corrective measures to prevent accident re-occurrence.

The investigations cover different needs, such as

Legal evidence (police, court, etc.)
Since the early days, investigations of traffic accidents have been performed by police to understand the legal responsibility of the accidents. During the 1950s, several methods were developed in many Western countries by police and local road authorities.

For such investigations, mostly performed by legal entities such as police, the purpose is mostly to collect evidence for the guiltiness of the accident. This evidence can be used in court or by any other means to clarify who was guilty of the accident. Insurance companies sometimes have a similar approach to in-depth investigations.

Learning and product development (original equipment manufacturer (OEM), etc.)
Many vehicle manufacturers have created in-depth accident investigating teams; as an example, Volvo created an Accident Research Team in 1970. In their existing accident database, there were more than 70 000 accidents in

[1]Chalmers Industriteknik, Sweden

2020 involving Volvo cars and/or trucks. Many other manufacturers in Europe and the US have similar teams. These accidents are normally used to understand the performance of existing safety measures and to provide feedback for the development of future safety systems.

Research (academia, manufacturer, insurance, etc.)
Some in-depth databases are also used for academic or industrial research on traffic safety. For example, databases such as GIDAS or iGLAD might be used for such research. Another resource that OEMs could use is the INTACT database, which contains objective data from on-scene investigations as well as deeper case analysis.

4.2 Methodologies

4.2.1 Different needs of methods and techniques

In-depth accident investigations are normally conducted on-scene at the accident sites, when evidence and pictures can be collected directly after the accident. If it is not possible to reach the location in time, there are also possibilities to perform investigations retrospectively, either by using already collected data by the emergency response team or by visiting the location at a later stage. Another type of investigation would be the special investigations on severe, unusual accidents such as bus accidents with many fatalities. Because of their unique conditions, these events are typically investigated individually at the national level by special accident investigation teams.

4.2.2 Background investigation

During the investigations, it is important to process findings methodologically. One often-used way to organize the findings is to use the Haddon matrix. The Haddon matrix is a framework used in product development and research in injury prevention and public health. It was developed by Dr. William Haddon, a public health expert and physician, in the 1960s. The matrix makes it easier to analyse and understand factors contributing to injuries, particularly in motor vehicle accidents. It can also be applied to other types of incidents. The matrix organizes factors influencing injury outcomes into a 3×3 grid based on two dimensions:

Phases of an event (columns):
Pre-event: Factors that contribute to the risk of an event occurring (e.g. speed limits, road conditions, driver behaviour).
Event: Factors that influence the event itself and how it unfolds (e.g. vehicle safety features like airbags, seatbelts or crash barriers).
Post-event: Factors affecting the outcomes after the event has occurred (e.g. emergency response, availability of medical care, quality of trauma centres).

Types of factors (rows):
Human factors: Characteristics of the people involved, such as driver behaviour, knowledge, skill, fatigue or alcohol use.

Vehicle (or equipment) factors: Characteristics of the physical objects involved, like vehicle safety features or road infrastructure.
Environmental factors: The external environment, including road conditions, weather and traffic laws.
Figure 4.1 shows NTHSA uses the Matrix in Report No. FHWA-SA-09-029.

Normally, all in-depth accidents use a similar methodology with a four-step approach:

(1) Background investigation to understand the circumstances during the accident.
(2) Site investigation, measurements and photos to collect objective data.
(3) Interviews of involved persons, as soon as possible, to get unbiased witness information from the accident scene.
(4) Analysis, conclusion and reporting to conclude the collected information and store information in a database for future analysis. In many cases, causation of the accident is also included in this stage. It is important that the conclusions are not influenced or biased by the investigator's early judgement. The future analysis could then be statistical or case studies for further conclusions.

4.2.3 On-scene investigation

With an accident investigation team, there is a possibility to reach the accident scene 24/7 directly after the accident; typically, together with the emergency team, such as the police and ambulance. In this case, the accident investigation team can observe the accident scene and take measurements, with the crashed vehicles still in their rest positions. Another advantage is that witnesses can be interviewed

PERIOD	HUMAN	VEHICLE/ EQUIPMENT	PHYSICAL ENVIRONMENT	SOCIO-ECONOMIC
PRE-CRASH	Poor vision or reaction time, alcohol, speeding, risk taking	Failed brakes, missing lights, lack of warning systems	Narrow shoulders, ill-timed signals	Cultural norms permitting speeding, red light running, DUI
CRASH	Failure to use occupant restraints	Malfunctioning safety belts, poorly engineered air bags	Poorly designed guardrails	Lack of vehicle design regulations
POST-CRASH	High susceptibility, alcohol	Poorly designed fuel tanks	Poor emergency communication systems	Lack of support for EMS and trauma systems

Figure 4.1 Haddon matrix for crashes in an urban area (ref. NHTSA Report No. FHWA-SA-09-029)

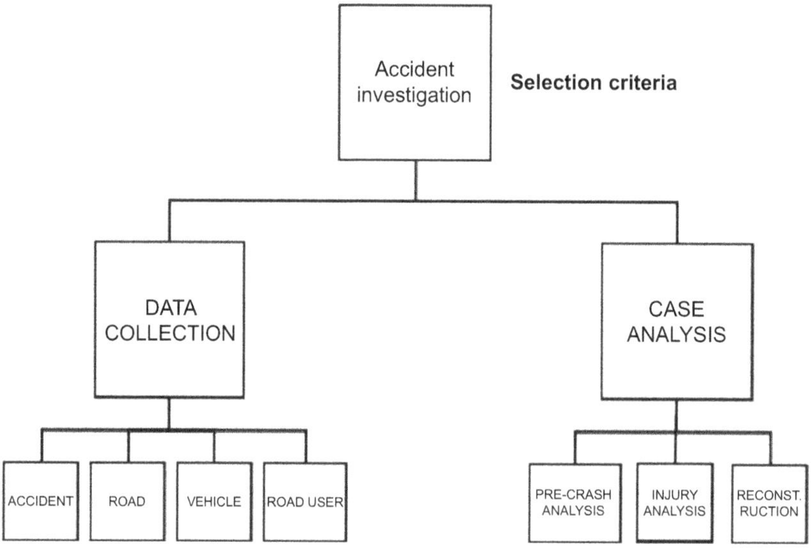

Figure 4.2 Two parts of an INTACT accident investigation

immediately, and their witnesses are less biased. An obvious disadvantage is of course the large cost for personal resources and equipment that need to be on stand-by.

As an example of on-scene investigations, one could look at the Swedish research project INTACT (Figure 4.2). In this research project, the accident investigation process consists of two parts: data collection and case analysis, which are conducted in a structured manner.

Site investigation, measurements and photos

On the scene, the investigators need to talk to police, rescue services, persons involved in the accident and witnesses. The investigator then enters the course of events, contact information and general accident information into the on-scene accident-vehicle protocol. The vehicle data will also be put into the on-scene accident-vehicle protocol.

The investigators also need to document photos and measurements of the vehicle rest position(s), collision point(s)/collision object(s) and other physical traces. On the scene is important to mark vehicle rest position(s), collision point(s) and vehicle/object traces and take photos of all marked areas. Then, a sketch of the accident site, vehicle paths and end positions will be done.

One example of a data collection form for objective data collection is from 'Uniform Crash Manual – Florida Department of Highway Safety and Motor Vehicles 2019', where vehicle data is being collected in the on-scene investigation (Figure 4.3).

Vehicle Page

Vehicle Section: This section is designed to identify vehicle information. The following are instructions for entering data into the vehicle section of the Florida Traffic Crash Report (HSMV 90010S). The vehicle data elements are used to describe the characteristics, events and consequences of the motor vehicles involved in the traffic crash.

VEHICLE # ☐ **Check if Commercial** ☐ | REPORTING AGENCY CASE NUMBER | HSMV CRASH REPORT NUMBER

1 Vehicle in Transport
2 Parked Motor Vehicle
3 Working Vehicle ☐ | VEHICLE LICENSE NUMBER | STATE | REGISTRATION EXPIRES | Check if Permanent Registration ☐ | VIN

Hit and Run
1 No
2 Yes
88 Unknown ☐ | YEAR | MAKE | MODEL | STYLE | COLOR | DAMAGE: 1 Disabling 4 Minor 2 Functional 88 Unknown 3 None ☐ | EST. AMOUNT

INSURANCE COMPANY | INSURANCE POLICY NUMBER | Towed due to Damage: 1 No 2 Yes ☐ | VEHICLE REMOVED BY | 1 Rotation 2 Owner Request 3 Driver 77 Other, Explain in Narrative ☐

NAME OF VEHICLE OWNER (Check if Business) ☐ | CURRENT ADDRESS | CITY & STATE | ZIP CODE

TRAILER # | LICENSE NUMBER | STATE | REGISTRATION EXPIRES | Check if Permanent Registration ☐ | VIN | YEAR | MAKE | LENGTH | AXLES

TRAILER # | LICENSE NUMBER | STATE | REGISTRATION EXPIRES | Check if Permanent Registration ☐ | VIN | YEAR | MAKE | LENGTH | AXLES

VEHICLE TRAVELING N ☐ S ☐ E ☐ W ☐ Off-Road ☐ Unknown ☐ | ON STREET, ROAD, HIGHWAY | AT EST. SPEED | POSTED SPEED | TOTAL LANES

HAZ. MAT. RELEASED 1 No 2 Yes 88 Unknown ☐ | HAZ. MAT PLACARD 1 No 2 Yes 88 Unknown ☐ | HAZ. MAT. NUMBER | HAZ. MAT. CLASS

Area of Initial Impact

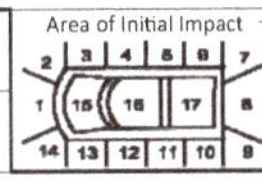

18 Undercarriage 18
19 Overturn 19
20 Windshield 20
21 Trailer 21

Most Damaged Area

MOTOR CARRIER NAME | US DOT NUMBER

MOTOR CARRIER ADDRESS | CITY & STATE | ZIP CODE | PHONE NUMBER

Vehicle Body Type ☐
1 Passenger Car
2 Passenger Van
3 Pickup
7 Motor Home
8 Bus
11 Motorcycle
12 Moped
13 All Terrain Vehicle (ATV)
15 Low Speed Vehicle
16 (Sport) Utility Vehicle
17 Cargo Van (10,000 lbs (4,536 kg) or less)
18 Motor Coach
19 Other Light Trucks (10,000 lbs (4,536 kg) or less)
20 Medium/Heavy Trucks (more than 10,000 lbs (4,536 kg))
21 Farm Labor Vehicle
77 Other, Explain in Narrative
88 Unknown

Trafficway ☐
1 Two-Way, Not Divided
2 Two-Way, Not Divided, with a Continuous Left Turn Lane
3 Two-Way, Divided, Unprotected (painted >4 feet) Median
4 Two-Way, Divided, Positive Median Barrier
5 One-Way Trafficway
88 Unknown

Commercial Motor Vehicle Configuration ☐
1 Vehicle 10,000 lbs or less Placarded for Hazardous Materials
2 Single-Unit Truck (2-axle and GVWR more than 10,000 lbs (4,536 kg))
3 Single-Unit Truck (3 or more axles)
4 Truck Pulling Trailer(s)
5 Truck Tractor (bobtail)
6 Truck Tractor/Semi-Trailer
7 Truck Tractor/Double
8 Truck Tractor/Triple
9 Truck more than 10,000 lbs (4,536 kg), Cannot Classify
10 Bus/Large Van (seats for 9-15 occupants, including driver)
11 Bus (seats for more than 15 occupants, including driver)
77 Other, Explain in Narrative
88 Unknown

Comm/Non-Commercial ☐
1 Interstate Carrier
2 Intrastate Carrier
3 Not in Commerce/Government
4 Not in Commerce/Other Truck

Trailer Type TRAILER 1 ☐ TRAILER 2 ☐
1 Single Semi Trailer
2 Tandem Semi Trailer
3 Tank Trailer
4 Saddle Mount/Trailer
5 Boat Trailer
6 Utility Trailer
7 House Trailer
8 Pole Trailer
9 Towed Vehicle
10 Auto Transport
77 Other, Explain in Narrative
88 Unknown

Cargo Body Type ☐
1 No Cargo
2 Bus
3 Van/Enclosed Box
4 Hopper
5 Pole-Trailer
6 Cargo Tank
7 Flatbed
8 Dump
9 Concrete Mixer
10 Auto Transport
11 Garbage/Refuse
12 Log
13 Intermodal Container Chassis
14 Vehicle Towing Another Vehicle
15 Not Applicable (vehicle 10,000 lbs (4,536kg) or less not displaying HM placard)
77 Other, Explain in Narrative
88 Unknown

Comm GVWR/GCWR ☐
1 10,000 lbs (4,536 kg) or less
2 10,001-26,000 lbs (4,536-11,793 kg)
3 More than 26,000 lbs (11,793 kg)
4 Not Applicable

Most Harmful Event ☐

Sequence of Events 1st ☐ 2nd ☐ 3rd ☐ 4th ☐

Non-Collision
1 Overturn/Rollover
2 Fire/Explosion
3 Immersion
4 Jackknife
5 Cargo/Equipment Loss or Shift
6 Fell/Jumped From Motor Vehicle
7 Thrown or Falling Object
8 Ran into Water/Canal
9 Other Non-Collision

[40-46 Sequence of Events only]
40 Equipment Failure (blown tire, brake failure, etc.)
41 Separation of Units
42 Ran Off Roadway, Right
43 Ran Off Roadway, Left
44 Cross Median
45 Cross Centerline
46 Downhill Runaway

Collision with Non-Fixed Object
10 Pedestrian
11 Pedalcycle
12 Railway Vehicle (train, engine)
13 Animal
14 Motor Vehicle in Transport
15 Parked Motor Vehicle
16 Work Zone/Maintenance Equipment
17 Struck By Falling, Shifting Cargo or Anything Set in Motion by Motor Vehicle
18 Other Non-Fixed Object

Collision Fixed Object
19 Impact Attenuator/Crash Cushion
20 Bridge Overhead Structure
21 Bridge Pier or Support
22 Bridge Rail
23 Culvert
24 Curb
25 Ditch
26 Embankment
27 Guardrail Face
28 Guardrail End
29 Cable Barrier
30 Concrete Traffic Barrier
31 Other Traffic Barrier
32 Tree (standing)
33 Utility Pole/Light Support
34 Traffic Sign Support
35 Traffic Signal Support
36 Other Post, Pole, or Support
37 Fence
38 Mailbox
39 Other Fixed Object (wall, building, tunnel, etc.)

Emergency Vehicle Use ☐
1 No
2 Yes
88 Unknown

Roadway Grade ☐
1 Level
2 Hillcrest
3 Uphill
4 Downhill
5 Sag (bottom)

Roadway Alignment ☐
1 Straight
2 Curve Right
3 Curve Left

Vehicle Maneuver Action ☐
1 Straight Ahead
3 Turning Left
4 Backing
5 Turning Right
6 Changing Lanes
8 Parked
10 Making U-Turn
11 Overtaking/Passing
13 Stopped in Traffic
14 Slowing
15 Negotiating a Curve
16 Leaving Traffic Lane
17 Entering Traffic Lane
77 Other, Explain in Narrative
88 Unknown

Traffic Control Device For This Vehicle ☐
1 No Controls
4 School Zone Sign/Device
5 Traffic Control Signal
6 Stop Sign
7 Yield Sign
8 Flashing Signal
9 Railway Crossing Device
10 Person (including Flagman, Officer, Guard, etc.)
13 Warning Sign
77 Other, Explain in Narrative
88 Unknown

Vehicle Defects ☐ ☐
1 None
2 Brakes
3 Tires
4 Lights (head, signal, tail)
6 Steering
7 Wipers
9 Exhaust System
10 Body, Doors
11 Power Train
12 Suspension
13 Wheels
14 Windows/Windshield
15 Mirrors
16 Truck Coupling/Trailer Hitch/Safety Chains
77 Other, Explain in Narrative
88 Unknown

Special Function of Motor Vehicle ☐
1 No Special Function
2 Farm Vehicle
3 Police
7 Taxi
8 Military
9 Ambulance
10 Fire Truck
11 Farm Labor Transport
12 School Bus
13 Transit/Commuter Bus
14 Intercity Bus
15 Charter/Tour Bus
16 Shuttle Bus
17 Farm Labor Bus
88 Unknown

VIOLATIONS

PERSON #	NAME OF VIOLATOR	FL STATUTE NUMBER	CHARGE	CITATION NUMBER
PERSON #	NAME OF VIOLATOR	FL STATUTE NUMBER	CHARGE	CITATION NUMBER
PERSON #	NAME OF VIOLATOR	FL STATUTE NUMBER	CHARGE	CITATION NUMBER

HSMV 90010 S (V/P) (rev 06/13) Page ___ of ___

Figure 4.3 Uniform Crash Manual – Florida Department of Highway Safety and Motor Vehicles 2019

4.2.4 Retrospective investigation

If on-site investigation is not possible directly after the accident, a retrospective investigation is many times efficient. It is often an effective way to collect data if the emergency team has taken objective measures and photo documentation. The need for high-quality collection of objective data is obvious if the investigation team shall be successful in understanding the accident circumstances. If the initial data collection is done with high quality, this could be a way to create an accident database with less personal cost and effort.

4.2.5 Special Crash Investigations (SCIs, severe accidents with special attendance)

These SCIs are normally run by governmental bodies for larger, severe accidents. Typically, bus accidents with many fatalities, air-bag investigations, etc. There are several different Special Crash Investigation teams, that is

- US: NHTSA SCI,
- Swe: Swedish Accident Investigation Authority (Statens haverikommission, SHK).

4.3 Selection/sampling criteria

When the database is created, it is of great importance to define the purpose of the database and its scope. The different types of investigations mentioned above normally will need different selection criteria. The selection criteria are normally not the same in different databases and therefore need to be understood. For example, if the purpose is to collect evidence for a lawsuit, the selection criteria will be different from the creation of a database for OEM developments. Some databases use tow-away as the selection criteria, whilst others might use severe injury or fatal injury as a selection criterion. These different types of criteria will create a bias in the content of the database.

4.4 Interviews of involved persons

To better understand the cause of the accident, interviews are conducted in a structured way. The questions should describe the accident and provide an objective understanding of the situation.

One example is from a manual, 'Possible Accident Investigation Questions' (Minnesota Department of Administration), when they use a template for interviewing persons involved in accidents:

HOW

- How does the injured person feel now?
- How did the injury occur?
- How could this accident have been prevented?

WHO

- Who was injured?
- Who witnessed the accident?
- Who was working with the injured person?
- Who had assigned the person to the work task?
- Who had trained the person on the hazards and protective measures for this task?
- Who else was involved?

WHAT

- What were the causal factors of the accident?
- What were the injuries?
- What was the person doing when injured?
- What had the person been instructed to do?
- What tools was the person using?
- What machinery/equipment was involved?
- What was the condition of the machinery/equipment involved?
- What training had been given?
- What specific precautions were necessary?
- What personal protective equipment was being used?
- What personal protective equipment should have been used?
- What will be done to prevent a recurrence?
- What safety rules were in place to prevent this type of accident?
- What safety rules were being followed?
- What were the environmental conditions (e.g. lighting, floor surface, etc.)?

WHEN

- When did the accident occur?
- When did the person start this task?
- When was the person assigned to this department?
- When had the supervisor last checked on the job progress?

WHY

- Why were the persons injured?
- Why did the person do what he/she did?
- Why wasn't protective equipment used?
- Why weren't specific instructions issued?

WHERE

- Where did the accident occur?
- Where was the person at the time of the accident?

4.5 Variables in the database

A comprehensive traffic accident database must normally include a variety of variables to accurately capture the details of each accident and enable thorough analysis for safety improvements and preventive measures. Dependent on the usage of the database for different reasons, i.e. product development, national statistics or research, the content might vary. In most cases, some essential variable groups are needed. These variables can be grouped into several categories: crash details, vehicle information, human factors, environmental factors and post-crash outcomes.

4.5.1 Crash details

- Date and Time of Accident: Including day, month, year and time of day. This helps analyse patterns related to traffic volume, daylight and time-specific behaviours.
- Location (Geographical Coordinates): Provides the exact location of the crash, allowing for analysis to identify high-risk areas and blackspots.
- Accident Type: Specifies the type of collision (e.g. rear-end, head-on, side-impact, rollover, pedestrian accidents).
- Collision Description: A brief narrative or code describing the sequence of events leading up to and during the crash.
- Number of Vehicles Involved: Identifies the number of vehicles or other road users involved in the crash.
- Accident Severity: Coded as minor, serious or fatal to assess the seriousness of the crash.
- Impact Points: Describes the point of impact on the vehicles to determine the nature of the collision.

4.5.2 Vehicle information

- Vehicle Type, which indicates the type of vehicle involved (e.g. car, truck, motorcycle, bicycle, bus).
- Make, Model and Year to provide specific information about the vehicles involved, which can be useful for identifying patterns with certain vehicle types or models.
- Vehicle Condition, which describes the condition of the vehicle before the accident
- Safety Features, which specify whether safety features like airbags, seatbelts or electronic stability control were present and deployed during the crash.

4.5.3 Human factors

- Driver Information: Includes the age, gender, license status and driving experience of the drivers involved.

- Occupant Information, with data on all vehicle occupants, such as whether they were passengers or drivers and whether they were using safety devices like seatbelts.
- Driver Behaviour, which describes driver actions leading up to the crash (e.g. speeding, distracted driving, drunk driving, etc.).
- Driver Condition, which will indicate whether the driver was impaired by alcohol, drugs, fatigue, or any medical condition at the time of the crash.
- Driver Reaction: How the driver reacted before the accident (e.g. braking, swerving or no reaction).
- Toxicology Reports: Results of alcohol or drug tests for drivers, which can indicate impaired driving.

4.5.4 Environmental factors

- Weather Conditions: Specifies weather conditions at the time of the crash (e.g. clear, rainy, foggy, snowy) that could have contributed to the accident.
- Lighting Conditions: Describes the lighting at the scene (e.g. daylight, dusk, night with or without streetlights).
- Road Conditions: Describes the state of the road surface (e.g. dry, wet, icy, gravel).
- Road Type: Specifies the type of road (e.g. highway, urban road, rural road) and number of lanes.
- Speed Limit: Indicates the posted speed limit at the crash location.
- Traffic Control Devices: Describes the presence of traffic lights, stop signs, pedestrian crossings or other traffic controls.
- Road Design and Geometry: Describes road design, such as intersections, curves, ramps or roundabouts, which could influence accident risk.
- Obstructions to Visibility: Any obstructions (e.g. trees, parked vehicles, construction) that may have affected the driver's ability to see.

4.5.5 Post-crash outcomes

- Injuries and Fatalities: Detailed records of the number and severity of injuries, including data on whether they were minor, serious or fatal, and the type of medical response required.
- Emergency Response Time: The time it took for emergency services to arrive on the scene after the crash.
- Hospitalization Data: Information on whether and where individuals were transported for medical treatment.
- Damage Assessment: Estimates of property damage, including the vehicles involved and any other damaged property (e.g. buildings, infrastructure).
- Insurance Information: Details about the insurance status of the vehicles and individuals involved for claims processing and liability determination.

4.5.6 Legal and regulatory variables

- Traffic Violations: Information on any traffic violations committed by drivers (e.g. speeding, failure to yield, running a red light).
- Fault Determination: A preliminary assessment of fault or legal liability, often made by law enforcement or insurance investigators.
- Accident Reporting Officer: Details about the officer or authority who reported or investigated the accident, including contact information for follow-up.

4.5.7 Other variables

- Time of day and traffic conditions that describe the general traffic flow at the time of the accident
- Statements from witnesses can provide additional context and details not captured in physical evidence.
- In cases where dash cams or traffic cameras recorded the event, this data can be crucial for reconstructing the crash.
- Roadwork in progress that may have affected traffic patterns or road safety.

4.6 Case analysis

Dependent on the reason for the investigation (legal guilt, product development, improvement of infrastructure, etc.). Case analysis in traffic accident investigations involves a systematic approach to understanding the causes, contributing factors and consequences of an accident. Investigators use various methods and tools to collect, analyze and interpret data to determine what happened, why it happened and how future accidents can be prevented.

A root cause analysis should be performed during the case analysis:

Primary Cause: The investigation identifies the primary cause of the accident, which could be human error, mechanical failure or environmental factors.

Contributing Factors: Investigators also identify secondary factors that contributed to the accident but were not the main cause, such as inadequate road maintenance or driver distraction.

Use of the Haddon Matrix: The Haddon Matrix (described earlier) may be applied here to analyse human, vehicle and environmental factors across the pre-event, event and post-event phases.Depending on the reason for the investigation, different legal or regulatory considerations might be analysed. Investigators will check if the involved vehicles comply with all relevant compliance safety regulations. The investigators will determine if any traffic regulations and laws were violated (e.g. running a red light) and whether these violations caused the accident. After the case analysis, and sometimes, specific analysis meetings, an Accident Report is prepared. The findings of the investigation are compiled into a formal report, which includes a detailed narrative of what happened, diagrams or reconstructions, and conclusions about the causes of the accident. If necessary, experts such as accident reconstruction specialists or forensic engineers may be consulted to provide additional analysis or verification of findings. The report most often includes recommendations for preventing similar accidents in the future, such as road improvements, changes to traffic laws or vehicle safety enhancements.

4.7 Representativity of collected data

The question of representativeness in a database is very delicate. An accident database only becomes representative when it accurately reflects the diversity, distribution, and characteristics of the population or events being studied. In the databases of traffic accidents, it is important to have a representative database, or at least understand the bias, for accurate decisions on the identification of trends and designing effective countermeasures to reduce accidents. Below are a few ways to try to understand how the accident databases can be made representative.

4.7.1 Comprehensive data collection

The database should include accidents of varying severity (e.g. minor, serious, and fatal accidents) and involve different types of participants (drivers, passengers, pedestrians, cyclists). This is typically done by OEMs, where the collected data is used to calculate risk curves and thereby understand how an introduced safety system affects the outcome of the number of accidents. Accidents from all areas, such as urban, suburban, and rural areas, should be included in this case. As a result, the database can represent different traffic environments and conditions. The data should cover a broad range of accident types (e.g. vehicle collisions, rollovers, pedestrian incidents) to provide a complete view.

4.7.2 Random sampling

To prevent over-representation of certain types of accidents (like high-profile or fatal ones), the database should use random sampling methods to capture a more accurate distribution of incidences. For example, scheduled time sampling, where data is collected over different times of day, days of the week and seasons of the year to ensure that the data reflects varying traffic volumes and environmental conditions.

4.7.3 Demographic representation

The collected data must include information, such as age, gender and socio-economic status, on people from all demographic groups who may be involved in accidents, ensuring that the database reflects the population. The investigators might need to include data for various vehicle types (cars, trucks, motorcycles, bicycles) to account for differences in accident risk.

4.7.4 Geographic and environmental factors

The database should include data from different regions, accounting for differences in road design, traffic laws, climate and terrain. Based on the purpose of the database, the sampling region might be different. In some cases, it might be interesting to cover only a small region or city if the user needs data for this region. In other cases, the data coverage might be global coverage, as in iGLAD, where many national databases are normalized and merged into a global accident database (Bakker *et al.*, 2017). Accidents occurring under different weather conditions (e.g. snow, rain, clear skies) and road conditions (e.g. highways, rural roads, city streets) should be captured.

4.7.5 Standardized data collection methods

Using consistent criteria over time for collecting and recording accident details ensures that the data can be compared across different regions, time periods and demographic groups. Data on factors like driver behaviour, vehicle characteristics and environmental conditions should be consistently recorded to allow for meaningful analysis.

References

Bakker, J., Jeppson, H., Hannawald, L., Spitzhüttl, F., Longton, A., and Tomasch, E. (2017). IGLAD – International Harmonized In-Depth Accident Data. In *ESV Conference Proceedings*. National Highwy Traffic Safety Administration.

Highway Safety Improvement Manual NHTSA fhwasa09029

INTACT Final report: Methodology Development for Advanced Accident Investigations – Ver. 1.0.

Safety Accident Investigation Questions and Needed Info – Minnesota Department of Administration tcm36-208795.

Uniform Crash Manual – Florida Department of Highway Safety and Motor Vehicles 2019.

Chapter 5
Healthcare data

Niels Bos[1], *Aliaksei Laureshyn*[2] *and Carl Johnsson*[2]

In this chapter, we will discuss the limitations of police accident data and the added value of health data as a complementary source. Several types of health organisations exist that cover a certain severity of traffic casualties. A number of medical indicators will be discussed, as well as methods to estimate the true number of serious injuries and the completeness of the police reporting. For two countries (the Netherlands and Sweden), the processes, research options and lessons learned will be discussed more comprehensively.

5.1 Limitations of police accident records

Traditionally, police records serve as the source for the official statistics on road traffic accidents. However, there are good reasons why they are not fully representative of the scale of the road safety problem and should be complemented with data from other sources, such as health records.

Since minimizing the health consequences of an accident is an obvious priority, medical help is often called first. Only secondary comes the question of traffic offences and liability, in which police officers play an important role – next to their role of guarding traffic along the accident site. Therefore, it is not surprising that medical registrations are generally more complete for (severe) injuries. Also, police officers tend to report the traffic rules that have been violated in order to facilitate the handling of the accident in court and by insurers. The juridical view of an accident with one main cause hinders finding other contributing factors such as errors in road design – relevant for accident prevention. For this type of information, it is often necessary to do in-depth accident investigations (see Chapter 4).

Furthermore, not all safety-related incidents in traffic are legally required to be reported at all (Figure 5.1). Usually, the definition of a road traffic accident includes criteria like: (i) it must happen on a public road, suddenly and unintentionally; (ii) it involves at least one moving vehicle; and (iii) the severity of consequences must be above a certain threshold (person injuries are more than

[1]Department Data and Analysis for Policy, SWOV Institute for Road Safety Research, Netherlands
[2]Department of Technology and Society, LTH, Lund University, Sweden

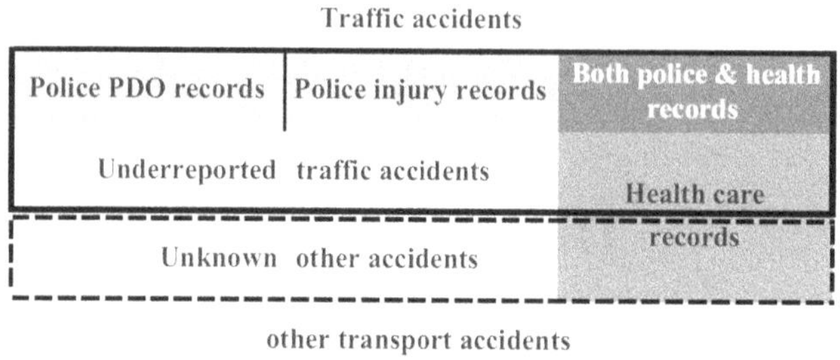

Figure 5.1 *Road traffic accidents and other road transport accidents and their coverage by police and health care records (not to scale): PDO = property damage only*

minor cuts and bruises or the property damage is sufficiently large) (Eurostat *et al.* 2019). In some countries accidents resulting in property damage only (PDO) are not reported by police; then, insurers completely cover the liability and financial handling of these crashes.

The traffic accident definition can be (and, at least within the research community, actually is) questioned (e.g. Methorst *et al.* 2017). Why are non-public (private) roads excluded? Why is a moving vehicle – in some countries even limited to a *motor* vehicle –a necessary requirement? And why, unlike in aviation and rail transport, are possibly serious near-crashes not considered?

Answers to these questions can be found in arguments such as that a government agency (public road owners) cannot be responsible for cases that happen outside of their jurisdictions; however, from a public health point of view, it does not really matter if someone crashes on their own carport or on a mountain bike trail. The moving vehicle criterion excludes injuries when departing or entering a stopped bus and getting on/off a bicycle or moped, even when that is occurring in traffic (e.g. at traffic lights); even though getting on/off a bicycle involves motion, the definitions in ICD10* are not clear about whether this is a 'traffic accident' or an 'other transport accident' (WHO 2019). It also excludes pedestrian falls that occur on slippery pavements or uneven surfaces while they are – just as all other traffic – moving from A to B; such accidents, however, contribute up to 70% of all pedestrian injuries in traffic (Eriksson *et al.* 2022; Methorst *et al.* 2017). The reporting of near-misses is not done systematically since it will greatly increase the number of cases to be investigated and reported. Unlike in the aviation and railway domains, the number of actual road accidents is still huge, while some doubts exist on whether the mechanisms to prevent near-misses would also prevent serious accidents (read more on that in Chapter 9 on 'Surrogate measures of safety: site-based observations').

Not all accidents that are reportable find their way into the police records either. Usually, the fatality numbers are the most reliable, even though minor deviations from the true count still occur. For example, according to the current

*See more about ICD classification in Section 5.2.3.

definition used in European Union countries, a traffic fatality occurs within 30 days after the accident. This requires a check-up at the end of this period before the accident record can be finalised, a procedure that might sometimes not be performed properly.

However, traffic fatalities represent only the top of the iceberg in all the negative health and economic impacts of traffic accidents. The non-fatal traffic injuries provide a significantly larger bulk of data, allowing for a more nuanced picture of the situation and trends within the safety domain. For example, while traffic fatalities are generally decreasing in Western countries, the number of serious injuries is not decreasing at the same rate, and in some countries, it even increases (ITF 2011). In a way, the fatalities have 'converted' into severe injuries, which can be seen only as a minor improvement as many people are still suffering greatly.

Attempts to compare police and hospital records clearly show that there is only a partial overlap between them (Figure 5.1). The share of accidents unreported by police increases as the accident severity decreases. The police reporting rate varies greatly between countries (ITF 2011). Such variation, aggravated by the differences in the definitions and the reporting systems used, complicates the international comparison of road safety performance. However, what is even more important, a significant variation in under-reporting is systematically found within individual countries, mainly related to accident types, road users involved, road categories and the type of road environment, type of police force involved, etc. (Amoros *et al.* 2006; Alsop and Langley 2001).

Figure 5.2 illustrates the last point with an extract from the Swedish accident database Strada that contains both police and hospital records. For injury accidents involving motor vehicles, the police reporting rate reaches 50%–70%. However, single bicycle accidents, constituting up to 70% of all bicyclists' severe injuries (Eriksson *et al.* 2022; Niska *et al.* 2013), are hardly known to the police. The pedestrian falls, as mentioned earlier, are not officially included in the accident definition and, despite being the major single cause of injuries in traffic (including severe injuries), are only known through hospital records (Eriksson *et al.* 2022; Berntman 2015).

There is also a difference in the data available in police and hospital records. Usually, a police report contains information about the time and location of the accident, drivers, vehicles and manoeuvres involved, weather and road conditions, etc. However, a police officer lacks proper expertise to judge the injuries and their consequences, providing only a very simple (yet still often found erroneous) classification of the injury severity (e.g. fatal, serious and slight injuries). On the opposite, a hospital record contains detailed information on the person (age, gender, etc.), a description of the injuries, their severity and potential consequences, the duration of the hospital admission and so on. The description of the accident circumstances, conversely, is limited to what the injured person has told (or sometimes is as simple as 'External cause' according to ICD10 classification). From this perspective, merging the two data sources contributes to a better understanding of the accident conditions and contributing factors as well its outcomes.

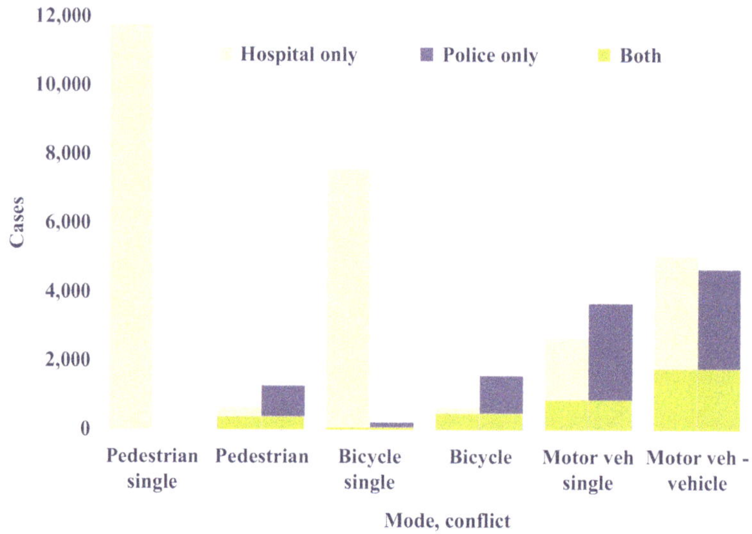

Figure 5.2 Reporting by police and health care for different accident types (all injury severity levels, based on data from the Swedish accident database Strada for 2019)

5.2 Healthcare data

Healthcare data is first collected to optimise the care, diagnose and treatment process. Hospitals use the data for the planning of medical staff throughout the week (but also seasonal fluctuations or events, such as fireworks accidents during New Year celebrations), for comparing different treatments, for financial handling with patients and care insurers, for benchmarking between different hospitals, etc. Examples are the Hospital Standardised Mortality Ratio, HSMR (Jarman *et al.* 1999), to compare the number of fatalities in a hospital by standardising over a number of characteristics such as diagnose, age and gender and the billing based on diagnoses (ICD, United States) or diagnose-treatment combinations (DBC, the Netherlands).

Vital patient parameters, such as body temperature, blood pressure, exact type and time at which certain medications were administered, are of great importance to follow and optimise the care process, however of less importance for road safety research. Conversely, the external cause that caused a patient to break his leg is of lesser importance for the treatment in a hospital where it is crucial information to select injuries by traffic accidents from other external causes.

In principle, healthcare ranges from the general practitioner, emergency medical services (ambulances), accident and emergency (A&E) departments, clinical admissions, intensive care to rehabilitation centres and hospices. Generally, the 'highest' care given is described as the 'severity' of a patient, however within that several nuances are possible as we will see later. For assessing the total care demand or cost of a traffic casualty, be aware that for example for a person hospitalised there is not just the hospital stay. There is likely also an ambulance trip involved, A&E treatment and several follow-up meetings with the medical specialist and general practitioner, physiotherapist or psychological help later in the process of cure.

Data harmonisation in the field of healthcare is an ongoing process. For Hospital Discharge Reports or Accident and Emergency visits, possibly only a few countries integrate the information from all hospitals into one register. Countries in the European Union report on the number of hospital discharges by a limited number of characteristics to Eurostat (n/d). Global information on the number of injuries is only roughly available. International organisations, such as WHO, take a lot of effort to estimate the global number of traffic injuries (WHO 2023) and its burden on society (GBD n/d, 2024), both partially based on models to deal with missing information and differences in definitions and reporting.

5.2.1 Potentials of healthcare data for road accident research

First, applications of health data for road traffic accident research were the verification of the number of casualties as reported by the police.

Police records can distinguish severity scales between killed, transported to hospital, treated on the spot and superficial or no injury. Depending on the regulations in the country, health personel is allowed to communicate to the police about to which hospital the casualty is transported and whether this resulted in hospitalisation or just treatment at the A&E department after which the person returned home. On the severe spectrum of outcome, it may also be possible that the casualty died (within 30 days). Communication about specific injuries from health to police is not often the case, even on request by the police. In the UK, police reports injuries from a specific list (Department of Transport n/d); however, the comparability between medical injuries and AIS-severity is not always consistent (Department of Transport 2020, pp. 30–35). In the US, a KABCO scale (NHTSA 2024; MMUCC 2012) is used to differentiate between (K) fatal injury, (A) suspected serious injury, (B) suspected minor injury, (C) possible injury, and (O) no apparent injury. This makes it difficult to compare the numbers of traffic casualties hospitalised from health data as this cannot easily be selected from the police data; the patients concerned may have a different severity in the police system. Only by extensive breakdown (e.g. comparing the number of car occupants between 18 and 24 years old in a certain region and month) or by data matching (one-to-one pairing of police and health registrations) the true overlap will become apparent and also the size of underreporting and misclassification of severity.

In police reporting, but also on hospital admission, the state of the casualty immediately after the crash is captured and registered as severity. It makes sense to look at the final state of a casualty at hospital discharge or even after 2 or 5 years (compare five-year survival after cancer diagnosis). Follow up on casualties to assess their severity 5 years later, makes a timely analysis of road safety problems quite difficult, so for effective road safety analysis and policy development of road safety measures, predictors for complete recovery and partial impairment are needed.

Healthcare data is rich and should not be reduced just to label casualties as 'not injured', 'slightly injured', 'seriously injured' and 'killed'. The way healthcare data in road safety is interpreted is sometimes different from 'classical' health care. Fatalities are counted at the time of the crash in road safety, whereas in classical health care it is at the time of death. The internationally agreed 30-day definition of road safety may not be applicable from a health perspective. Therefore, the number of traffic fatalities in a year may be somewhat different from the number of people who died due to traffic accidents in a year.

A broader use of healthcare data includes

- assessment of the type, location and severity of injuries sustained (section 5.2.3);
- impairment and burden of injury (section 5.2.3);
- care demand;
- medical costs of traffic injuries;
- completeness of accident data (police and hospital) (section 5.2.4);
- bias of police data (section 5.2.4).

5.2.2 Types of hospital records, availability

By 'hospital data', we usually mean that it concerns information on traffic casualties that were hospitalised for at least one night as in-patients. However, before a casualty reaches hospital, several routes can be taken as illustrated in the flowchart from the Dutch Healthcare Authority (Figure 5.3) (NZA 2022). The exact scheme and naming are specific to the Netherlands; however, similar complexity could be expected in most countries.

When a person requires acute medical care, most patients (88%, Dutch numbers in 2020) visit their General Practitioner (GP), who can treat the illness/injury or refer them to appropriate care at a hospital's A&E department. In more severe cases, the patient is taken directly to the A&E, either by ambulance services or by own means (12%), to be amended with the 14% of the patients that took the route via the GP. At the A&E department, 26% of all patients are diagnosed and treated, and about one-third of them are hospitalised. In principle, each of the institutions mentioned (GP, ambulance, A&E, hospital) could collect information on traffic casualties and even pass traceable information to the next care giver. As for the GDPR and other laws of medical data protection, the information hardly ever has personal identifiers enabling data linking. Medical status information is handed over in case reports or orally, and for example information of the circumstances or location of the crash is of no importance for the treatment of injury, so this information is hardly ever stored in medical files being passed over.

The number of slight cases is much larger than the severe ones. The severity of an accident is called after the most severe injury in it. In a fatal accident, there is at least one person deceased, but maybe others are injured, and there will probably be damage to the vehicles involved. Some of the people involved may have no injuries at all. In a PDO accident, there are no persons injured, although that assessment

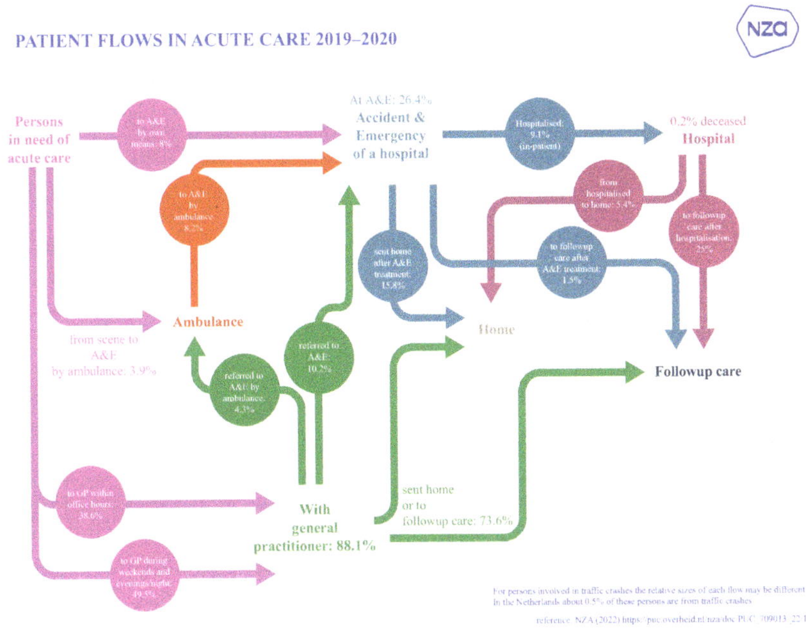

Figure 5.3 Patient flows in acute care in the Netherlands (adopted from NZA 2022)

may appear wrong when some consequences become visible later. It may also be subjective as the severity of bruises and scratches might be judged differently. Psychological consequences, such as fear of driving or posttraumatic anxiety/stress disorders may develop later.

The number of persons entering the healthcare system through a GP is huge, so is the number of GPs as well as the variety of problems they handle – majority of which are not traffic-related. This makes it difficult to collect harmonised data at the GP level if the goal is to cover all traffic injuries, though data collection from a representative sample of GPs is possible (this is done, for example, by the Netherlands Institute for Health Services Research – NIVEL (n/d)).

The number of A&Es in the country is more manageable; however, the stream of patients they handle is still seen as too high to collect all the data on the incoming injuries. This is done only in 12% of the A&Es in the Dutch Injury Surveillance System (DISS) operated by VeiligheidNL (n/d).

When a patient is hospitalised after being treated at an A&E (or when the person dies at the A&E), the data are passed to the Trauma Register, TR, operated by the National Network of Acute Care (LNAZ n/d-a) and collecting data from all the hospitals.

Within the hospital, most casualties are treated in the Trauma department, some of them temporarily in the Intensive Care Unit (ICU). Hospitals have an obligation to report about the patients they treat in the Hospital Discharge Register (HDR) (DHD n/d). Information on the medical process at the ICU, which in some

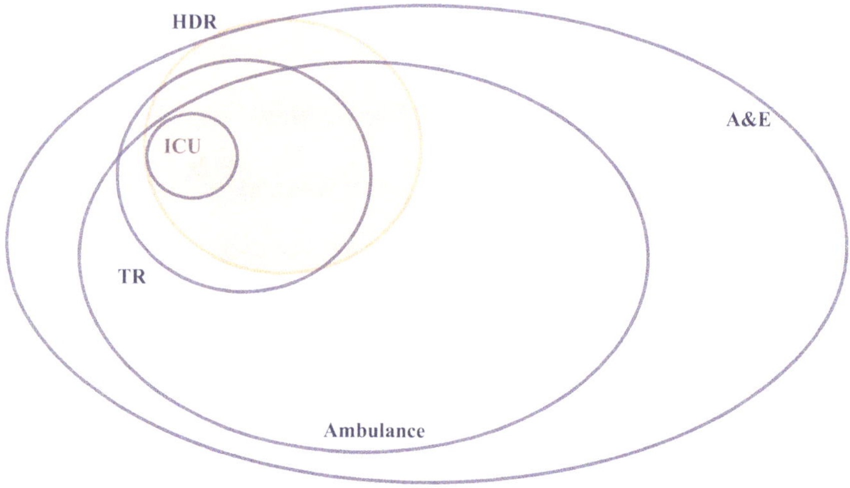

Figure 5.4 *Datasets containing information about traffic injuries (not to scale): HDR, Hospital Discharge Register (hospitalised in-patients); A&E, Accident and Emergency hospital department; ICU, Intensive Care Units; TR, Trauma Register; Ambulances (note that only part of the injured patients requiring hospitalisation arrives by ambulance)*

cases may also be of relevance, is collected by the National Intensive Care Evaluation (NICE n/d).

After discharge from the hospital, a process of rehabilitation starts where the focus is on regaining mobility as before the accident or developing physical activity that is still possible considering the impairment a casualty may have. Also, casualties may suffer from mental problems such as post-traumatic stress or anxiety which may need proper psychological treatment.

To sum up, there are several points along the patients' routes through diagnosis and treatment in the healthcare system where the data is registered; some registers are samples, and some overlap with others, as visualised in Figure 5.4. As most countries in the European Union are to report on the number of in-patient patients to Eurostat (EUR-Lex 2008, 2022), the HDR is often the most harmonised and complete data source for severe injuries.

5.2.3 Injury classifications

Injuries obtained in an accident can be characterised by their type, organs and body parts injured, the probability of immediate death, invalidity or long-term health consequences, etc. Moreover, an accident may cause multiple injuries and there is a need to assess their combined effect. Several measurement scales have been developed to classify injuries and reflect these aspects.

International Classification of Diseases (ICD)

In most countries, hospitals do not define injury severity as such; instead, based on the medical diagnosis, each injury is classified using the International Classification of Diseases (ICD). ICD is a globally used tool maintained by the World Health Organisation (WHO 2019). It is periodically revised and the latest, 11th version, was officially released in 2022. However the earlier versions, ICD9 and ICD10 are still widely used.

The injuries that are recorded for each patient, are detailed and enable the grouping and prioritising of injuries in order to make them meaningful for epidemiological and clinical analyses. For example, Barell developed her matrix (Barell *et al.* 2002) displaying 12 nature of injury columns and 36 body regions rows. Each ICD9 or ICD10 injury is assigned uniquely to one cell in the matrix. For example, traumatic brain injury (TBI) was separated into three types and injuries to the eye have been separated from other facial injuries. Other head injuries such as open wounds and burns were categorised separately. Injuries to the spinal cord and spinal column were categorised differently as are the abdomen and pelvis. Extremities have been divided into upper and lower with a further subdivision into more specific regions. Hip fractures were separated from other lower extremity fractures.

Based on a grouping of ICD-codes, the Integris project proposed a certain hierarchy for casualties with multiple injuries, so that the injury group with most impact in terms of disability and health impact and costs was preferred. They defined 39 so-called Eurocost groups (Belt *et al.* 2009) and derived disability weights for each of the groups (Haagsma *et al.* 2012).

Abbreviated Injury Scale (AIS) and Maximum AIS (MAIS)

The AIS© is a coding system developed by the Association for the Advancement of Automotive Medicine (AAAM) to classify the severity of injuries for nine regions of the body (head, face, neck, thorax, abdomen/pelvis, spine, upper extremities, lower extremities, external). First published in 1969, it has been updated on multiple occasions with the latest version issued in 2015 (AAAM 2015). Earlier versions AIS2005/update08 and AIS1990 are also still in use. With each new version, the number of injuries covered has been increased while the severity scores have been refined with regard to more recent medical treatment possibilities.

The AIS scale is a measurement tool for a single injury, and it reflects the threat to life associated with it rather than severity of the injury itself. AIS severity is measured with a six-point ordinal scale 1 to 6, 1 being a minor and 6 being the maximal (untreatable) injury. The AIS code 9 is reserved for occasions in which no sufficient information is available to assess the severity of the injury.

Typically AIS is used by ambulances and Trauma Registers, also as a component in the triage process to prioritise patients with the most critical injury. Although there are some limitations, it is possible to convert ICD codes into AIS-severity, such as 'AIS ICD ISS Map' software developed by the AAAM (AAAM

2018). A more general overview of the software products and mapping tables can be found in Pérez *et al.* (2016). Direct coding of AIS is preferred over (ICD-) derived AIS as conversions are always limited. Some studies show that the number of serious injuries resulting from ICD10-mapping is lower than direct coding of AIS (Airaksinen *et al.* 2019).

In case a person suffers from multiple injuries, the overall severity is then measured by MAIS, the Maximum of the AIS scores of all reported injuries.

Counting Serious Injuries

It is not easy to count the number of serious injuries in a way that it is comparable to other countries. Not in all countries access is possible to hospital data, and if it is available, the characteristics to select traffic casualties and to assess the severity of the injuries sustained may be difficult. In the project SafetyCube (Pérez *et al.* 2018; Weijermars *et al.* 2018), a harmonised way to select serious injuries (defined as MAIS3+) from hospital data using ICD-to-AIS transformations and to estimate the total number of serious injuries was set in practical guidelines. Three methods were suggested: (1) linking police and hospital data and apply capture/recapture, (2) selection from HDR only, (3) from police data using factors derived from a sample of hospitals (HDR) and/or linking study in order to correct for underreporting and to harmonise severity. Depending on some national characteristics of HDR data availability (e.g. the number of diagnoses per patient), some additional corrections were suggested.

Injury Severity Score (ISS) and New Injury Severity Score (NISS)

The ISS aims at representing complex information on the multiple injuries with a single number. The ISS utilises the highest AIS scores from the three most severely injured regions of the body. The original AIS nine body region are first regrouped into six ISS body regions (head/neck, face, chest, abdominal & pelvic contents, extremities & pelvic girdle, external). Then, the AIS scores are squared and added, producing the ISS score.

The NISS is computed as the sum of squares of the three most severe injuries, regardless of the body region in which they occur.

Both indicators are assigned a value of 75 when the patient has an injury of AIS = 6, so the ISS and NISS range from 1 to 75. Medically, physicians consider an ISS of 16 and more as polytrauma. This can be achieved by one injury of AIS = 4 or multiple injuries of AIS 3 + 3 or 3 + 2 + 2.

Other indicators

The AIS-related scores described above are all reflecting the threat-to-life aspect of the injuries. There are, however, other aspects of relevance, such as the duration of hospital admission and rehabilitation in general, probability of sustaining permanent damage and invalidation, reduced quality of life, inability to work, etc. There is a plethora of indicators developed aiming to reflect these consequences of an injury. In practice, however, they are seldom collected in a systematic way so that they can be used on a large scale in traffic safety work. Some work has been done on quality of life of persons treated at A&E (Haagsma *et al.* 2012; Belt *et al.* 2009) and this has been applied to road traffic casualties in the SafetyCube project (Pérez *et al.* 2016; Weijermars *et al.* 2016c). The risk of permanent medical impairment (RPMI) has been investigated on the Folksam insurance data (Malm *et al.* 2008) and applied to road traffic casualties (Berg *et al.* 2016).

Years of Life Lost and Burden of Injuries

For fatalities, it is possible to calculate the number of life years lost (YLL) by comparing the age at death with the life expectancy at that age. In a simple approximation, one could just multiply the number of fatalities by the average expected remaining life (which is the life expectancy minus the average age of a fatality, say 80 − 45); $\mathrm{YLL} = \mathrm{N} * (\mathrm{LE} - \langle \mathrm{age} \rangle)$

A more sophisticated calculation takes into account each fatality by age and gender and the remaining life expectancy at that age. For example, for a male traffic fatality of 30 years old the expected age at death would be 80 years, so the remaining 50 years are lost by the traffic crash.

For persons injured something similar can be developed by accounting the loss of quality of life for the remaining life of the casualty. These burden of injury calculations take into account an *average* disability weight and the remaining life expectancy to estimate the Years Lived with Disability (YLD), as it is impossible to follow traffic casualties for the rest of their lives to determine the impact of their injury. In the simple approximation, one could just multiply the number of hospitalised or serious injuries with the fraction of casualties that has permanent impairment (say $f = 25\%$), the average disability weight (say dw=0.1) and the average expected remaining life (say 35 years): $\mathrm{YLD} = \mathrm{N} \cdot f \cdot \mathrm{dw} \cdot (\mathrm{LE} - \langle \mathrm{age} \rangle)$.

The more sophisticated calculation takes into account fractions and disability weights that depend on the injury sustained in the accident, the actual age/gender of the casualty and the remaining life expectancy at that age. If the 30-year-old male did not die, but was hindered by a wrist fracture (disability weight 0.21) for the rest of his life this would lead to $50 \cdot 0.21 = 10.5$ YLD. Not every traffic casualty with a wrist fracture will have a permanent impairment as many will completely recover, so only a fraction of 18% of the persons with wrist fractures will be taken into account[†].

Together the fatalities and injuries make one new indicator: the Disability Adjusted Life Years (DALY = YLL + YLD). This indicator is useful in comparing the impact of traffic crashes with other types of accidents

[†]Disability Weights and Fractions are taken from Haagsma *et al.* (2012)

or diseases. This is done, e.g. in the Global Burden of Diseases studies (GBD 2020), where traffic injuries were estimated to account for 3% of world-wide DALY.

In another work-package of the SafetyCube project, the burden of injury was calculated for six European countries/regions (AT, BE, ES, FR, GB, NL) (Weijermars *et al.* 2016c).

In the method to estimate the burden of injury, the *fraction of casualties that has permanent impairment* plays an important role. In the research by Haagsma *et al.* (2012), this was assessed for persons treated at A&E departments and separated 38 groups of injuries (Belt *et al.* 2009) (Eurocost groups) based on ICD9 or ICD10.

In Sweden, based on the Folksam car occupant casualty data and scores by insurance doctors, Malm derived percentages of lifelong disability for each combination of 5 AIS-levels and 10 body regions (Malm *et al.* 2008). Doctors used the method of Grading Medical Impairment (Sveriges Försäkringsförbund 2004) so that they could assess the impact of the disability at 1%, 5% and 10% level. As an example, losing a finger would be impairing at a 2% level. So only at the Risk of Permanent Medical Impairment (RPMI) level of 1%, this injury counts as a lifelong impairment. At the 5% and 10% level, the lower boundary for taking injuries into account is higher, so this injury will drop out, resulting in lower probabilities of impairment at the 10% level than with the 1% level.

Berg *et al.* (2016) applied Malm's method to Swedish data from Strada for 2013. Applying the RPMI matrix to 14.623 road traffic casualties with at least one injury of severity AIS = 2 or higher (MAIS2+) resulted in 6179 casualties (42%) with a permanent impairment at the 1% level. For the higher impairment levels, the number (and proportion) of casualties facing permanent impairment is lower. Limiting cases to more serious injuries (MAIS3+) results in a higher proportion of permanent impairment.

Weijermars *et al.* (2016a) applied the Integris method to hospital data from the Netherlands. For MAIS2+ traffic casualties in the years 2000–09 about 20% of the cases were estimated to encounter lifelong consequences.

The same method was also applied to several European countries in the SafetyCube project (Weijermars *et al.* 2016c, 2018). For Serious Injuries (MAIS3+), the proportion of casualties with lifelong disabilities averaged 25%, but varied between 19% and 33%.

MAIS3+ as definition of serious injury

Ideally, all traffic injuries should be described and reported. From the pragmatic perspective, however, a priority should be given assuring the quality of the data on serious injuries. Such limitation both narrows the scope of the task (thus making its completion more feasible) and is in line with the Safe System paradigm that prioritises elimination of the fatalities and serious injuries in traffic yet tolerating the existence of minor accidents. To define a serious injury in practical terms is the first step in that direction.

To be able to internationally compare road safety data, the definition of serious injury should not be affected by the differences in demographic and economic factors, clinical practices, availability and access to the medical services, etc. This rules out, for example, indicators such as duration of hospital admission as richer countries may offer more complex treatments compared to countries with less resources. We have also seen a reduction of the length of hospital stay (LoS) during the last decades by a factor of 3, in spite of a more or less constant number of serious injuries (van Kampen 2007). A low or decreasing LoS is also seen as an indicator of efficiency for inpatient care (Brasel *et al.* 2007; McMahon 1985).

From this perspective, the immediate risk to life reflected by MAIS seems to be the most universal and transferable measure of severity. An IRTAD working group on Serious Road Traffic Casualties discussed a range of options using hospital data to consider the 'best' option for a practical international definition of a serious casualty, in terms of threat to life. They considered various injury severity scales, the length of stay in hospital, and lists of particular injury diagnoses and analysed a set of six criteria to choose the most suitable indicator. The working group recommended using MAIS level 3 and above (MAIS3+) as a suitable definition of a serious injury (ITF 2011). Similar conclusions were reached in the European research projects SafetyCube (Pérez *et al.* 2018) and SafetyNet (Broughton *et al.* 2007). While MAIS level 2 (MAIS2, moderate injury) could have also served as a good cut-off point, in many countries it is not practically possible to separate between data on MAIS levels 1 and 2, thus making the lowest feasible threshold at MAIS3. This is also in line with the AIS classification which defines *serious injury* starting from AIS level 3.

The polytrauma definition of ISS $\geq$ 16 is not considered suitable since it is mostly useful for emergency doctors and requires accurate registration of more than one injury, which is difficult to obtain in road safety research.

Focus on MAIS3+ or ISS16+ patients (approximately 25% of all hospitalised road traffic casualties) makes sense when viewed from within the hospital. However, from the perspective of healthy people traveling from A to B, this does not reflect the general need for safe traveling without being involved in a traffic crash, let there be a need to attend hospital for an injury and even stay there as an in-patient with only 'moderate' injury.

5.2.4 Linking police and hospital records

Linking records in police and hospital databases has several important purposes. First, for the accidents which records have been found in both databases, more detailed information becomes available. The total number of accidents found in at least one database is a more accurate estimate of the *true* accident number. Linking also reveals which accidents are missing in each of the databases and if there is some systematic pattern for some accident types being under-reported by each data source. Finally, with two linked data sources, it is possible to estimate the number of accidents that are not reported by any of the databases (the method for that is described in the section 'Estimation of the true number of injuries').

Data linkage methods

There are three main methods for linking the records from two databases – manual, deterministic and probabilistic. The manual method is, probably, the most accurate and, as the name suggests, is done manually by a human. Because of its labour intensity, this method is only feasible when the number of records and the amount of data in each record is small.

The remaining deterministic and probabilistic methods both rely on computer algorithms that 'decide' whether two records belong to the same accident. This allows working with large databases, however, at the cost of higher frequency of erroneous linkages (false positives and false negatives).

In the deterministic approach, the decision is taken based on the exact matching of one or several unique identifiers, such as personal identification number, vehicle number plate, etc. If such data is missing in at least one of the databases, the linking becomes impossible.

This problem is addressed by the probabilistic method. It is based on estimating the two probabilities – that the two records belong to the same person in the same accident and that they are matched by chance. Unlike in the deterministic method, the variables that are not unique identifiers (e.g. age/date of birth, gender, region of accident or hospital) still provide useful information if used together. The individual variables can be weighed in the calculations of the final probabilities based on how likely their values are (higher weights being assigned to more rare values).

There is a practical limitation on how many records could be handled with the probabilistic method in combination with the 'key' of identifying variables. When the total number of records in both databases approaches the number of combinations of the variables used as the identifiers, finding the true pairs is no longer possible.

Various software tools, both freely available and commercial, can perform data linkage. In most cases, however, the quality of the original data is much more important that the choice of the software. It is also to be expected that the matching performed by a computer will yield some dubious pairs that have to be checked manually afterwards.

Estimation of the true number of injuries

In a strict sense, the number of traffic injuries that are registered neither by the police nor by the healthcare system will remain unknown since no data about them is available. However, some estimates could be made.

The method used for estimating the under-reporting comes from biology and is referred to as 'capture-recapture' (ITF 2011). For example, one wants to estimate the total number of fishes in a pond. The method involves the following steps: (i) catching and tagging a certain number of fishes – capture; (ii) releasing them back; (iii) making the second catch – recapture – and measuring the share of fishes that are tagged. The general idea is that there is a relation between the probability of catching the same fish twice and how big the entire fish population is.

The two databases (police and hospital) can be seen as an analogy for two independent 'catches' from the entire traffic injury population. However, the capture-recapture method relies on several assumptions that do not strictly hold in case of traffic injuries:

- **Perfect identification of the subjects of interest.** When the police file contains many slight injuries for which no hospital admission is needed or when in-hospital traffic injuries can not be selected from other causes, then it will be impossible to estimate the number of relevant cases in each dataset and additional cases that are not present in the other.
- **Closed population.** This means that no injuries can deliberately 'escape' registration in both rounds. However, people may deliberately want to avoid involving police (e.g. if not being sober) or choose to seek medical help outside the region for which the data is being collected. This also plays a role where police registrations report on cases occurring in one year, while HDR concerns discharges also from the previous year and does not include hospitalisations that have not yet resulted in discharge.
- **Perfect identification of subjects common in both registrations.** As discussed earlier, none of the data linkage methods is perfect and some of the matched pairs would be erroneous and some will be missing.
- **The registrations are truly independent.** Arriving at an injury accident scene, police is likely to call for the medical care service; on the opposite, medical personal is not likely to call police when treating a person who approached them after a traffic accident.
- **Homogeneity of capture.** This means that all injuries should have the same chance of being registered. However, there is a systematic pattern in how the chances of an accident to be registered are influenced by its type, severity, and road users involved.

It is an ongoing debate whether the capture-recapture method can be relied on in case of traffic injuries. Some of the issue can be addressed by disaggregating the data (e.g. making estimates for each severity level and accident type separately), while other are quite impossible to compensate for.

Capture/Recapture example

The capture–recapture method states that the proportion of cases in the second catch that was also captured in the first catch reflects the proportion of the total number of cases that was captured in the first catch. The police dataset (first catch) consists of the injuries found in both dataset (B) and the injuries found only in the police register (P). Similarly, the hospital dataset consists of the injuries found in both datasets and in hospital records only (H). Thus,

$$\frac{B}{B+H} = \frac{B+P}{\text{Total}} \quad \text{and} \quad \text{Total} = \frac{(B+P)\cdot(B+H)}{B}.$$

The number of injuries not captured by any of the datasets N could be estimated, using the knowledge that $Total = B + P + H + N$, as

$$N = \frac{P \cdot H}{B}.$$

both (*B*)	**hospital only (*H*)**	**Hospital**
police only (*P*)	**in neither *N* = *P* * *H* / *B***	
Police		***Total***

An example. Police registers 1200 serious casualties being hospitalised, and the hospital has 1500 casualties of which 1000 (66%) are also in the police files. The total number of injuries could be estimated as

$$Total = \frac{(1000+200)\cdot(1000+500)}{1000} = 1800.$$

The number of non-reported injuries is

$$N = \frac{200 \cdot 500}{1000} = 100.$$

Some study examples and important insights

Police reporting is biased for motor vehicle involvement and no good material to study the group of bicycle crashes (i.e. single, with pedestrian or obstacle or among each other). In the Netherlands, the number of MAIS2+ casualties from crashes without involvement of a motor vehicle (i.e. bicycles) is increasing and since 2012 larger than the number from crashes with involvement of a motor vehicle (i.e. all motor vehicle occupants including powered two wheelers and bicyclists in a crash with a motor vehicle). The completeness of the police reporting for crashes without motor vehicles or with just one vehicle is much lower than for crashes with motor vehicles.

Groups targeted with prevention strategies for fatalities also target MAIS3+ casualties, however, to prevent MAIS2 injuries, a different approach is needed.

Studies international:

- PENDANT (Martin *et al.* 2005), an in-depth study of motor vehicle crashes combining vehicle register information with police and hospital information. For example, the difference in impact for drivers and passengers was investigated using accident manoeuvre and injury severity. It includes the analysis of injuries and risk factors in car to car crashes, whiplash in car to car and pedestrian injuries in car to pedestrian collisions.
- SafetyNet (Broughton *et al.* 2007), the relation between the number of police-reported casualties and hospitalised casualties was investigated for eight European countries/regions (AT, CZ, EL, ES, FR, HU, NL, UK) by linking. Different severity indicators were suggested to harmonise severity, such as length of stay and AIS. A severity of MAIS3+ was recommended as the best way to harmonise data and methodology throughout the European Union.
- Sustain (Aarts *et al.* 2016), the largest groups of serious injury casualties were clustered in terms of mode, age and gender. Measures to bring down these numbers were suggested. While the EU Member States were proceeding to estimate the total number of serious injuries using the new MAIS3+ definition, there was a need to know more about main crash circumstances of MAIS3+ casualties in order to make a start with strategies and measures that are effective in the prevention of these injuries.

  The study was directed at providing an understanding of the main circumstances and factors that affect the emergence of serious road traffic injuries, for the road traffic modes with many casualties in the EU: pedestrians, bicyclists, motorcyclists and car occupants. Data were gathered for nine countries from in-depth sources, hospital discharges, trauma registers and police records linked to medical registers. For each database, the available main crash characteristics have been extracted and used to define common scenarios for each traffic mode. Also the most affected body regions of the casualties have been gathered per traffic mode and per database. Furthermore, differences in injury patterns per crash scenario have been studied in order to find first clues for effective measures. For each of the four modes, most common characteristics of crashes were listed.
- Analysis of Swedish and Dutch accident data on cyclist injuries in cyclist-car collisions (Leo *et al.* 2019). To reduce the number of serious injuries sustained by cyclists in crashes with vehicles, it is important to understand which kinds of injuries are occurring to identify what should be assessed by means of virtual testing. A detailed analysis of injuries was made based on Swedish and Dutch accident data. The most

frequently injured body regions and the most frequent single injuries of these body regions were analysed. Cyclists most frequently injured their heads, upper and lower extremities, and bone fractures as well as brain injuries were identified as one of the most important injuries. The study concludes that for the virtual assessment of cyclist protection, injury predictors for long bone, skull and pelvic fractures as well as brain injuries are required in Human Body Models.

5.2.5 Hinders for access and use of hospital data

Healthcare records include sensitive personal information and need high protection. Frequently these data are not easily accessible, even when the intentions are good. Moreover, even if access is granted, the data need to be anonymised, i.e. the personal identifiers removed and stored in a protected environment with access only for authorised researchers with a confidentiality statement. Even so, however, matching the two datasets (police and healthcare) creates new knowledge about the individuals, consequences for which must be considered. In most cases, ethical approval for accessing and matching the data would be necessary. When reporting the results it should not be possible to trace any personal information of casualties involved. In Europe, the General Data Protection Regulation (GDPR 2016) came active in 2016 to arrange for that, see Chapters 16 and 17.

While we now have a harmonised definition and procedure to estimate the true number of Serious Injuries (SafetyCube 2016), this does not mean that it is easy to apply this method in all countries.

Many countries experience problems in accessing hospital data (either national or regional), to derive an AIS-severity, to select traffic casualties and to use the right inclusion and exclusion criteria (or correct the result for deviations from those criteria, for example, to compensate for the inability to select crashes on public roads only). So at the moment estimates of the number of serious injuries are not yet harmonised, resulting in large differences between countries, for example in the factor of Serious Injuries to Fatalities, ranging from 1 to 12 (ETSC 2023).

This makes it difficult to compare countries or developments between countries in a reliable way. It is currently investigated how to improve this approach in order to further harmonise injury data. Maybe a different approach is needed that can also be applied to low- and middle-income countries.

5.3 Case example: the Netherlands

5.3.1 History

As a check on completeness and bias of police-data, hospital data was analysed already in the early 1980s (Maas and Harris 1984), leading to a pilot linkage study in 1988 between HDR and the police registration (Blokpoel and Polak 1991). An

Table 5.1 Variables in police and hospital datasets, variables in italic are used for the linking

Police	Hospital
Date and time of crash	*Date and time of admission*
Municipality of crash	–
Region of hospital	*Region of hospital*
Date of birth	*Date of birth*
Gender	*Gender*
Severity	–
–	*External cause (ICD)*
Mode of transport	External cause type
Opponent	External cause type
Person class	External cause type
Location	–
Circumstances	–
Manoeuvre	–
–	Injuries (ICD)
–	Length of stay
–	Treatment

Italic variables that are in one dataset only are used to set a preference, e.g. to prefer a case that was actually sent to hospital over a case that died on the spot or was only slightly injured.

existing Injury Surveillance System on personal injuries at some A&E departments was extended with traffic accidents (van Kampen 1994). Initially, linkage studies were carried out with the medical partners, from 2003 onwards it became easier to obtain HDR data for annual linkage studies. Historic HDR data became available starting from 1978 including relevant key variables, see Table 5.1, such as date of birth, enabling linking to police data.

5.3.2 The linked database

Both police and hospital datasets consist of anonymised records, containing detailed data, suitable for linking. In the Netherlands, the size of the police file is about 30,000 casualties, of which some 10,000 are hospitalised according to the police. The HDR that is obtained for linking is larger than just traffic casualties. All patients with an injury and/or an external cause are compared to allow for misclassification (e.g. falls are sometimes misinterpreted as pedestrians instead of cyclists). The HDR contains typically 100,000 patients of which 25,000 are transport accidents.

Just by comparing these numbers, it is evident that police data is underreported for road traffic casualties that are hospitalised.

Procedures for linking police and hospital data

A probabilistic, distance-based linking method was developed to match casualties with patients. In this method it is not required that pairs are exactly equal on all relevant variables. Initially, pairs with hospital admissions were allowed between −1 and +4 days compared to the date and time of the crash. A recent study by Bos *et al.* (2019) showed that for serious/acute patients this time window could be narrowed to −3 h to +36 h. For less severe injuries, however, especially those to the extremities, the window needed to be extended to 18 days. Each hospital record is compared to all police records within the specified time window.

Missing values or slight differences lead to a small distance between a pair. A distance is added as well when according to the police the casualty was not hospitalised or when according to the hospital the external cause was not a traffic crash. Pairs with the lowest distances are interpreted as matches when this distance is below a certain threshold. Each record in either dataset can only match with one record in the other dataset.

Using hospital data and severity indicators

Patients in hospital are diagnosed by codes from the International Classification of Diseases (WHO 2019). The injuries are coded as well as the external cause of the injury.

Early 2000, when severity mappings were introduced, it was realised that hospitalised not always meant serious: admissions for observation only or with slight injury (AIS = 1) occur. Application of ICD-AIS mappings enabled the filtering of serious injuries. Initially, MAIS2+ was used as a cut-off, still leaving the possibility to also estimate the number of MAIS3+ casualties.

Initially, in the Netherlands external causes and injuries were coded in ICD9cm (1993–2012). Since 2011 hospitals gradually implemented ICD10.

For AIS-mapping, the application ICDmap90 (ICDMAP 2002) was used (from ICD9cm to AIS1990). Mapping from ICD10 appeared to be difficult as AAAM's tables were designed for the clinical modification ICD10cm. Instead ICD10 injuries are transformed back to ICD9cm by applying a conversion from RIVM (WHO-FIC 2011), followed by the ICD-ISS-map (AAAM 2018) to AIS2005/update08.

ICD10 external causes do not facilitate a difference between mopeds and motorcycles, nor between bicycles and electric bicycles. Since 2015 hospitals in the Netherlands used a Dutch modification of ICD10 to enable that.

Procedures to estimate the total number of serious injuries

Linking two databases does not yet give a clear picture of the number of serious road injuries. As from police data, it is unknown if a casualty is seriously injured (MAIS3+), a police-only file cannot be used. Dutch data conflict with the 1st condition (see section 'Estimation of the true number of injuries'). However, matches between police records and hospital records can be separated into matches with traffic casualties and matches with patients with other external causes. As the police was present at the scene and as they said it was a traffic accident, we assume that the external cause for the casualty was miscoded in the hospital.

The completeness of the police file is quite different for crashes with motor vehicles (M) and for crashes without the involvement of a motor vehicle (N). Therefore, the data should be separated, as shown in Table 5.2.

Table 5.2 Typical annual number of MAIS3+ casualties in the Netherlands (average for 2014–2021)

	Traffic M in HDR	**Traffic N in HDR**	**Not a traffic accident in HDR**	**Sum**
Linked M	1600	50	60	1700
Linked N	30	350	20	400
Not in police	1220	3000	*Unobserved*	*4450*
Sum	2850	3400	*300*	*6550*

Numbers are rounded. M = casualty in a crash with a motor vehicle involved, N = casualty in a crash without any motor vehicle involved.

Using a capture–recapture like approach and allowing for misclassification of crashes with or without involvement of a motor vehicle – we believe the police file here – the number of unobserved cases and the total number of serious injuries is estimated (Reurings and Stipdonk 2011).

Compared to the number of fatalities, the number of serious injuries (MAIS3+) is about 10 times higher. For injuries from crashes involving motor vehicles (M), the factor is about 5, and for crashes without motor vehicle involvement (N), it is much higher (crashes between bicycles and pedestrians, other cyclists, objects and single), mainly because the number of fatalities is low for N crashes.

Until 2009, the linked dataset made up a big part of the total number of serious injuries. The reporting rate, defined as the number of matches divided by the total number was estimated at 70%. This enabled estimates of the number of serious injuries by mode, age, gender, region and police records could be filtered to just include serious injuries. A completeness drop of the police reporting in 2010 caused that the linked databases became much smaller. Also, police no longer reported on the (region of the) hospital where the casualty was transported to, and information on whether the casualty was hospitalised was omitted. As a proxy, the region of the crash was used in the linking procedure. With larger uncertainty, the total number of serious injuries can still be estimated however disaggregates are not possible in a reliable manner. After 2014, the completeness improves; however, case-information on hospitalisation is lacking so disaggregates are also difficult.

In the Netherlands, the linked database does not really exist for privacy reasons, and as from the linking methodology, there is uncertainty about the matched pairs. Statistically it is sufficient to be able to estimate the true number of Serious Injuries from the intersection and remainder files, including a capture/recapture estimate of unobserved cases.

Since 2014, privacy regulations have become more strict; also the implementation of the GDPR in 2018 required stricter procedures for the work with health and criminal information. At the moment, police datasets may contain violations of the traffic law and the fact that someone got injured is considered as medical information. Therefore, data are brought together in a protected environment, and only information that cannot reveal any personal information is allowed to leave that environment.

Some national studies

Annually the number of serious injuries is estimated using linked HDR and police data. Information on MAIS3+ and MAIS2 is provided. The most recent study is: Serious injuries 2022, estimate of the number serious road traffic injuries in the Netherlands (Bos *et al.* 2023).

- Serious road injuries in the Netherlands dissected (Weijermars *et al.* 2016b). In this study, the distribution of serious road injuries over (1) travel mode and gender and (2) crash type and age are compared for the years 2000–2011. Moreover, the distribution of the injuries over the body regions is illustrated using coloured injury body profiles for different travel modes and age groups (Figure 5.5).

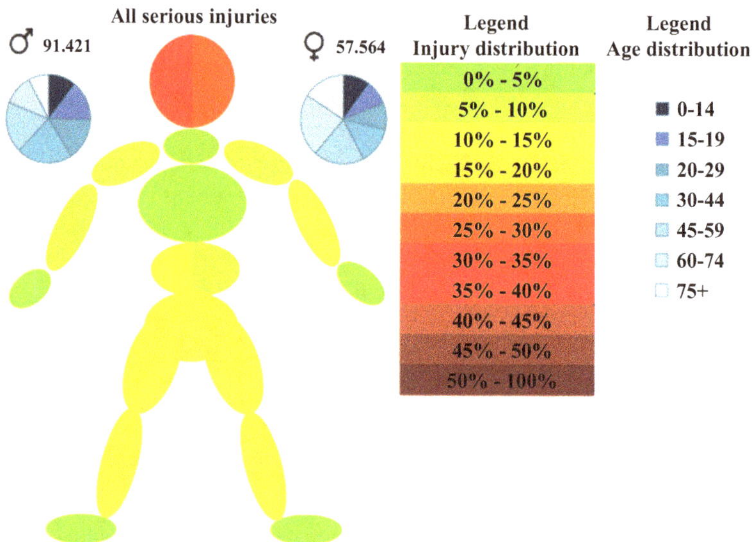

Figure 5.5 Injury body profile showing the distribution of injuries over the Barell body regions for all serious road injuries (MAIS2+) among men (left) and women (right) (based on hospital discharge data, LMR2000-2009.

- Burden of road traffic injuries: Disability-Adjusted Life Years in relation to hospitalisation and the Maximum Abbreviated Injury Scale (MAIS) (Polinder *et al.* 2014). When Years of Life Lost (YLL) for fatalities are compared with the Burden of Injury (YLD) for survivors, the following picture resulted from their analysis (Figure 5.6): 80% of injuries only need treatment at the A&E department, making less than 20% of the burden. Most burden can be avoided by preventing MAIS2+ injuries.
- Health burden of serious road injuries in the Netherlands (Weijermars *et al.* 2016a). In this study, the disability weights of Haagsma were used to calculate the burden of injury for traffic casualties. For casualties with at least one MAIS2+ injury, the average burden amounted to 2,1 YLD (Years Lived with Disability) (Figure 5.7).

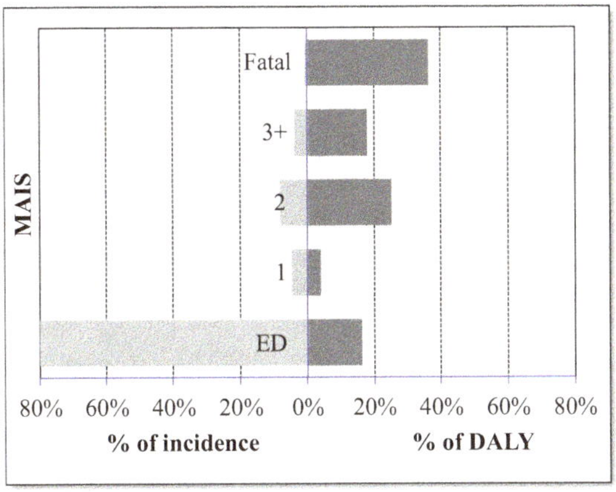

Figure 5.6 *Contribution of MAIS severity levels to incidence and disease burden for Road traffic injuries (the Netherlands, 2007–09). Incidence and DALYs of admitted, ED treated and fatal casualties.*

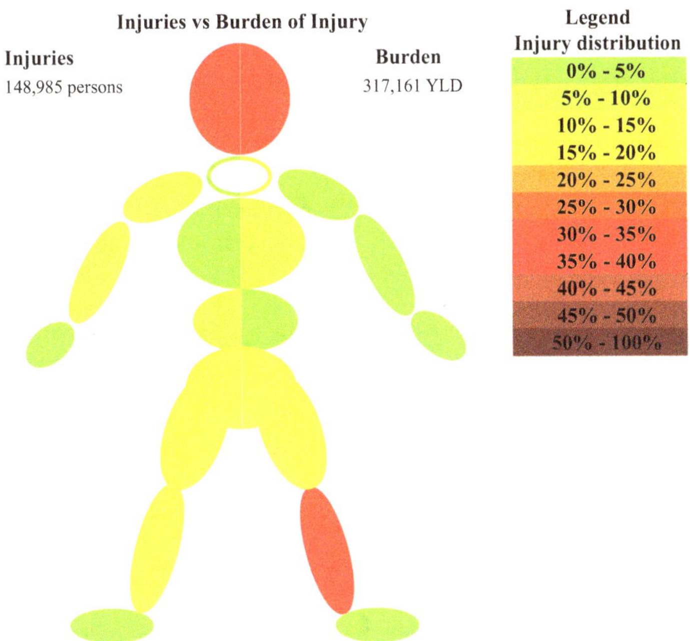

Figure 5.7 *Distribution of injuries and burden of injury over the human body. Hospital discharge data, LMR2000-2009, MAIS2+.*

- Characteristics and injury patterns of seriously injured older road users compared to 30–45 year-olds (Boele *et al.* 2019). Differences between older (70+) and younger (30–45) road users were analysed by mode of transport based on Hospital Discharge Data 2011–17, MAIS2+. More than half of the 70+ age group had a severity score of MAIS3+, whereas the majority of the 30–45-year-olds sustained MAIS2 injuries to the upper and lower extremities. The highest absolute numbers of seriously injured road users concerned older cyclists who were involved in a crash without involvement of a motor vehicle. They primarily sustained injuries to the lower extremities. Although the share of serious head injuries is relatively low in these crashes compared to bicycle crashes with motor vehicles, the absolute number of serious head injuries is at least twice as many. The differences found in injury mechanism, severity and patterns may be a starting point to develop measures aimed at injury prevention.

Other medical data

Apart from hospital in-patients (Hospital Discharge Register), there are also other medical data in the Netherlands:

- Ambulance trips (collected by AZN, RIVM), are seen as promising intermediate data between hospital admissions and police data as from the HDR the accident location is unknown (Aarts *et al.* 2020). Also by itself ambulance data are relevant as they are probably more complete on injury crashes than police data. Privacy regulations and sometimes missing information on the type of accident (transport or other) hinder broader dissemination of these data. As ambulances for most of the trips pick up patients from their home addresses, the data collecting organisation AZN is reluctant to provide information on unknown accident types for privacy reasons. Datasets exist for some regions and a national pilot delivered 2009–12 trip-data. Efforts are being made to make data available for road safety research (Nijman *et al.* 2020).
- A&E department of hospitals. Since 1998, the Injury Surveillance System (LIS) is a representative sample of 8% of hospitals in the Netherlands reporting all A&E visits for injuries, among which traffic injuries (VeiligheidNL 2019). It separates A&E visits after traffic accidents by follow-up (sent home or hospitalised), their mode of transport, opponent, age, gender and if they arrived by ambulance. For coding the injury-location and nature of injuries a simple scheme is used, that was internationally agreed upon (EuroSafe n/d). It is possible to approach the persons afterwards for further inquiries via the hospital. This was done, for example, for casualties of bicycle and moped crashes (Krul *et al.* 2022).
- Trauma Register, national dataset (LTR) on persons hospitalised after visit to the A&E-department, includes persons that died at the A&E. The data collection started in 2007 Injuries are coded in AIS, recognising traffic injuries is

possible since 2014 (LNAZ n/d-b). The number of (serious) traffic injuries roughly corresponds to the number cases from the HDR.

- National Intensive Care Evaluation (NICE n/d). The data collection started in 1996 and annually some 75,000 cases are added. Unfortunately it is not possible to select traffic accidents.

5.4 Case example: Sweden

Strada (Swedish Traffic Accident Data Acquisition) is the national information system containing data on traffic injury accidents occurring in the Swedish road transport system. Strada is managed and maintained by the Swedish Transport Agency (*Transportstyrelsen*) that gathers traffic accident reports from both police and healthcare systems and merges them into a single database.

5.4.1 Historical note

The decision to create Strada was taken by the national government in 1996. The Swedish Road Administration[‡] (*Vägverket*) received the task to create an information system that would support traffic safety improvement efforts on a central, regional and local levels. In 1999, Strada was first launched as a pilot project covering only few geographical areas. Starting from 2003, the Strada's police reports cover the entire country. Connecting all the emergency hospitals was a much longer process which was completed only in 2016. In 2009, the responsibility for Strada maintenance and development was transferred to the newly created Swedish Transport Agency (*Transportstyrelsen*) (Transportstyrelsen 2024).

Strada is an important tool within the national traffic safety strategy Vision Zero, adopted by the Swedish government in 1997 (Johansson 2009). For example, one of the most important tools within the Vision Zero is the management-by-objectives, meaning that road safety work is steered towards intermediate goals and that prompt adjustments are taken if the development is not going in the right direction or at the sufficient pace. Both the fatality and serious injury numbers are included in the list of safety performance indicators monitored and reported annually on the national and regional levels, with Strada being the primary data source (Hurtig *et al.* 2023). Moreover, the current interim traffic safety goals for the period 2020–30 include the 25% reduction of serious injuries due to pedestrian falls, which are technically not included in the official definition of traffic accidents. Pedestrian falls are normally not registered by the police, thus all the data for monitoring the progress in this area comes into Strada from the healthcare system.

Over the years, Strada brought many new insights regarding both the scale and the nature of the road safety problems (see the textbox on page 87).

[‡]Swedish Road Administration (*Vägverket*) was merged together with the Swedish Rail Administration (*Banverket*) in 2010, forming the current Swedish Transport Administration (*Trafikverket*).

5.4.2 Strada database

The general structure of the Strada system is shown in Figure 5.8.

First, the original reports are submitted by the police (a report per accident) and emergency hospitals (a report per person who sought care for an injury in traffic and who agreed to fill in such a report). The records are matched, where possible, using the personal identification numbers and other variables like accident date, location, accident type, etc. (Strada 1 in Figure 5.8). Up to this point, all the data in the records are preserved, including the personal information considered sensitive, and access to the database is strictly limited. At the next step, the reports are anonymised and transferred to a secondary database (Strada 2 in Figure 5.8), which can be accessed online by safety professionals.

Police data

The police reports to Strada traffic accidents with at least one person injured (accidents without injuries are not submitted, even if the police were present at the accident site). The accident must have occurred on a public road and must have involved at least one vehicle in motion. One report (see Figure 5.9(a)) is created per accident, and everyone involved is included in the report regardless of whether they were injured or not. The report includes severity rating for the injured persons (lightly injured, severely injured or dead). The severity is determined at the accident site, therefore it is based only on observable injuries.

There are several pre-defined accident types, such as single accident, rear-end collision, and accident with wildlife. This typology is not particularly detailed if, for example, compared to the CADaS taxonomy (CARE 2018). The decision to use a simplified classification came from the reasoning that while creating extra burden for the police officers, such disaggregation brings little additional value, particularly due to very low numbers collected in many sub-categories. The possibility for

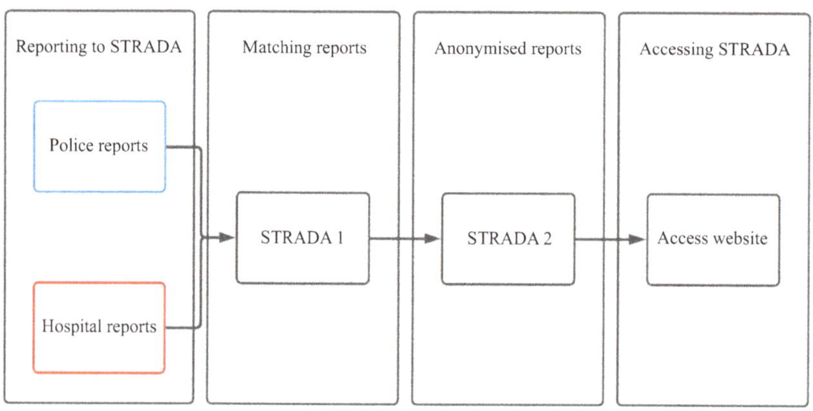

Figure 5.8 Strada database – general structure

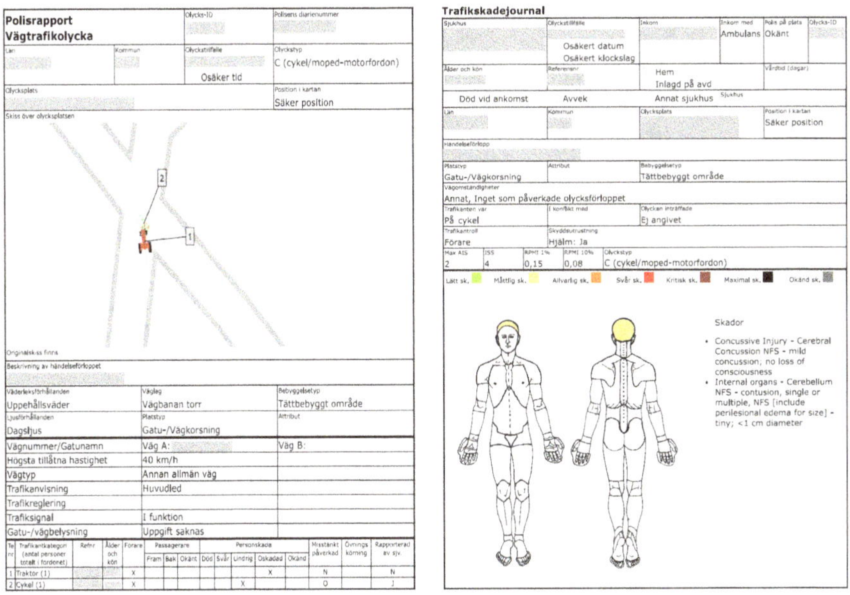

(a) (b)

Figure 5.9 Report examples from the police (a) and hospitals (b) in Strada

such in-detail analyses is not fully eliminated; however, a researcher needs to do own classification based on textual descriptions and sketches found in the report.

On the average, the police submit 15,000 accident reports per year (20,000 persons injured or killed).

Hospital data

A separate emergency hospitals report is created for each person seeking medical care for an injury received in traffic (on public roads). Unlike the police, the hospital system reports traffic injuries even if there was no vehicle involved in the accident. Thanks to that, most of the knowledge about single falls of pedestrians (and even cyclists, even though formally they should be reported by police) is coming from the hospital data.

Reporting to Strada is voluntary both for the hospitals and for the injured individuals. Each hospital makes its own agreement with the Swedish Transport Agency and receives financial compensation for delivering data to the Strada. In the case of individual reports, the informed consent of each injured person must be given before the registration process starts. It is estimated, however, that the drop-out rate at this stage is very low.

Each report contains general information about the person such as age and gender combined with a brief explanation of what caused the event, usually in just a short sentence (Figure 5.9(b)). The injury description is more detailed and includes a list of individual injuries with respective AIS, as well as the combined ISS severity rating.

On the average, emergency hospitals submit 40,000 reports each year.

Data matching in Strada

When possible, the reports from police and hospitals related to the same accident are merged into a joint Strada report. Since the police reports accidents while the hospitals report injured persons, multiple hospital records can be linked with a single police record. The criteria for linking the two types of records are that the persons involved have the same personal identification numbers, accident date and time do not differ more than 24 h, and that the reported accident locations are within 1,000 m from each other. There are also checks performed within Strada to remove duplicate hospital records (caused by patients being moved between hospitals). Approximately 30% of all hospital records get connected to corresponding police records, and 40% of all police reports get connected to at least one hospital report (Howard and Linder 2014).

In the reports that has been linked, the injury severity estimated by the hospital is always prioritised. However, if the only existing report is from the police, the serious injuries (ISS 9+) are downgraded to a moderate degree (ISS 4–8). This adjustment was introduced after a study that compared police and hospital severity ratings in linked records and concluded that the police officers generally tend to overestimate the severity of injuries (Howard and Linder 2014).

It takes different time for the police and the hospitals to deliver their reports before the data linkage can be initiated. About 90% of the accident reports made by the police are delivered within 24 days. Reaching the same 90% takes the hospitals up to 90 days.

Privacy in STRADA

Over the years, the rules for access to the online database have been made more and more restrictive. It has to do both with the general evolution of the personal data protection legislation and with some negative experiences of data being used by people not really understanding it (e.g. journalists attempted to make own methodologically flawed 'analyses' that later received public attention). As of year 2024, access is granted on a personal basis (through the electronic ID to prevent sharing of the logins) and only to professionals employed by few selected institutions and whose work duties are directly related to traffic safety data analysis. It is also obligatory to take a one-day course about the database structure and operation, the data quality and limitations, as well as the rules of conduct when using the Strada.

5.4.2.1 Accessing STRADA

Strada can be accessed by the users through a website hosted by the Swedish Transport Agency (*Transportstyrelsen*). The website provides a graphical user interface with accidents located on a geographic map, each containing one or several reports. The query tool allows filtering by the geographical area, time period, accident type, road category, injury severity, etc. The query results, as well as individual reports for injured persons/accidents can be extracted and saved offline for further analysis.

Researchers without a direct access to Strada may apply for an extract from the database to be performed by the Swedish Transport Agency. Each application should describe the purpose of the study, specify which data need to be extracted, how they will be analysed and stored after the analysis, etc. Each application is treated on individual bases, so, for example, if a researcher decides to extend the period for which the data is needed, a new application must be submitted.

Understanding single bicycle and pedestrian accidents

The awareness about single bicycle and pedestrian accidents was created by the first explorative studies performed in 1980s–1990s (Berntman *et al.* 1997; Nilsson 1986). Using healthcare records, it was clearly shown that falls in public space were common, could lead to serious (and sometimes fatal) injuries, and yet they were virtually non-existing in the official traffic accident register that was based on the police data. Naturally, since the researcher had to manually go through the hospital records, such studies were only feasible for relatively small geographical areas and short time periods.

Introduction of Strada allowed for making analysis on a much larger scale. Thulin and Niska (2009) analysed bicyclist injuries during 2003–06 and reported that 72% of bicyclist injuries were caused by single accidents, with slippery surfaces being a contributing factor in 23% of cases. Other contributing factors mentioned were kerbstones, uneven road surfaces, collisions with temporary or permanent objects, as well as cycling under alcohol influence.

Niska *et al.* (2013) continued the analysis for the period 2007–11, complementing Strada with in-depth investigation reports of the fatal accidents as well as deep interviews with some seriously injured bicyclists. Similarly to the previous report, it was estimated that up to 80% of injured cyclists who needed emergency treatment were injured in single accidents. It was also estimated that single accidents contributed 20% of cyclist in traffic fatalities. More exact shares of the accident contributing factors were established, too. Out of 1,274 cyclists seriously injured in single accidents, 27% could be related to road operation and maintenance, 20% to road design, 27% to cyclist-bicycle interaction, 15% to the behaviour and state of the cyclist, and 11% to the interaction of the cyclist with other road users. It was common that several factors were at play. Also, the problem of single accidents taking place near road works areas was highlighted. This led to several research projects on how the temporary bicycling facilities around road works/construction areas should be designed, marked, and equipped [e.g. some barriers meant to protect cyclists were actually found to be death traps when subjected to crash tests (Niska *et al.* 2022)].

Berntman (2015) performed a similar analysis with focus on pedestrian falls (data from 2009 to 2013). About 13% of pedestrian fatalities occurred as a result of a fall. Majority of them were men older than 55 years; nearly all deceased women were 75 years or older. When it comes to severe injuries, it was estimated that more than 30 times more pedestrians are injured in falls than in collisions with vehicles. Women suffered most, contributing to 70% of fall-related injuries; most of them were 55 years or older. The age pattern is quite different for men, with the age group of 0–34 being the largest. In winter months, the slippery conditions were the most dominating contributor. For the rest of the year, deficient maintenance of the walking infrastructure contributed most to the falls.

It is worth mentioning that in 2003 only 29 (40%) of emergency hospitals were connected to Strada. By 2013, five (though one being a major regional hospital) hospitals were still not connected.

The most recent study by Eriksson *et al.* (2022) (data for 2014–19) once again confirmed the importance of single accidents, found to contribute 94% of serious pedestrian injuries and 79% of serious bicyclist injuries. What was new this time is that the authors added the analysis of the free text descriptions of accidents. A special method was developed to systematically code important information that otherwise could be formulated in many different ways in its textual form. Since the coding was done manually, only a representative subset of the records was processed in this way. In particular, the already known accident contributor categories such as 'road maintenance' could be further disaggregated into finer groups of 'slippery: ice/snow', 'slippery: water', 'slippery: leaves', 'slippery: mud', 'uneven surface/pothole', 'stumble over temporary object', 'loose gravel', etc. Particularly in case of cyclists, it became transparent that while for a large share of accidents the countermeasures were obvious (remove snow/leaves/gravel, fix potholes, etc.), there was a significant group of accidents which could not be explained just by the presence of the contributing factors and better understanding of the mechanisms behind the accidents was necessary. Again, this initiated several (currently ongoing) research projects examining how the infrastructure may 'surprise' cyclists affecting their stability and eventually fall risk – for example, combination of a downhill with a sharp curve in front of a narrow tunnel would likely to require adjustments of both the speed and the path, which could result in a fall or a collision if done too sharply.

5.5 Conclusion

Healthcare data provide a rich source of information on traffic injuries that is not included (or not described in sufficient detail) in the police accident records. Complementing police-reported accidents with hospital records helps to get a better estimate of the true size of the traffic safety problem. This is particularly important for certain types of accidents that traditionally have not been prioritised by the police (e.g. pedestrian falls or single bicycle accidents), though unreported-by-police accidents can be found in virtually every category.

Naturally, the medical records provide much better insight in the types and severity of the injuries since the injured person is examined by medical professionals with access to equipment that normally is not available at the accident scene (e.g. X-rays). If injury classification is available both from the police and the healthcare, the latter is more reliable and is to be preferred.

It is important to realise that 'healthcare data' actually relates to data from many different sources as there are quite many medical facility types through which a patient may enter the system. Some decisions need to be made as for where the data is to be collected. Also, some relevant injuries still might not be captured by the healthcare registers (or might be not ear-marked as traffic injuries), thus there is always a number of 'unknown accidents/injuries' remaining, even if the police and hospital data are combined.

To estimate the amount of the 'unknown injuries', methods such as capture-recapture could be used. The reasoning behind these methods is based on certain assumptions, which quite often are violated in case of police and hospital injury registration. This does not mean that such estimates should not be attempted, but rather that the results have limited accuracy.

Pilots on combining police and healthcare data have been attempted in many countries, and in some of them it has become a regular routine in preparing the traffic accident statistics. However, it is not unusual to hear that the healthcare data is so well protected by some country's laws and regulations that there is no way for traffic safety specialists to get access to them. Such arguments are rather an explanation than an excuse. The personal data protection legislation in Sweden or the Netherlands, which we used as the two case examples, is by no means less strict compared to the rest of the world. The difference is in the political will to look for practical solutions that would not infringe too much on personal privacy yet benefits the entire society by reducing the citizen risks of being killed or injured in traffic. We hope that by now the reader is well equipped with knowledge and positive examples to become an advocate of including healthcare data analyses into the traffic safety practitioner's toolbox.

References

AAAM (2015), Abbreviated Injury Scale (AIS) (Association for the Advancement of Automotive Medicine), https://www.aaam.org/abbreviated-injury-scale-ais/, accessed 8 November 2022.

AAAM (2018), '*The Abbreviated Injury Scale (AIS) ICD-ISS Map v.2.0*' (Chicago, IL, USA: Association for the Advancement of Automotive Medicine), https://www.aaam.org/abbreviated-injury-scale-ais/licensing/ais-icd-iss-map/.

Aarts, L., J. Commandeur, R. Welsh, S. Niesen, M. Lerner, P. Thomas, N. Bos, R. Davidse (2016), *Study on Serious Road Traffic Injuries in the EU* (Brussels, Belgium: Directorate-General for Mobility and Transport, European Commission), MI-01-16-994-EN-N, https://doi.org/10.2832/29647.

Aarts, L. T., G. J. Wijlhuizen, F. Hermens, N. M. Bos (2020), 'Koppelmogelijkheden van ambulancedata met andere bronnen [Options for linking ambulance data with other sources]' (the Hague, the Netherlands: SWOV), R-2020-15, https://swov.nl/nl/publicatie/koppelmogelijkheden-van-ambulancedata-met-andere-bronnen.

Airaksinen, N. K., M. T. Heinänen, L. E. Handolin (2019), 'The reliability of the ICD-AIS map in identifying serious road traffic injuries from the Helsinki Trauma Registry', *Injury*, 50 (9), 1545–1551, https://doi.org/10.1016/j.injury.2019.07.030.

Alsop, J., J. Langley (2001), 'Under-reporting of motor vehicle traffic crash victims in New Zealand', *Accident Analysis & Prevention*, 33 (3), 353–359, https://doi.org/10.1016/S0001-4575(00)00049-X.

Amoros, E., J.-L. Martin, B. Laumon (2006), 'Under-reporting of road crash casualties in France', *Accident Analysis & Prevention*, 38 (4), 627–635, https://doi.org/10.1016/j.aap.2005.11.006.

Barell, V., L. Aharonson-Daniel, L. A. Fingerhut, et al. (2002), 'An introduction to the Barell body region by nature of injury diagnosis matrix', *Injury Prevention*, 8 (2), 91–96, https://doi.org/10.1136/ip.8.2.91.

Belt, E., S. Polinder, J. Haagsma, E. van Beeck, R. Lyons, S. Macey, M. Atkinson, J. Lund (2009), 'WP5 Injury disability indicators' (INTEGRIS – Integration of European Injury Statistics: FP7 project), https://citeseerx.ist.psu.edu/document?repid=rep1&type=pdf&doi=16370b624fe4e3ceff14b545e125d4bc63a76bbc.

Berg, H. Y., J. Ifver, M. Hasselberg (2016), 'Public health consequences of road traffic injuries – Estimation of seriously injured persons based on risk for permanent medical impairment', *Transportation Research Part F: Traffic Psychology and Behaviour*, 38, 1–6, https://doi.org/10.1016/j.trf.2015.12.007.

Berntman, M., K. Brundell-Freij, C. Hydén, A. Ståhl (1997), 'Fotgängares och cyklisters singelolyckor [Pedetrian and bicyclist single accidents]' (Sweden: Lund University, Faculty of Engineering, LTH, Department of Technology and Society, Transport and Roads), Bulletin 140/3000.

Berntman, M. (2015), 'Fotgängares olyckor och skador i trafikmiljö med fokus på fallolyckor [Pedestrian accidents and injuries in traffic envirobment with focus on fall accidents]' (Lund, Sweden: Lund Univerity, LTH, Department of Technology & Society), Bulletin 295, https://lup.lub.lu.se/search/files/3934697/8194817.pdf.

Blokpoel, A., P. H. Polak (1991), 'Koppeling tussen de landelijke medische registratie LMR en de verkeersongevallenregistratie VOR van in ziekenhuizen opgenomen verkeersgewonden: resultaten van een proefkoppeling [Link between the national medical registration LMR and the road accident registration VOR of hospitalized road injuries: results of a pilot linkage]' (Leidschendam, the Netherlands: SWOV), R-91-79, https://swov.nl/en/publicatie/koppeling-tussen-de-landelijke-medische-registratie-lmr-en-de.

Boele, M. J., N. M. Bos, R. J. Davidse (2019), 'Characteristics and injury patterns of seriously injured older road users compared to 30–45 year-olds', presented at *International Cycling Safety Conference* (Brisbane, Australia: 18–20 November 2019).

Bos, N. M., R. J. Decae, F. D. Bijleveld, F. Hermens, J. J. F. Commandeur, L. T. Aarts (2019), 'Ernstig verkeersgewonden 2018: Schatting van het aantal ernstig verkeersgewonden in 2018 [Serious road injuries 2018: Estimate of the number of serious road injuries in 2018]' (the Hague, the Netherlands: SWOV), R-2019-23, https://swov.nl/nl/publicatie/ernstig-verkeersgewonden-2018.

Bos, N. M., F. D. Bijleveld, L. T. Aarts, R. J. Decae (2023), 'Ernstig verkeersgewonden 2022: Schatting van het aantal ernstig verkeersgewonden in 2022 [Serious road injuries 2022: Estimate of the number of serious road injuries in 2022]' (the Hague, the Netherlands: SWOV), https://swov.nl/en/publicatie/ernstig-verkeersgewonden-2022.

Brasel, K. J., H. J. Lim, R. Nirula, J. A. Weigelt (2007), 'Length of stay: An appropriate quality measure?', *Archives of Surgery*, 142, 461–466, https://doi.org/10.1001/archsurg.142.5.461.

Broughton, J., E. Amoros, N. M. Bos, *et al.* (2007), 'Estimation real number of road accident casualties' (European Commission, Directorate-General Transport and Energy: SafetyNet – Building the European Road Safety Observatory), Deliverable D.1.15, http://www.dacota-project.eu/Links/erso/safetynet/fixed/WP1/D1.15_Estimation_real_number_of_road_accident_casualties_final%20report_3.pdf.

CARE (2018), 'CaDaS – Common accident data set', Version 3.7, https://road-safety.transport.ec.europa.eu/system/files/2021-07/cadas_glossary_v_3_7.pdf.

Department of Transport (2020), 'Reported road casualties in Great Britain: 2019 annual report', https://assets.publishing.service.gov.uk/media/5f735c2fd3bf7f286abc0748/reported-road-casualties-annual-report-2019.pdf.

Department of Transport (n/d), 'Classification of injury severity using the CRaSH reporting system', https://www.gov.uk/government/publications/guide-to-severity-adjustments-for-reported-road-casualty-statistics/guide-to-severity-adjustments-for-reported-road-casualties-great-britain#classification-of-injury-severity-using-the-crash-reporting-system, accessed 5 September 2024.

DHD (n/d), 'Dutch hospital data', https://www.dhd.nl/, accessed 7 September 2023.

Eriksson, J., P. Henriksson, M. Rizzi (2022), *'Oskyddade trafikanters inblandning i olyckor och deras skadeutfall: En jämförande studie mellan fotgängare, cyklister, mopedister och motorcyklister [Vulnerable road users involvement in accidents and their injury outcome: a comparative study between pedestrians, cyclists, mopedists and motorcyclists]'* (Linköping, Sweden: Swedish National Road and Transport Research Institute), VTI rapport 1133, http://vti.diva-portal.org/smash/get/diva2:1669879/FULLTEXT01.pdf.

ETSC (2023), *Ranking EU Progress on Road Safety: 17th Road Safety Performance Index Report* (Brussels, Belgium: European Transport Safety Council), https://etsc.eu/wp-content/uploads/ETSC-17th-PIN-Annual-Report-DIGITAL-1.pdf.

EUR-Lex (2008), 'Regulation No 1338/2008 of the European Parliament and of the Council of 16 December 2008 on Community statistics on public health and health and safety at work', *Official Journal of the European Union, L* 354 (1338), https://eur-lex.europa.eu/eli/reg/2008/1338.

EUR-Lex (2022), 'Commission Regulation 2022/2294 of 23 November 2022 implementing Regulation No 1338/2008 of the European Parliament and of the Council as regards statistics on healthcare facilities, healthcare human resources and healthcare utilisation', *Official Journal of the European Union*, L 304 (2294), https://eur-lex.europa.eu/eli/reg/2022/2294/oj.

EuroSafe (n/d), 'European Association for Injury Prevention and Safety Promotion', https://www.eurosafe.eu.com/home, accessed 7 September 2023.

Eurostat, ITF, UN (2019), 'Glossary for transport statistics' (Eurostat, International Transport Forum, United Nations), 5th edn, https://ec.europa.eu/eurostat/documents/3859598/10013293/KS-GQ-19-004-EN-N.pdf.

Eurostat (n/d), 'Discharges from hospitals', https://doi.org/10.2908/TPS00048, accessed 5 September 2024.

Global Burden of Disease Study (GBD) (2020), 'Global burden of 369 diseases and injuries in 204 countries and territories, 1990–2019: a systematic analysis for the global burden of disease study 2019', *Lancet*, 396 (10258), 1204–1222, https://doi.org/10.1016/S0140-6736(20)30925-9.

GBD (n/d), GBD Results Tool, https://vizhub.healthdata.org/gbd-results/, accessed 5 September 2024.

GBD (2024), 'Global burden of 288 causes of death and life expectancy decomposition in 204 countries and territories and 811 subnational locations, 1990–2021: a systematic analysis for the Global Burden of Disease Study 2021', *Lancet*, 403 (10440), 2100–2132, https://doi.org/10.1016/S0140-6736(24)00367-2.

GDPR (2016), 'General Data Protection Regulation 2016/679 of the European Parlament and the Council', https://eur-lex.europa.eu/legal-content/en/TXT/?uri=CELEX:32016R0679.

Haagsma, J. A., S. Polinder, R. A. Lyons, *et al.* (2012), 'Improved and standardized method for assessing years lived with disability after injury', *Bulletin of the World Health Organization*, 90 (7), 513–521, https://doi.org/10.2471/BLT.11.095109.

Howard, C., A. Linder (2014), 'Review of Swedish experiences concerning analysis of people injured in traffic accidents' (Linköping, Sweden: Swedish National Road and Transport Research Institute, VTI), VTI notat 7A-2014, http://vti.diva-portal.org/smash/get/diva2:699198/FULLTEXT01.pdf.

Hurtig, P., P. Larsson, M. Lindholm, *et al.* (2023), 'Analysis of road safety trends 2022' (Borlänge, Sweden: Swedish Transport Administration), https://trafikverket.diva-portal.org/smash/get/diva2:1779579/FULLTEXT01.pdf.

ICDMAP (2002), *ICDMAP-90 User's Guide* (Baltimore, MD: The Johns Hopkins University and Tri Analytics, Inc.).

ITF (2011), *Reporting on Serious Road Traffic Casualties: Combining and Using Different Data Sources to Improve Understanding of Non-Fatal Road Traffic Crashes* (Paris, France: International Transport Forum), https://www.itf-oecd.org/sites/default/files/docs/road-casualties-web.pdf.

Jarman, B., S. Gault, B. Alves, *et al.* (1999), 'Explaining differences in English hospital death rates using routinely collected data', *BMJ*, 318, 1515–1520, https://doi.org/10.1136/bmj.318.7197.1515.

Johansson, R. (2009), 'Vision Zero – Implementing a policy for traffic safety', *Safety Science*, 47, 826–831, https://doi.org/10.1016/j.ssci.2008.10.023.

Krul, I., H. Valkenberg, S. Asscherman, C. Stam, and K. Klein Wolt (2022), *'Fietsongevallen en snor-/bromfietsongevallen in Nederland [Bicycle crashes and light moped crashes in the Netherlands]'* (Amsterdam: VeiligheidNL), Rapport 934, https://www.veiligheid.nl/kennisaanbod/cijferrapportage/fietsongevallen-en-snor-bromfietsongevallen-nederland.

Leo, C., C. Klug, M. Ohlin, N. M. Bos, R. J. Davidse, and A. Linder (2019), 'Analysis of Swedish and Dutch accident data on cyclist injuries in cyclist-car collisions', *Traffic Injury Prevention*, 20, 160–162, https://doi.org/10.1080/15389588.2019.1679551.

LNAZ (n/d-a), 'National network of acute care', https://www.lnaz.nl/, accessed 7 September 2023.

LNAZ (n/d-b), 'LTR Landelijke Traumaregistratie', https://www.lnaz.nl/trauma/landelijke-traumaregistratie, accessed 5 September 2024.

Maas, M. W., S. Harris (1984), 'Police recording of road accident in-patients: Investigation into the completeness, representativity and reliability of police records of hospitalized traffic victims', *Accident Analysis & Prevention*, 16 (3), 167–184, https://doi.org/10.1016/0001-4575(84)90011-3.

Malm, S., M. Krafft, A. Kullgren, A. Ydenius, and C. Tingvall (2008), 'Risk of permanent medical impairment (RPMI) in road traffic accidents', *Annals of Advances in Automotive Medicine*, 52, 93–100, https://www.ncbi.nlm.nih.gov/pmc/articles/PMC3256772/.

Martin, J.-L., L. T. B. van Kampen, C. Pérez (2005), 'Hospital data analysis' (FP5 EC Research Programme: PENDANT – Pan-European Co-ordinated Accident and Injury Databases), Deliverable D9, https://swov.nl/en/publicatie/hospital-data-analysis-pendant-pan-european-co-ordinated-accident-and-injury-databases.

McMahon, S. M. (1985), 'The efficiency of inpatient medical care in a hospital secure unit', *Medical Care*, 23 (10), 1139–1147, https://doi.org/10.1097/00005650-198510000-00002.

Methorst, R., P. Schepers, N. Christie, *et al.* (2017), '"Pedestrian falls" as necessary addition to the current definition of traffic crashes for improved public health policies', *Journal of Transport & Health*, 6, 10–12, https://doi.org/10.1016/j.jth.2017.02.005.

MMUCC (2012), Model Minimum Uniform Crash Criteria, 4th edition definitions, https://safety.fhwa.dot.gov/hsip/spm/conversion_tbl/pdfs/kabco_definitions.pdf.

NHTSA (2024), *MMUCC Guideline: Model Minimum Uniform Crash Criteria*, 6th edn (National Highway Traffic Safety Administration), DOT HS 813 525, https://crashstats.nhtsa.dot.gov/Api/Public/ViewPublication/813525.

NICE (n/d), 'Nationale intensive care evaluatie', https://www.stichting-nice.nl/index.jsp, accessed 7 September 2023.

Nijman, S., K. Klein Wolt, B. Olij, and B. Blatter (2020), *'Ambulancedata verkeersslachtoffers: ontwikkeltraject landelijke ontsluiting ambulancedata voor verkeersveiligheidsbeleid [Ambulance data on road casualties: development process for national access to ambulance data for road safety policy]'* (Amsterdam, the Netherlands: VeiligheidNL), https://swov.nl/nl/publicatie/ambulancedata-verkeersslachtoffers-ontwikkeltraject-landelijke-ontsluiting-ambulancedata.

Nilsson, G. (1986), 'Halkolyckor: Förekomst och konsekvenser [Slip accidents: Occurrence and consequences]' (Linköping, Sweden: Swedish National Road and Transport Research Institute), VTI rapport 291.

Niska, A., S. Gustafsson, J. Nyberg, and J. Eriksson (2013), 'Cyklisters singelolyckor: Analys av olycks- och skadedata samt djupintervjuer [Bicycle single accidents: Analysis of accident- and injury data and deep interviews]' (VTI, Swedish National Road and Transport Research Institute), Rapport 779, https://www.diva-portal.org/smash/get/diva2:670651/FULLTEXT01.pdf.

Niska, A., J. Wenäll, and J. Karlström (2022), 'Crash tests to evaluate the design of temporary traffic control devices for increased safety of cyclists at road works', *Accident Analysis & Prevention*, 166, 106529, https://doi.org/10.1016/j.aap.2021.106529.

NIVEL (n/d), 'Netherlands Institute for Health Services Research', https://www.nivel.nl/en, accessed 7 September 2023.

NZA (2022), 'Patiëntenstromen acute zorg 2019–2020 [Patient flows in acute care 2019–2020]' (Nederlandse Zorgautoriteit), https://puc.overheid.nl/nza/doc/PUC_709013_22/1/.

Pérez, K., W. Weijermars, E. Amoros, *et al.* (2016), 'Practical guidelines for the registration and monitoring of serious traffic injuries' (SafetyCube: Horizon 2020 project), Deliverable D7.1, https://www.safetycube-project.eu/wp-content/uploads/SafetyCube-D7.1-Practical-guidelines-for-the-registration-and-monitoring-of-serious-traffic-injuries.pdf.

Pérez, K., W. Weijermars, N. Bos, *et al.* (2018), 'Implications of estimating road traffic serious injuries from hospital data', *Accident Analysis & Prevention*, 130, 125–135, https://doi.org/10.1016/j.aap.2018.04.005.

Polinder, S., J. A. Haagsma, N. M. Bos, *et al.* (2014), 'Burden of road traffic injuries: Disability-adjusted life years in relation to hospitalization and the maximum abbreviated injury scale (MAIS)', *Accident Analysis & Prevention*, 80, 193–200, https://doi.org/10.1016/j.aap.2015.04.013.

Reurings, M. C. B., and H. L. Stipdonk (2011), 'Estimating the number of serious road injuries in the Netherlands', *Annals of Epidemiology*, 21 (9), 648–653, https://doi.org/10.1016/j.annepidem.2011.05.007.

SafetyCube (2016), 'Practical guidelines for determining the number of serious road injuries (MAIS3+)' (Horizon 2020 project 633485), https://www.safetycube-project.eu/wp-content/uploads/SafetyCube-D7.1-Leaflet.pdf.

Sveriges Försäkringsförbund (2004), 'Medicinsk invaliditet. Gradering av medicinsk invaliditet 2004', https://www.svenskforsakring.se/globalassets/medicinska-tabellverk/aldre-medicinska-tabellverk—arkiv/gradering-av-medicinsk-invaliditet-2004_rev080812.pdf/.

Thulin, H., and A. Niska (2009), 'Tema cykel – skadade cyklister: Analys baserad på sjukvårdsregistrerade skadade i Strada [Tema Cycle – injured cyclists. Analysis based on hospital registered injury information from Strada]' (Linköping, Sweden: Swedish National Road and Transport Research Institute), VTI rapport 644, https://www.diva-portal.org/smash/get/diva2:675381/FULLTEXT01.pdf.

Transportstyrelsen (2024), 'STRADA: Swedish Traffic Accident Data Acquisition', https://www.transportstyrelsen.se/STRADA, accessed 27 February 2024.

van Kampen, L. T. B. (1994), 'Verkeersslachtoffers in Privé Ongevallen Registratie Systeem PORS VIPORS: voorbereidingsactiviteiten 1993 [Road Casualties in Private Accident Registration System PORS VIPORS: Preparatory Activities 1993]' (The Netherlands: SWOV), R-94-26, https://swov.nl/en/publicatie/verkeersslachtoffers-prive-ongevallen-registratie-systeem-pors-vipors.

van Kampen, L. T. B. (2007), 'Verkeersgewonden in het ziekenhuis [Traffic injuries in Hospital]' (Leidschendam, the Netherlands: SWOV), R-2007-02, https://swov.nl/system/files/publication-downloads/r-2007-02.pdf.

Veiligheid NL (2019), 'Verkeersongevallen 2018', https://www.veiligheid.nl/verkeer/feiten-cijfers/rapport-verkeersongevallen-2018.

Veiligheid NL (n/d), 'Knowledge Centre for Injury Prevention', https://www.veiligheid.nl/, accessed 7 September 2023.

Weijermars, W., N. Bos, and H. Stipdonk (2016a), 'Health burden of serious road injuries in the Netherlands', *Traffic Injury Prevention*, 17 (8), 863–869, https://doi.org/10.1080/15389588.2016.1157591.

Weijermars, W., N. Bos, and H. Stipdonk (2016b), 'Serious road injuries in the Netherlands dissected', *Traffic Injury Prevention*, 17 (1), 73–79, http://dx.doi.org/10.1080/15389588.2015.1042577.

Weijermars, W., J.-C. Meunier, N. Bos, *et al.* (2016c), 'Physical and psychological consequences of serious road traffic injuries' (SafetyCube: Horizon 2020 project), Deliverable D7.2, https://www.safetycube-project.eu/wp-content/uploads/SafetyCube-D7.2-Physical-and-psychological-consequences-of-serious-road-traffic-injuries.pdf.

Weijermars, W., N. Bos, A. Filtness, *et al.* (2018), 'Burden of injury of serious road injuries in six EU countries', *Accident Analysis & Prevention*, 111, 184–192, https://doi.org/10.1016/j.aap.2017.11.040.

WHO-FIC (2011), 'Conversie ICD-10 2006 – CVZ'80' (Collaborating Centre for the Family of International Classifications in the Netherlands), https://www.whofic.nl/downloads-en-links/icd-10.

WHO (2019), *International Statistical Classification of Diseases and Related Health Problems (ICD)* (World Health Organization), https://www.who.int/standards/classifications/classification-of-diseases, accessed 8 November 2022.

WHO (2023), *Global Status Report on Road Safety* (World Health Organization), https://iris.who.int/bitstream/handle/10665/375016/9789240086517-eng.pdf.

Chapter 6

The use of self-reported data in road safety research

Wouter Van den Berghe[1] *and Uta Meesmann*[2]

6.1 Introduction

6.1.1 Using self-reported data in research

Collecting and analysing data provided by interviewees or questionnaire survey respondents is a method used in many research areas and for collecting information within organisations and jurisdictions. Examples of applications are attitude surveys, censuses, market research surveys, psychological assessments, reporting on health behaviour and polls on political preferences.

Within road safety research, these so-called 'self-report' or 'self-declared' approaches involve collecting information directly from individuals about, among others, their experiences, characteristics, mobility, behaviour in traffic, attitudes, or perceptions. These methods can be based on various survey techniques, including interviews (face-to-face, by telephone or online video), written surveys (using mailings or handing them out to groups of people) and online surveys. Each technique offers unique advantages and disadvantages, influencing the quality and reliability of the data collected.

This chapter discusses how self-reported data can be used within road safety research. Particular attention is given to self-declared behaviour in traffic by road users and to the use of online panel surveys. Both advantages and limitations are discussed. Although we briefly discuss interviews in Section 1.3, the remainder of the chapter covers quantitative research methods only.

6.1.2 Sources and references

A large number of sources have been used for drafting this chapter. This includes a number of publications based on the authors' own professional experience in using questionnaire surveys at international level (Meesmann *et al.*, 2021; Meesmann and Wardenier, 2024; Pires *et al.*, 2020; Van den Berghe, 2022; Vollrath *et al.*, 2019). Other sources used throughout this document, in particular for the summary table in

[1]Tilkon Research & Consulting, Belgium
[2]Knowledge Centre, Vias Institute, Belgium

section 1.3 and the bullet lists in sections 2 and 3 of this chapter are af Wåhlberg *et al.* (2010), Choi and Pak (2005), Goldenbeld and De Craen (2013), Lajunen *et al.* (1997), Lajunen and Özkan (2011a), Lajunen and Summala (2003), Meade and Craig (2012), Nießen *et al.* (2019), Tellis and Chandrasekaran (2010), Ward and Meade (2018, 2023).

Useful general reference works on interviews and survey research are Baker *et al.* (2010), Bryman (2012a), De Leeuw *et al.* (2008a), Groves *et al.* (2009), Krosnick and Presser (2010), Rea and Parker (2014), Sudman and Bradburn (1982), Tourangeau and Smith (1996), N. J. Ward *et al.* (2019), Wolf *et al.* (2016). The ESRA-website (https://www.esranet.eu/) is also a rich source of information and results based on self-reported data covering a very wide range of road safety topics (see also Section 5.1).

Some of these sources discuss the use of questionnaire surveys in road safety research, while others are more general in nature. When a statement in this chapter is only based on one particular publication, this publication is explicitly referred to in the text.

6.1.3 Overview of main data collection methods for self-reported data

There are different possibilities for collecting self-reported data in road safety research. All these methods have specific advantages and limitations, which can vary according to the topic being researched. Table 6.1 (based on the sources listed

Table 6.1 Survey methods to collect self-reported data and their main advantages and limitations

	Method	**Main advantages**	**Main limitations**
Interviews	Face-to-face interview	• High response rates. • Possibility to clarify questions, reducing misinterpretation. • Suitable for complex questions and sensitive topics. • Richer information through open and follow-up questions.	• Time-consuming. • Needs recording, transcribing and codification. • Costly. • Potential interviewer bias. • May not be feasible for large geographical areas.
	Telephone interview	• More personal than online surveys. • Higher response rates than paper-based and online surveys. • Allows to clarify questions.	• Potential interviewer bias. • Potential for distraction. • Questions cannot be complex. • Limited to those with telephone access. • Not suitable for lengthy questionnaires.

(Continues)

Table 6.1 (*Continued*)

	Method	Main advantages	Main limitations
			• Needs recording, transcribing and codification of answers.
	Online interview	• Similar advantages as face-to-face (see above). • More possibility for visual aids. • Flexibility in scheduling.	• Requires internet access. • Potential for distraction. • Potential interviewer bias. • Less personal compared to face-to-face. • Needs recording, transcribing and codification of answers.
Questionnaire surveys	Paper-based survey	• Anonymity (can encourage honesty). • Avoids interviewer bias. • Can reach people without internet access.	• Can be costly to process data. • Low response rates. • Slow turnaround time. • Limited ability to clarify questions.
	Web-based survey	• Anonymity (can encourage honesty). • No interviewer bias. • Low cost. • Direct access to a specific group. • Design flexibility. • Easy to implement. • Filtering/ randomisation of questions • Possibility to use a smartphone.	• Low response rates. • Slow turnaround time. • Limited ability to clarify questions. • Requires internet access. • Self-selection bias. • Risk of manipulation. • May not represent target population. • High drop-out with long surveys.
	Online panels survey	• Quick access to a large pre-recruited, diverse pool of respondents. • Anonymity (can encourage honesty). • No interviewer bias. • Cost-effectiveness. • Speed and efficiency. • International comparisons. • Filtering/randomisation of questions • Possibility to use a smartphone.	• Composition of the panel (representativity, size). • Panel conditioning. • Careless respondents. • Membership attrition over time. • Limited ability to clarify questions.

in Section 1.2) lists the main survey methods, showing the utility of each survey method. A differentiation between interviews and questionnaire surveys is being applied. In the next sections, we will dive deeper into the benefits and limitations of questionnaire surveys, in particular online panel surveys, compared to other data collection approaches. A common advantage of all self-report methods is that they allow us to **measure unobservable constructs**, such as attitudes, beliefs, or feelings. Common limitations of self-reported data are that they are based on the **subjective perception** of the respondents, cannot reveal **unconscious behaviour** and may be subject **to recall** bias (these common advantages and limitations are not listed in the table).

This table illustrates that no single method is the best in all circumstances and that both scientific, ethical and economic criteria need to be considered.

6.2 Advantages and limitations of using questionnaire surveys in road safety research

This section starts with listing the advantages of using questionnaires surveys in general. Most of these advantages are applicable to all types of questionnaires, whether they are paper-based or conducted online, and whether they address the general population or a specific target group. Second, the specific additional advantages of surveys based on internet panels are outlined, since this method is becoming an increasingly used approach when a large population needs to be reached or when cross-national surveys are conducted. Thereafter, we discuss the limitations of questionnaires, first generally and subsequently for surveys using online panels.

6.2.1 Advantages of using questionnaire surveys

When needing to collect data for road safety research, it is often advantageous to use questionnaire surveys, whatever format they take. The most important advantages are as follows:

- **Flexibility and broad application.** There are several survey methods that can be used with questionnaires, each adaptable to different research settings and populations. This flexibility makes it a versatile tool for researchers and organisations across different areas – from research to HR management. Unlike interviews, participation is independent of the time of day and (when using online surveys) geographical location. For online surveys, it is often possible to use a smartphone.
- **Reducing costs.** Compared to interviews, questionnaires can keep costs down. They are cheaper to implement, and often, a larger number of participants can be reached than through interviews and in many cases also more than through observation (a counterexample is speeding, where high numbers of vehicles can be observed through cameras or radars).

- **Measuring unobservable constructs.** With questionnaires, it is possible to measure subjective constructs such as attitudes, beliefs, feelings, experiences, or past behaviour, which are difficult or impossible to observe directly. Self-reporting in surveys is a direct way to access an individual's thoughts, feelings and perceptions. This is particularly useful in psychology, market research and social sciences.
- **Enriching the data collected.** Questionnaire surveys can be used to collect data on many additional factors, which can be linked to the self-reported information – for example, linking self-reported behaviour on distraction or seatbelt use with variables such as education level, family composition trip characteristics and professional experience. Such surveys can also provide insights into socio-cognitive determinants of behaviour such attitudes, perceived social norms, risk perception, or existing habits. Socio-cognitive concepts can help to understand the underlying motivations of certain behaviour.
- **Reducing social desirability bias.** Compared to interviews, self-reported data are less prone to social desirability bias. Respondents are more likely to provide honest answers when they do not have to interact directly with an interviewer. (Goldenbeld and De Craen, 2013). Through an elaborate design of the questionnaire, such as including a social desirability scale, it is even possible to control for the effect of social desirability of other variables in further analysis (Nießen *et al.*, 2019).
- **Eliminating interviewer effects.** Self-administered questionnaires do not suffer from variability across interviewers asking questions in different ways or in a different order. Furthermore, as mentioned above, such surveys can offer anonymity, potentially leading to more honest and less socially desirable responses, especially on sensitive topics. All of this increases the data quality.
- **Facilitating longitudinal studies.** The same group of participants can be surveyed multiple times, facilitating longitudinal studies that track changes in attitudes, behaviours and experiences related to road safety over time.
- **Identifying careless respondents.** Because completion of a questionnaire survey is in general not monitored, some respondents may act carelessly and provide untrue responses. Questionnaire design (e.g. including instructed response questions) and data cleaning techniques help to identify such respondents and remove their data from the database – and hence increase the data quality (Goldammer *et al.*, 2020; Ward and Meade, 2023).
- **Design flexibility.** Questionnaire surveys, in particular online surveys, can be designed with filters and randomised response items to make them more relevant to the participant, avoiding unnecessary questions and reducing the risk of order effects.

6.2.2 Additional advantages of using online panels for questionnaire surveys

An online access panel consists of a very large number of people (for most countries, often over 100,000 people) who have indicated that they are willing to participate in online services. They receive a small reward every time they complete a survey. Panel providers strive to have a large representative part of the population. The panels can be used as a sampling frame for web surveys. The quality of online panel survey relies primarily on the representativity of the panel (sampling frame) and the sampling procedure which is used to reach survey respondents. Sampling could be based on a probability sample or a convenience sample. In most cases, a probability sample is to be preferred. The additional advantages of submitting web-based questionnaire to such panels are the following:

- **Representativity.** A requirement to achieve a representative sample is probability sampling (also called random sampling; see also section 4.1.3). Online access panels provide a vast database of potential respondents, making it easier than in other survey methods to achieve a large sample, even for specific target groups. It should be noted, however, that the representativity of the sample is restrained by the composition of the online panel (sample frame).
- **Greater reach.** Online panels can provide access to a wide range of demographics, geographic locations (even across countries) and specialised groups (e.g. specific driver categories).
- **Cost-effectiveness.** Compared to other data collection methods such as face-to-face interviews or postal surveys, online panel-based surveys are generally less expensive because there are almost no costs for physical materials, travel and personnel time. This allows for the collection of data from a larger number of participants across diverse geographic locations.
- **Speed and efficiency.** Online panels allow for rapid data collection – often only a week or so – because surveys and studies can be distributed to thousands of participants simultaneously. This leads to quicker turnaround times compared to traditional methods. Additionally, a well-designed survey with online error correction leads to lower needs for data cleaning and removal of incomplete data.
- **International comparisons.** Using online panels allows for a systematic and uniform data collection across different countries, because it can be programmed as one survey for all the countries, and subsequently translated. This is especially beneficial for studies with an international scope, as it allows for the inclusion of a wide range of participants from different countries. It should be noted, however, that the comparability across countries also depends on the quality of the translation or cultural and international differences (see Section 2.3).

6.2.3 Limitations of questionnaire surveys

Despite the obvious advantages of questionnaire surveys, one should also consider that there are several limitations and disadvantages. These include

- **Reduced representativity.** For surveys based on a non-probability sample (also called non-random sampling), the representativity of the sample for the targeted population is limited (see Section 4.1.3). Furthermore, when using certain survey methods (telephone-, postal- or email-based survey) contact information of part of the target group might be missing or be incorrect. If people without correct contact information differ significantly from those where contact information is available, survey results will be biased.

- **Risk of non-response.** The representativeness of the survey results may be lower if certain groups of people do not respond to surveys, either due to lack of access, interest or time. Security and privacy concerns might also deter the participation of certain groups. This non-response can be problematic when the individuals who do not participate differ significantly from those who do. This can be particularly problematic in telephone surveys and website-based surveys, which are accessible to a large number of possible respondents.
- **Slow return time** (for written questionnaires) because of the logistics involved and efforts needed to get a higher response rate (such as reminders to participants).
- **Questionnaire design limits.** There are restrictions on the type of questions that can be used in questionnaire surveys. Open questions are to be avoided, since they are often not answered, and/or it is challenging coding large numbers of responses to open questions. In general, closed questions are recommended in questionnaires.
- **No control of respondents.** When using questionnaire surveys, most often it is not possible to motivate participants. Especially with long surveys, there is a risk of a high drop-out and careless respondents (see also below). There is also potential for distraction or other interruptions while respondents complete the questionnaire.
- **Misunderstanding of questions.** Questions with difficult words or jargon, long questions and questions that are formulated in a negative way can be misunderstood by participants. Poorly designed surveys with complex questions lead to misunderstandings and inaccurate responses. Unlike in interviews, there is in general no mechanism for clarifying questions or checking whether the questions are well understood.
- **Subjectivity and perception.** Whilst subjectivity is inherent for constructs such as attitudes and feelings, for other topics, the subjective views may not correspond to reality, because of recall errors or social desirability. For instance, when asked about trips undertaken, survey respondents often underestimate the length of the trips undertaken on foot. Or their unsafety feeling on the street may not correspond with the real unsafety level of that street.
    - **Social desirability bias**. This is the tendency of respondents to provide answers which present a favourable image of themselves. This may be less of an issue in questionnaire surveys than when people are interviewed, but there is anyhow also a potential bias in surveys.
    - **Recall bias**. People may unintentionally provide faulty answers due to memory errors, for example, failure to correctly remember past behaviour. The bigger the time period between the phenomenon and the completion of the survey, the higher the risk of recall bias.
    - **Unconscious behaviour bias**. When driving a vehicle, one may not be aware of particular behaviour (e.g. distracted driving, driving unconscious

over the speed limit, drowsiness). Consequently, when asked about risky driving behaviour afterwards, one cannot report it. This bias might even be more important than the recall bias.

  - **Careless responses**. The anonymity of questionnaire surveys can sometimes encourage responses or dishonesty. Careless respondents are participants who fail to read or give sufficient attention to item content, resulting in data that may not accurately reflect respondents' true levels of the constructs being measured. This applies in particular if participants are more motivated by incentives than contributing meaningful data.
- **Manipulation**. When the response sample is uncontrolled and anonymous, some people may be tempted to answer a survey several times in order to increase the frequency of certain answers.
- **Cultural and international differences**. In cross-national studies, differences in social values, personality, status of persons, laws, road safety culture, infrastructure, traffic composition, enforcement, interpretation of scale values (e.g. whether a score of 6 on a scale from 0 to 10 is a pass or a fail), culture related response tendencies, etc., can influence responses. All such factors potentially lead to non-comparable results across countries and possibly erroneous conclusions.

6.2.4 Specific limitations of using online panels

In addition to the 'generic' limitations for all categories of questionnaire surveys, there are some additional concerns and possible risks when using online panels:

- **Composition** of the panel. Despite efforts undertaken by panel providers, it is not possible to have a distribution of panel members that is a completely representative of the total population. This is caused by factors such as variations in internet penetration and use (which differs by age, gender, education level and socio-economic status), the lack of interest in being a panel member by certain types of people (e.g. older, busy, rich or introvert people, not interested or willing to give their opinion) and the economic interest in being a panel member.
- **Panel conditioning.** Regular participation in surveys may lead to 'panel conditioning' where participants become overly familiar with survey techniques, potentially biasing their responses.
- **Careless respondents.** Some panel members are mainly motivated by the collection of rewards and tend to complete the survey as fast as possible. The experience in E-Survey of Road users' Attitudes (ESRA) suggests that careless responding is increasing in panels of market research agencies.
- **Low number of panel members.** Especially in small countries and in low- and middle-income countries, the panels available may be too small for large-scale sampling, leading to a higher risk of non-representativity, panel conditioning and often also to higher costs.

- **Manipulation.** Two types of manipulation could occur respondents answering several times for maximising incentives and polling agencies adding duplicate entries to reach the targeted number of respondents. Data should always be checked for duplicates.

6.2.5 Suitability of self-reported data collection for road safety research

Within road safety research, questionnaire surveys are particularly well-suited for:

- **Attitudinal and behavioural studies**. Such studies explore drivers' attitudes towards road safety regulation, driving habits, the social norm in society and self-reported behaviours (e.g. adherence to speed limits, use of mobile phones while driving). Such information is critical for assessing the traffic safety culture of a group.
- **Assessments of acceptance and support for policy measures**. This covers the understanding of public support for, and gauging public reaction to, proposed, or recently implemented road safety policies and interventions.
- **Awareness and education evaluations**. This includes studies assessing the effectiveness of road safety campaigns and educational programs in changing knowledge, attitudes and behaviour.
- **Perception of the effectiveness of road safety interventions.** When measures have been implemented (e.g. road infrastructure, new regulations, enforcement, etc.) the perceived effectiveness and safety level can be measured through questionnaire surveys.

For other road safety topics, alternative methods may provide more reliable or detailed information. Examples are accident analysis and reconstruction, infrastructure safety, traffic flows and vehicle safety. For driving behaviour, both direct observations as well as self-declared behaviour can be used to determine the nature and extent of risky behaviour in traffic. This will be discussed in Section 3.

In conclusion, while self-reported data offer valuable insights into areas that are otherwise challenging to study, such as personal attitudes and behaviours related to road safety, researchers must carefully consider these advantages and limitations when designing studies and interpreting data. To mitigate some of the limitations, employing mixed methods or triangulating data with observational or objective measures where possible can enhance the robustness of research findings.

6.3 Measuring road user behaviour through self-reporting

6.3.1 Methods for measuring road user behaviour

As was illustrated in Section 2, it is possible to use self-reported data for measuring behaviour on the road. Traditionally behaviour in traffic used to be measured using direct observation based on data collected (see, e.g. European Commission, 2022):

- through roadside surveys, with observations made and coded by trained field researchers
- through roadside surveys, with observations made by cameras or radars along the road
- by the police through their law enforcement activities
- by the police when participating in an experimental design
- by in-vehicle sensors (GNSS, odometry, ABS, etc.), tracing and GPS systems
- by mobile phones of road users
- through induction loops under the roads

Direct observation can provide accurate, real-time data on specific behaviours, free from the biases inherent in self-report methods. It is especially useful for quantifying behaviours that can be directly measured, such as vehicle speed, seat-belt usage and the presence of alcohol through breath tests. Direct observation is the predominant measurement method for road safety performance indicators in Europe and the European Trendline programme (see Chapter 12).

However, there are also limitations and disadvantages of using observational methods, including

- They may not capture the context or motivation behind behaviours.
- They are limited to actions that are observable.
- Applying the methodology at the national level can be resource-intensive and costly.
- It can be challenging to use a representative sample of observation locations across a country and weight the observations in an appropriate way.
- The approach may raise privacy concerns (e.g. when using cameras), limiting their use in certain contexts.
- There may be legal limitations for random measurements, in particular, for driving under the influence of alcohol and drugs.
- Observed behaviours may not represent typical driving patterns due to the Hawthorne effect (McCarney *et al.*, 2007), whereby individuals modify their behaviour because they know they are being observed.
- For the reliability and repeatability of behaviour measurements, often certain behaviours are excluded from the observations (e.g. congested traffic for speed measurements, or using mobile phones at night), creating a bias in the observation results.

When collecting data on road safety behaviour through different methods (i.e. self-reported versus observed behaviour), often discrepancies in values emerge. Such discrepancies can be attributed to various factors, including social desirability bias, recall bias and the limitations inherent in each data collection method, but one of the key differences is that different types of prevalence are being compared (period-prevalence versus. point-prevalence). This is a result of different ways of scaling the occurrence of the behaviour. For example, behaviour measurements based on self-reporting typically ask whether a certain behaviour has occurred at least once within a certain period (e.g. the last 30 days; this is an example of period-prevalence). Roadside surveys, on the other hand, typically provide a point estimate (i.e. which percentage of drivers at any given time is not respecting certain traffic regulations; this is an example of point-prevalence). Given the differences in the formulation of the dimensions and scale of the risky behaviour, values from self-reported data are likely to be much higher than those obtained through direct observation.

A specific limitation of measuring behaviour based on self-report data is that it cannot capture unconscious behaviour – which adds to the recall bias discussed earlier: the recall bias refers to behaviour road users were conscious of when it took place, but they have forgotten since; the unconscious behaviour bias refers to behaviour (e.g. speeding, distracted driving, etc.) road users were not aware of at the moment itself.

6.3.2 Example comparing self-reported and observed behaviour

When comparing self-reported and observed behavioural data, it is important to understand that these methods measure different constructs. Very often self-reported data is presented as a binary variable (yes/no), which shows the percentage of respondents admitting to a behaviour in a certain period (period prevalence). This says nothing about the frequency with which the respondents show this behaviour within that period. Observed data provides a precise estimate of the portion of traffic driven by drivers exhibiting a certain behaviour (point prevalence), but it does not indicate whether this is due to a small number of drivers who consistently exhibit this behaviour or to a larger number of drivers who exhibit this behaviour less frequently. Therefore, comparisons of these data should always be interpreted with caution and mainly have an explorative tentative nature.

Despite these methodological challenges, this section presents an example of a tentative comparison of self-reported and observed behaviour data. The example focuses on the topic of driving above the legal alcohol limit. This topic was assessed within the Baseline project (Yannis and Folla, 2023) and the ESRA2 survey (see Section 5.1). In the ESRA2 survey, car drivers were asked if they had driven a car when they may have been above the legal limit for drink driving at least once in the past 30 days [score 2–5 on a 5-point scale from 1 'never' to 5 '(almost) always'] (Achermann Stürmer *et al.*, 2019). Within the Baseline project, eight countries conducted roadside surveys based on random breath testing at random locations by the police. For six of these countries, ESRA2 data are available. The data in Baseline were collected between 2021 and 2022, and the data in ESRA2 were collected in 2018, except for Luxembourg, where the survey was conducted in 2020. Figure 6.1 shows the correlation between observed and self-reported drink-driving behaviour for these six countries.

Although these results must be interpreted with caution due to the limited number of countries and other methodological issues, the data show a strong correlation (0.81) between self-reported and observed behaviour. Apart from Spain, most countries aligned closely on the linear regression line, suggesting that

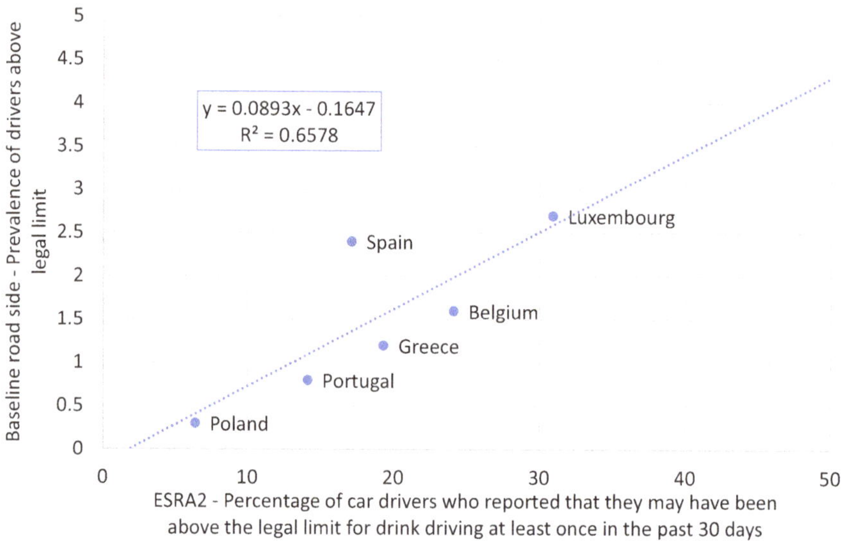

Figure 6.1 Observed drink driving above the legal alcohol limit (Baseline) as a function of self-reported drink driving above the legal alcohol limit (ESRA2)

self-reported data could predict 66% of the international variation in observed drink driving. Even though the self-reported data does not account for the frequency of infractions, or the distance driven while exceeding the legal limit, it still enables relatively accurate prediction of the proportion of kilometres driven over the legal limit in these six countries. An analysis of the correlations of observed behaviour with the concept of 'habits'*, which is also a form of self-reported past behaviour, revealed even higher correlations than those shown in Figure 6.1, which is based on a direct question about their behaviour in the last 30 days.

Not all road safety topics that can be compared between Baseline and ESRA2, show such a strong correlation. For instance, the observed use of a hand-held mobile phone while driving does not seem to correlate with self-reported behaviour, whereas the use of a seatbelt as a car driver or the use of a helmet as a cyclist does.

At the moment of writing this chapter, two groups of experts within the Trendline project (the follow-up project of Baseline) are assessing in more detail the potential of self-reported data as an alternative or complementary indicator for observed road safety performance indicators, for example, in cases where conducting a roadside survey is not feasible or too expensive. The focus of these two

*In ESRA2 'habit' for driving under the influence of alcohol was based on 3 items: 'I often drive after drinking alcohol', 'Even when I am a little drunk after a party, I drive', and 'It sometimes happens that I drive after consuming a large amount of alcohol (e.g. a litre of beer or half litre of wine)'; answered on a 5-point scale from 1 'disagree' to 5 'agree').

groups is on self-reported behaviour (Silverans and Meesmann, 2024) and road safety attitudes, including self-reported habits (Meesmann and Silverans, 2024). In the thematic reports of the ESRA project the Baseline results are compared with the most recent ESRA3 data, which was collected in 2023. They are published on: www.esranet.eu/en/publications/esra3-publications/.

6.3.3 International acceptance of data on self-reported behaviour

Despite the limitations of using self-reported data (see sections 2.3 and 2.4), mainly because of the inherent subjectivity, the risk of recall bias and the social desirability bias, over the last decades there has been a growing acceptance within the research and policy world on the usefulness of road safety data based on questionnaire surveys.

In Europe, this started with the SARTRE surveys (see Cestac and Delhomme, 2010) which included a range of questions on self-reported behaviour. The surveys were predominantly undertaken by using face to face interviews. Only in SARTRE4, the last version of SARTRE, researchers in the Netherlands used an online survey in parallel to the face-to-face interviews. The international ESRA initiative (see Section 5.1) built further on the foundations of SARTRE but used a new approach that was fully based on online panels. This also facilitated its extension outside Europe. A similar evolution from paper-based to online questionnaires can be noted for other road safety surveys, such as those based on the Driver Behaviour Questionnaire (DBQ) items (see, e.g. Winter and Dodou, 2016) and the Traffic Safety Culture Index survey (AAA Foundation for Traffic Safety, 2023).

Recently, using self-reported data also became an accepted methodology for data collection for certain road safety performance indicators in the Baseline and Trendline programmes (see Chapter 12). ESRA2 data on self-reported behaviour was fully integrated into the most recent version of the WHO Global Status Report on Road Safety (World Health Organization 2023).

It should be noted, however, that using online panels is often much more expensive in low- and middle-income countries than in high-income countries; it might be even cheaper to undertake road-side surveys with human observers along the road. Thus, from a cost-effectiveness perspective, it may be disadvantageous for such countries to use surveys with online panels. Furthermore, the representativity of online access panels in these countries for the national population may be limited, i.e. when countries have low internet penetration and certain groups of the population have no access to the internet.

6.3.4 Future directions for self-reported data on behaviour in traffic

With new types of sensors and measurement tools, such as sensors detecting fatigue (see, e.g. Kassem *et al.*, 2021), automatic detection of mobile phone use (see, e.g. Jain *et al.*, 2023) or detection of helmet use by motorcyclists (see, e.g. Siebert *et al.*,

2021), observations of several types of behaviour may become more scalable and less costly – and hence preferable to self-reported behaviour.

Yet, technologies like mobile apps and wearables could also bring a new dimension to self-reported data collection. Such tools enable real-time data collection, offering insights into behaviours and conditions with unprecedented detail and accuracy. For example, apps can prompt users to report their behaviours or experiences shortly after driving, reducing recall bias. Wearables can provide objective data on physiological responses, complementing self-reported measures. The challenge and opportunity lie in integrating these technologies in a way that respects participant privacy and data security.

The future of self-reported data in road safety research is likely to be shaped by advancements in technology and methodology. The increasing ubiquity of internet-connected devices opens new avenues for real-time and context-aware data collection. Machine learning and AI techniques could offer sophisticated tools for analysing large datasets, identifying patterns, and predicting behaviours more accurately. The challenge will be to balance these technological advancements with ethical considerations, ensuring that research continues to be conducted responsibly and with respect for participants' rights and privacy.

6.4 Good practice in self-reported data collection

6.4.1 Survey design

Designing survey questions

In designing the survey, good practice has been followed to increase the validity and relevance of the data (Bryman, 2012b; Bryman and Cramer, 2011; De Leeuw *et al.*, 2008b; Lajunen and Özkan, 2011b):

- Questions should be easy to understand.
- Questions should be short.
- Questions should be asked about one issue at a time.
- Questions should be non-leading.
- It should be obvious how to respond to the question.
- The lists of possible response items to a question should not be too long.
- Appropriate filters should be used so that respondents only need to answer questions relevant to them (this is often impractical in paper-based surveys but can be done in online surveys and computer-assisted interviews).
- Open questions should be avoided because codification and analysis are very time consuming.
- The sequence of response items should be randomised to avoid order effects (this cannot be done in paper-based surveys).
- The list of items used for closed questions should be complete, and the items should not overlap.
- Where possible, questions and their scales should have been validated in other surveys. This holds in particular for international, multilingual surveys.

- For online surveys, the 'length of interview' should not be longer than 20 minutes (to avoid 'respondent fatigue').
- Include measures to reduce careless respondents, (e.g. instructed response questions, see Goldammer *et al.*, 2020; Ward and Meade, 2023).

Two types of questionnaire surveys for measuring behaviour

Data on self-reported behaviour, as for example drink driving behaviour, can be collected through two types of questionnaire surveys, referred to as the 'period-prevalence' and the 'trip-based prevalence' survey (Boets *et al.*, 2021; Vollrath *et al.*, 2019):

- **Trip-based** prevalence surveys measure the prevalence of a certain risky behaviour (e.g. driving above the legal alcohol limit, using the mobile phone, driving while being tired, etc.) during a specific randomly chosen recent trip. This method is used most in connection with the analysis of mobility patterns (see, e.g. Commission *et al.*, 2022) but has also been used in other contexts, such as measurement of driving under the influence of alcohol (see, e.g. Schrauth and Funk, 2022) or drowsiness (see, e.g. Diependaele, 2015). One of the advantages of using trip-based surveys is that the values obtained are, or should be, comparable to those obtained through observational studies, because these also aim at estimating a point-prevalence.

Three types of data collection methods can be used for trip-based prevalence surveys:

- Roadside interviews: traffic participants are approached directly after a trip or during a trip and asked about risky behaviour (e.g. mobile phone use) in a phase of their last/current trip.
- Telephone interviews: participants are asked about their last made trip and then about their risky behaviour during that trip.
- Online survey: participants are asked about their last trip or about all their trips made in the last 24 hours and then about their risky behaviour during one (randomly selected) trip.
- **Period-prevalence** surveys measure the prevalence of risky driving over a specific period (e.g. the last 30 days). This appears currently the most common method in road safety research; it has been used in SARTRE (Cestac and Delhomme, 2010) and is still being used in ESRA (Meesmann *et al.*, 2018; Meesmann and Wardenier, 2024). One of the advantages is that the values are in general much higher than in trip-based surveys, and a lower sample can be used to obtain meaningful results.

Sampling approach

The choice of sampling approach in questionnaire surveys depends on various factors, including research objectives, available resources, the availability and quality of a sampling frame (the population from which the sample is extracted), as well as ethical and statistical considerations (see, e.g. Groves *et al.*, 2009). Two main approaches can be distinguished: probability sampling (also called random sampling) and non-probability sampling (non-random sampling).

If the self-report data collection is undertaken in view of estimating national means, the data should come from a representative sample of the national population. To achieve this, probability sampling must be used to ensure every element in the population has an equal chance of being included in the sample. Commonly used probability sampling techniques are simple random sampling (each member has an equal chance of selection), stratified sampling (population divided into homogeneous groups), systematic sampling (selecting every nth element), cluster sampling (randomly selecting groups), or multistage sampling (combining multiple techniques) (Lohr, 2019).

For financial or practical considerations, it is not always possible to extract a random sample directly from the national population (e.g. random selection of respondents based on national identification number in national registers). Therefore, many online surveys make use of an online access panel (sampling frame). The quality of the representativity of the sample depends on the quality of the sampling frame. An example of how respondents are sampled in the ESRA survey can be found in Section 5.1.

Compensation of recall bias

Failure to remember correctly what happened in the recent past ('recall bias') is an important source of error or bias in self-reported data collection. In a survey, the respective driving, cycling or walking episode should be not too far in the past. The bigger the time lag between the survey moment, the higher the risk of recall bias. Self-reports in telephone interviews and online surveys are usually more delayed than in roadside interviews or trip-based online surveys, since the latter are usually carried out immediately after the trip (Vollrath *et al.*, 2019).

In order to minimise this recall bias, interviews or surveys should be conducted as soon as possible after the respective trip. Roadside interviews for car drivers can be carried out by doing the interviews at parking areas or by conducting roadside interviews where cars are stopped. Pedestrians and cyclists can be stopped along their route, during their trip (if they do not go too fast) or can be approached when stopped or standing at road crossings. In online trip-based surveys or telephone interviews, one would ask the participants to remember their last trip as vividly as possible. In period prevalence surveys, the period considered should be as recent as possible (e.g. the last 30 days rather than a particular month).

Compensation of social desirability bias

A frequently occurring bias in collecting data from interviewees and survey participants is the 'social desirability bias', i.e. the willingness to report truthfully. This is mainly an issue for roadside (and telephone) interviews.

To reduce the social desirability bias, respondents should be assured in advance that all answers will be kept confidential and anonymous. No personal data that can link directly to the individual (e.g. name, address etc.) should be collected. Interviewers should avoid any negative comment about behaviours reported but rather focus on encouraging interviewees to report anything they can remember. To minimise interviewer bias, thorough interviewer selection, training and control are required to maximise accuracy of the interviews.

In online surveys, the answers can also be affected by social desirability bias, but to a lesser extent (Goldenbeld and De Craen, 2013) because there is no risk of an interviewer effect. Nonetheless, a social desirability bias in online surveys cannot be completely eliminated. Predefined texts can be included in online questionnaires to encourage respondents to answer honestly. It is also possible to include questions in the survey that can be used to control for social desirability. For example, in the ESRA survey, a standard social desirability scale question is included, which can be used to control the effect of other variables in further analysis (Meesmannand Wardenier, 2024).

6.4.2 Data analysis

Analysing data from questionnaire surveys is a multi-step process that involves several stages from data collection to data interpretation. The depth and coverage of these processes depend on the research objective and the selected study design. Some general elements of data processing are presented below:

- **Data Cleaning**. This is a crucial step before data analysis should start. Data cleaning includes:
  - removing duplicates (ensuring each respondent's data is entered only once)
  - handling missing data (deciding how to deal with responses that are incomplete; options include imputation, using mean/mode substitution, or simply excluding missing data points or the whole record)
  - filtering outliers (identifying and deciding how to handle data points that fall outside the expected range)
  - correcting errors (addressing inconsistencies or obvious errors in data entry or measurement)
  - remove careless respondents (e.g. in online surveys remove respondents who provide answers without having read the questions attentively; using, e.g. instructed response items, check of straight lining on matrix questions or monitor the response time).
- **Data coding, recoding and transformation**. Responses, especially from open-ended questions, need to be coded into numerically interpretable formats. This may involve:
  - assigning numerical values to ordinal responses (e.g. Likert scale responses)
  - categorising nominal data into groups
  - normalising data to ensure comparability, especially if scales differ among questions or variables.
- **Exploratory data analysis** (EDA). Before diving into complex analyses, performing an EDA can provide insights into the basic properties of the data: This involves calculating descriptive statistics such as mean, median, standard deviation, etc., to get a sense of the data distribution, and visualisation

techniques such as histograms, bar charts, box plots and scatter plots to visually inspect the distribution of responses and identify patterns or outliers.

- **Statistical analysis**. One can use descriptive statistics to summarise data (see previous point) and inferential statistics to draw conclusions about the population based on sample data. Common techniques include:
  - *t*-tests/ANOVA for comparing group means
  - Chi-square tests for examining relationships between categorical variables
  - correlation and regression analysis to investigate relationships between variables
  - factor analysis or principal component analysis (PCA) for reducing data dimensionality or identifying underlying variables
  - cluster analysis to group data into clusters that exhibit similar characteristics without predefined categories.
- **Validation of results**. Validating findings involves several methods such as checking for consistency of results across different sub-samples, testing the robustness of the results by applying alternative statistical methods or using different subsets of data, and/or compare the results with findings from other studies.

6.4.3 Ethical considerations when questioning people in surveys

Ethical considerations in self-reported research underscore the importance of informed consent, ensuring participants understand the study's purpose, their role and their rights, including withdrawal without penalty. Data privacy and protection are paramount, especially with sensitive information. In Europe, GDPR requirements should be adhered to, in order to guarantee privacy. Researchers must adhere to protocols that anonymise data and secure storage and transmission methods to protect participant confidentiality.

When using online panels provided by market research companies and other panel providers, in general the burden for protection privacy is on them. When selecting such companies, researchers should check whether the companies have implemented thorough procedures to guarantee the privacy of the respondents.

6.5 Some examples of studies using self-report data

6.5.1 The ESRA initiative

ESRA is a global initiative, bringing together road safety organisations, research centres and regional sponsors from all over the world. The aim is to collect and analyse comparable data on road safety performance and, in particular self-reported data on road safety culture and behaviour by the road users in each participating country. The ESRA data are used as a basis for a large set of road safety indicators and empower policymakers at all levels to make informed decisions based on scientific evidence.

The ESRA initiative has been initiated by Vias Institute in Brussels, Belgium, which coordinates ESRA together with ten steering group partners: BASt (Germany), DTU (Denmark), IATSS (Japan), ITS (Poland), KFV (Austria), NTUA (Greece), PRP (Portugal), SWOV (the Netherlands), TIRF (Canada) and University Gustave Eiffel (France).

The core of ESRA lies in a jointly developed questionnaire survey, which is subsequently translated into national language versions. This survey explores self-reported behaviours, attitudes and opinions regarding unsafe traffic behaviour, along with experiences with law enforcement and support for policy measures. It covers several road safety topics, such as driving under the influence, speeding, distractions, and seatbelt and helmet use. The survey targets various road users, including car occupants, moped and motorcycle riders, cyclists, pedestrians and e-scooter riders. In the latest edition, the focus on vulnerable road users was broadened, and for the first-time questions on e-scooters and infrastructure were included.

In ESRA, the self-reported data is collected through online panel surveys. The national online panels are provided by subcontracted polling agencies (sampling frame). From these national panels, the respondents are randomly selected based on predefined quota for gender and age (interlaced, gender*six age groups between 18 and 75) and regional distribution of the national population (United Nations Statistics Division 2023). The goal is to have at least 1000 respondents per country. The fieldwork is conducted simultaneously in all participating countries (Meesmann and Wardenier, 2024).

Three editions of the ESRA survey have already been conducted since 2015. For ESRA1, the data was collected in 2015–2017 (Torfs *et al.*, 2016), for ESRA2 in 2018–2020 (Meesmann *et al.*, 2022) and for ESRA3 in 2023 (Meesmann and Wardenier, 2024). Overall, by 2023, the ESRA survey had been conducted in 68 countries across six continents, translated into numerous national languages and programmed in nine different character sets: Arabic, Cyrillic, Devanagari, Greek, Hebrew, Japanese, Korean, Latin and Thai. In the latest edition (ESRA3), self-reported data from more than 37,000 road users in 39 countries, across five continents was collected simultaneously in 2023. The intention is to repeat this survey every 3 to 4 years, and the fieldwork of the next edition (ESRA4) is planned for 2026. An overview of the ESRA initiative and the project results is available on: www.esranet.eu.

The ESRA data provides the opportunity to compare information across countries and benchmark national road safety performance. Key results are publicly available on the ESRA website; this includes national country fact sheets (some in national language), the ESRA dashboard (Vias Institute, 2024), synthesis reports of each edition, thematic reports and detailed methodology reports. The outputs of the ESRA project have become building blocks of national and international road safety monitoring systems such as the Global Status Report on Road Safety (GSRRS) of the World Health Organization (2023) and the country profiles and thematic reports of the European Road Safety Observatory (ERSO) of the European Commission.

The ESRA initiative has shown the feasibility and the added value of a jointly developed approach for a centrally organised collection of self-reported data on road safety culture and behaviour. It improves the comparability of the data and enhances the communication between organisations by facilitating the exchange of expertise from partners of countries with various levels of experience in road safety. Many high-income countries participate in ESRA, but also middle- or low-income countries from other regions such as Africa or South-East and Central Asia.

6.5.2 Public support for policy measures

An important factor influencing the willingness of policymakers to implement road safety measures is the expected level of public support for the interventions envisaged. The more a decision appears to be unpopular, the less policymakers will be inclined to decide to implement such an intervention.

It is important for policymakers to understand the factors that influence the resistance to particular measures (Smith *et al.*, 2014). A high level of public support for a measure is also a requirement for 'success' of the measure, with people actually adapting their behaviour in the desired direction (Goldenbeld, 2002). If a measure meets with resistance from a large group of road users, this group is likely to organise itself and start a movement against the measure. An example is the massive protest in France against the reduction of speed limits on rural roads from 90 to 80 km/h, which even led to a partial reversion of the measure after it had been voted and implemented (Ligue de Défense des Conducteurs, 2020).

Arguments for supporting or opposing policy measures are often linked to the perceived consequences and effects of the measures – and these can only be understood by questioning people about this. Eby *et al.* (2017) found that support for lowering the legal blood alcohol concentration (BAC) limit for drunk driving was partially tied to beliefs about the impacts of a change in the legal BAC limit. In relation to making alcohol interlock systems mandatory in the vehicles of people convicted of driving under the influence of alcohol, McCartt *et al.* (2010) found that most people favouring the measure thought that it would prevent alcohol-impaired driving, save lives, or prevent crashes. Molin and Brookhuis (2007) showed that the understanding of the effects of speeding, the support for the policy goal and the perceived effectiveness were among the factors increasing the acceptability of ISA. Vlassenroot (2011) concluded that factors such as the perceived effectiveness of ISA, as well as equity of penetration, effectiveness of ITS and personal and social aims had the largest effect on the acceptability of ISA. Van den Berghe (2020) showed that the expected consequences of several measures had a major impact on the willingness to support these. These examples suggest that when surveying public opinion on road safety measures, it is important to include questions about the expected effects and consequences of the policy measure. This will allow to understand what arguments – even if these are incorrect – lead people to support or oppose policy measures.

Different survey methods can be used, but currently online surveys are predominant. With a dedicated sampling strategy that ensures representativity of the target population and using a dedicated scale – typically a Likert scale with 5 or 7 values is used – it is relatively easy to measure the acceptability and public support for a specific measure. Often such polls are conducted in a particular country only, but in the field of road safety, there are also international surveys that look at a range of measures across a large number of countries – see e.g. Cestac and Delhomme (2010), Goldenbeld and De Craen (2013), Van den Berghe (2023), Van den Berghe *et al.*(2022) and Vias Institute (2024).

When analysing the survey results, it is often interesting to relate the degree of public support to other variables, such as behaviour on the road, subjective safety and socioeconomic situation. It is also important to distinguish between those affected and those not affected by the measures: because many road safety measures are perceived as restrictive, often those not using a particular transport mode (e.g. cycling) are more in favour of new regulation (e.g. cyclists having to wear a helmet) than those who are affected (e.g. people riding a bike regularly) (Van den Berghe, 2020).

If the level of public support is averaged across a country (or at a subnational level), it can be considered a national indicator and used in international comparisons. It can also be used in modelling and correlation analysis. For example, it has been shown that support for the introduction of ISA systems is strongly correlated to cultural differences (Van den Berghe *et al.*, 2020).

6.5.3 *Traffic Safety Culture Index*

Since 2008, the AAA Foundation for Traffic Safety annually produces a 'Traffic Safety Culture Index' (TSCI), which is based on an online questionnaire. At the time of writing this chapter, the most recent version of the TSCI (2022) surveyed five core areas pertaining to (1) people's perceived danger, (2) perceived risk of apprehension, (3) social disapproval, (4) self-reported behaviour and (5) support for safety countermeasures (AAA Foundation for Traffic Safety, 2023). The TSCI currently is based on a representative sample of 2500 adult motorists in the USA and over 900 teenagers aged 15–18 (more information on the sampling strategy can be found in the reference cited above. English and Spanish versions of the survey are available. Since 2008, the TCSI has been produced annually; although there have been changes in the methodology and the formulation of questions, the TSCI allows to identify trends over time.

The TCSI report is meant to be a useful reference for researchers, practitioners and traffic safety advocates to gain better understanding of people's perceptions and attitudes towards risky driving behaviours, to identify relevant issues and to develop corresponding countermeasures. The rich dataset can be used for a broad range of analyses, in particular when combining different variables. For example, data from the 2022 survey has been used to describe driving profiles in our sample of drivers. Six unique driving profiles were identified in the data and the percentage are: (1) safe drivers (41.2%), distracted drivers (15.0%), speeding drivers (22.7%), impaired drivers (1.3%), distracted and aggressive drivers (17.3%) and most dangerous drivers (2.4%). Such an analysis would be very difficult to conduct based on roadside observations of behaviour.

6.5.4 Measuring drowsy driving

Measuring the prevalence of drowsy driving and its role in crash causation is not self-evident. It is highly challenging to estimate sleepiness purely based on physical characteristics (certainly after a crash occurred). When drivers that caused a crash can be interviewed, there is a general bias not to report sleepiness – either by unawareness or by unwillingness. As a result, crash reports are usually not accurate with respect to driver sleepiness, and few data exist on the overall prevalence of sleepiness at the wheel.

A study in Belgium managed to obtain informative results using a self-report approach (Diependaele, 2015). Its main aim was to obtain a trip-based estimate of the prevalence of sleepiness at the wheel among Belgian car drivers. In an online questionnaire, drivers were asked to reflect on a single journey they had made during the last 24 h and to indicate the level of sleepiness they experienced while driving. The scale used was the Karolinska Sleepiness Scale (KSS) (Kaida *et al.*, 2006). Sleepiness was measured with respect to an actual driving episode. This methodology reduces the risk of response bias (not admitting sleepiness at the wheel) because it avoids direct contact between respondents and researchers (as in a road-side set-up, for instance). At the same time, it was feasible to reach a large sample of drivers that was representative of the Belgian population of car drivers. The study found that overall, 4.8% of the journeys by car drivers in Belgium involved a driver who is showing signs of sleepiness.

Because many additional variables were collected, it was also possible to consider the impact of a wide range of contextual variables that had been found in the literature to influence the sleepiness of drivers. Various circumstances may result in a prevalence of sleepiness that is considerably higher than the overall estimate of 4.8%, such as; spending more than 4 hours a day at the wheel (31%); having caught less than 8 hours of sleep (25%); having an irregular sleep-wake pattern with frequent shifts of more than 2 h (15%); being an adolescent/young adult (18–30 years: 11%); having consumed 2 or more standard units of alcohol prior to driving (2–4 units: 11%), etc. (Diependaele, 2015). All results were consistent with findings from other international studies. The advantage of the method was that, unlike in most other studies, a very large number of possible contributory factors could be included in a single study.

6.6 Conclusion

Collecting and analysing self-reported data can be very valuable for obtaining a wide range of information across various disciplines, including road safety. The choice of survey technique plays a critical role in balancing the advantages and disadvantages inherent to self-reported data. While biases and accuracy concerns are common challenges across all methods, strategic design and implementation of surveys can often mitigate these issues.

In road safety research, self-reported data are particularly useful for exploring drivers' attitudes towards road safety regulation, subjective safety, driving habits, the social norm in society, self-reported behaviour and public support for policy measures. The methods can also be used for evaluating the effects of campaigns and education interventions and for assessing the (perceived) effect of road safety measures.

Different methods can be used for collecting self-report, each with their advantages and limitations. The effectiveness of self-reported data in capturing accurate and reliable information hinges on careful consideration of survey technique, question design and population characteristics.

Online surveys based on online panels are increasingly used because the method is flexible, efficient and relatively cheap. Yet, care should be taken to design an appropriate questionnaire, to ensure representativeness of the sample and to mitigate the effects of recall bias, social desirability bias and careless responders.

References

AAA Foundation for Traffic Safety. (2023). *2022 Traffic Safety Culture Index*. https://aaafoundation.org/wp-content/uploads/2023/09/202311-AAAFTS-Traffic-Safety-Culture-Index-2022.pdf.

Achermann Stürmer, Y., Meesmann, U., and Berbatovci, H. (2019). Driving under the influence of alcohol and drugs. ESRA2 Thematic report Nr. 5. ESRA project (E-Survey of Road users' Attitudes) (Issue 5). Swiss Council for Accident Prevention. https://www.esranet.eu/storage/minisites/esra2018thematicreportno5drivingunderinfluence.pdf.

af Wåhlberg, A. E., Dorn, L., and Kline, T. (2010). The effect of social desirability on self reported and recorded road traffic accidents. *Transportation Research Part F: Traffic Psychology and Behaviour*, 13(2), 106–114. https://doi.org/10.1016/j.trf.2009.11.004.

Baker, R., Blumberg, S. J., Brick, J. M., *et al.* (2010). Research synthesis. AAPOR report on online panels. *Public Opinion Quarterly*, 74(4), 711–781. https://doi.org/10.1093/poq/nfq048.

Boets, S., Silverans, P., Houwing, S., Forsman, Å., Klipp, S., and Folla, K. (2021). *Methodological Guidelines – KPI Driving under the Influence of Alcohol*. Brussels: Vias Institute. https://www.baseline.vias.be/storage/minisites/methodological-guidelines-kpi-alcohol.pdf.

Bryman, A. (2012a). Social Research Methods. 4th edition. In *Social Research Methods* (4th ed.). Oxford: Oxford University Press. https://doi.org/10.4135/9781849209939.

Bryman, A. (2012b). Social Research Methods. 4th edn. In *Social Research Methods* (4th ed.). Oxford: Oxford University Press. https://doi.org/10.4135/9781849209939.

Bryman, A., and Cramer, D. (2011). *Quantitative Data Analysis with IBM SPSS 17, 18 and 19: A Guide for Social Scientists*. London: Routledge.

Cestac, J., Delhomme, P., Antov, D., *et al.* (2010). European road users' risk perception and mobility. The SARTRE 4 survey.

Choi, B. C. K., and Pak, A. W. P. (2005). A catalog of biases in questionnaires. *Preventing Chronic Disease*, 2(1), 1–13.

European Commission: Directorate-General for Mobility and Transport, Armoogum, J., Garcia, C., Gopal, Y., *et al.* (2022). *Study on New Mobility Patterns in European Cities – Final Report. Task A, EU Wide Passenger Mobility Survey*. Publications Office of the European Union. https://doi.org/doi/10.2832/728583.

De Leeuw, E. D., Hox, J. J., and Dilman, D. A. (2008a). *International Handbook of Survey Methodology*. Milton Park: Taylor & Francis.

De Leeuw, E. D., Hox, J. J., and Dilman, D. A. (2008b). *International Handbook of Survey Methodology*. Milton Park: Taylor & Francis.

Diependaele, K. (2015). Sleepy at the wheel. Analysis of the extent and characteristics of sleepiness among Belgian car drivers. https://www.vias.be/publications/Slaperig%20achter%20het%20stuur/Sleepy%20at%20the%20wheel%20-%20Analysis%20of%20the%20extent%20and%20characteristics%20of%20sleepiness%20among%20Belgian%20car%20drivers.pdf.

Eby, D. W., Molnar, L. J., Kostyniuk, L. P., *et al.* (2017). Perceptions of alcohol-impaired driving and the blood alcohol concentration standard in the United States. *Journal of Safety Research*, 63, 73–81. https://doi.org/10.1016/j.jsr.2017.08.013.

European Commission. (2022). *Road Safety Thematic Report – Road Safety Performance Indicators (RSPIs). European Road Safety Observatory*. European Commission, Directorate General for Transport. https://road-safety.transport.ec.europa.eu/system/files/2022-10/Road_Safety_Thematic_Report-Road_Safety_Performance_Indicators.pdf.

Goldammer, P., Annen, H., Stöckli, P. L., and Jonas, K. (2020). Careless responding in questionnaire measures: Detection, impact, and remedies. *Leadership Quarterly*, 31(4). https://doi.org/10.1016/j.leaqua.2020.101384.

Goldenbeld, C. (2002). *Publiek draagvlak voor verkeersveiligheid en verkeersveiligheidsmaatregelen. Overzicht van bevindingen en mogelijkheden voor onderzoek*. (Issue D-2002-2). SWOV Stichting Wetenschappelijk Onderzoek Verkeersveiligheid.

Goldenbeld, C., and De Craen, S. (2013). The comparison of road safety survey answers between web-panel and face-to-face; Dutch results of SARTRE-4 survey. *Journal of Safety Research*, 46, 13–20. http://doi.org/10.1016/j.jsr.2013.03.004.

Groves, R. M., Fowler, F. J., Couper, M. P., Lepkowski, J. M., Singer, E., and Tourangeau, R. (2009). *Survey Methodology*, 2nd edn. New York: Wiley. https://www.wiley.com/en-be/Survey+Methodology%2C+2nd+Edition-p-9780470465462.

Jain, H., Zope, H., Jain, H., Rakte, I., and Raut, S. (2023). Mobile phone detection and alcohol detection with engine locking for the prevention of car accidents. *2023 International Conference on Digital Applications, Transformation and Economy, ICDATE 2023*. https://doi.org/10.1109/ICDATE58146.2023.10248639.

Kaida, K., Takahashi, M., Åkerstedt, T., *et al.* (2006). Validation of the Karolinska sleepiness scale against performance and EEG variables. *Clinical Neurophysiology*, 117(7), 1574–1581. https://doi.org/10.1016/j.clinph.2006.03.011.

Kassem, H. A., Chowdhury, M., and Abawajy, J. H. (2021). Drivers fatigue level prediction using facial, and head behavior information. *IEEE Access*, 9, 121686–121697. https://doi.org/10.1109/ACCESS.2021.3108561.

Krosnick, J., and Presser, S. (2010). Question and questionnaire design. In P. Marsden and J. Wright (eds.), *Handbook of Survey Research* (2nd edn) Leeds: Emerald Group Publishing. http://www.stanford.edu/dept/communication/faculty/krosnick/docs/2010/2010Handbook of Survey Research.pdf.

Lajunen, T., Corry, A., Summala, H., and Hartley, L. (1997). Impression management and self-deception in traffic behaviour inventories. *Personality and Individual Differences*, 22(3), 341–353. https://doi.org/10.1016/S0191-8869(96)00221-8.

Lajunen, T., and Özkan, T. (2011a). Self-report instruments and methods. In B. E. Porter (eds.) (Ed.), *Handbook of Traffic Psychology* (pp. 43–59). New York: Academic Press. https://doi.org/10.1016/B978-0-12-381984-0.10004-9.

Lajunen, T., and Özkan, T. (2011b). Self-report instruments and methods. In B. E. Porter (Ed.), *Handbook of Traffic Psychology* (pp. 43–59). New York: Academic Press. https://doi.org/10.1016/B978-0-12-381984-0.10004-9.

Lajunen, T., and Summala, H. (2003). Can we trust self-reports of driving? Effects of impression management on driver behaviour questionnaire responses. *Transportation Research Part F: Traffic Psychology and Behaviour*, 6(2), 97–107. https://doi.org/10.1016/S1369-8478(03)00008-1.

Ligue de Défense des Conducteurs. (2020). *Le 80 km/h perd du terrain: plus de 31 500 km repassés à 90 en 2020!* Ligue de Défense des Conducteurs. https://www.liguedesconducteurs.org/vitesse-infos-et-intox/le-80-km-h-fiasco-de-l-annee-pour-le-gouvernement.

Lohr, S. L. (2019). *Sampling: Design and Analysis.* 2nd edn. Boca Raton, FL: Chapman and Hall/CRC. https://doi.org/10.1201/9780429296284.

McCarney, R., Warner, J., Iliffe, S., van Haselen, R., Griffin, M., and Fisher, P. (2007). The Hawthorne effect: A randomised, controlled trial. *BMC Medical Research Methodology*, 7(1), 30. https://doi.org/10.1186/1471-2288-7-30.

McCartt, A. T., Wells, J. K., and Teoh, E. R. (2010). Attitudes toward in-vehicle advanced alcohol detection technology. *Traffic Injury Prevention*, 11(2), 156–164. https://doi.org/10.1080/15389580903515419.

Meade, A. W., and Craig, S. B. (2012). Identifying careless responses in survey data. *Psychological Methods*, 17(3), 437–455. https://doi.org/10.1037/a0028085.

Meesmann, U., and Silverans, P. (2024). KPI road safety attitudes. Results of first stage data exploration. *Report produced as part the Trendline project, supported by the European Union (manuscript under review).*

Meesmann, U., Torfs, K., Nguyen, H., and Van den Berghe, W. (2018). *Do We Care about Road Safety? Key Findings from the ESRA1 Project in 38 Countries*. Brussels: Vias Institute – Knowledge Centre Road Safety.

https://www.vias.be/publications/ESRA2018 – Key findings ESRA1/ERSA2018_Key_findings _from_the_ESRA1_project_in_38_countries.pdf.

Meesmann, U., Torfs, K., Wardenier, N., and Van den Berghe, W. (2021). *ESRA2 methodology. ESRA2 report Nr. 1 (updated version). ESRA project (E-Survey of Road users' Attitudes)* (Vol. 1). Brussels: Vias Institute. https://www.esranet.eu/storage/minisites/esra2-methodology-report-updatewave2-def.pdf.

Meesmann, U., and Wardenier, N. (2024). ESRA3 *Methodology* ESRA3 Thematic *Report* Nr. 1. *Version 1.0. ESRA Project (E-survey of Road users' Attitudes).* https://www.esranet.eu/storage/minisites/esra3-methodology-report.pdf.

Meesmann, U., Wardenier, N., Torfs, K., Pires, C., Delannoy, S., and Van den Berghe, W. (2022). A global look at road safety. *Synthesis from the ESRA2 Survey in 48 Countries. ESRA Project (E-Survey of Road Users' Attitudes)*. Brussels: Vias Institute. www.esranet.eu.

Molin, E. J. E., and Brookhuis, K. A. (2007). Modelling acceptability of the intelligent speed adapter. *Transportation Research Part F: Traffic Psychology and Behaviour*, 10(2), 99–108. https://doi.org/10.1016/j.trf.2006.06.008.

Nießen, D., Partsch, M. V., Kemper, C. J., and Rammstedt, B. (2019). An English-language adaptation of the social desirability – Gamma Short Scale (KSE-G). *Measurement Instruments for the Social Sciences*, 1(1), 2. https://doi.org/10.1186/s42409-018-0005-1.

Pires, C., Torfs, K., Areal, A., *et al.* (2020). Car drivers' road safety performance: A benchmark across 32 countries. *IATSS Research*, 44(3), 166–179. https://doi.org/10.1016/j.iatssr.2020.08.002.

Rea, L. M., and Parker, R. A. (2014). *Designing and Conducting Survey Research: A Comprehensive Guide*, 4th edn. New York: Wiley. https://www.wiley.com/en-ie/Designing+and+Conducting+Survey+Research:+A+Comprehensive+Guide,+4th+Edition-p-9781118767023.

Schrauth, B., and Funk, W. (2022). *Key Performance Indicator für Alkohol – Entwicklung einer Methodik und Ersterhebung.*

Siebert, F. W., Hellmann, L., Pant, P. R., Lin, H., and Trimpop, R. (2021). Disparity of motorcycle helmet use in Nepal – Weak law enforcement or riders' reluctance? *Transportation Research Part F: Traffic Psychology and Behaviour*, 79, 72–83. https://doi.org/10.1016/J.TRF.2021.04.005.

Silverans, P., and Meesmann, U. (2024). KPI self-reported behaviour. Results of first stage data exploration. Report produced as part of the Trendline project, supported by the European Union.

Smith, K. C., Debinski, B., Pollack, K., *et al.* (2014). Research-informed evidence and support for road safety legislation: Findings from a national survey. *Accident Analysis and Prevention*, 73, 109–115. https://doi.org/10.1016/j.aap.2014.08.016.

Sudman, S., and Bradburn, N. M. (1982). *Asking Questions: A Practical Guide to Questionnaire Design*. San Francisco, CA: Jossey Bass.

Tellis, G. J., and Chandrasekaran, D. (2010). Does culture matter? Assessing response biases in cross-national survey research. *International Journal of Research in Marketing*, 27(4).

Torfs, K., Meesmann, U., Van den Berghe, W., and Trotta, M. (2016). *ESRA 2015 – The results. Synthesis of the Main Findings from the ESRA Survey in 17 Countries*. Belgian Road Safety Institute. https://www.vias.be/publications/ESRA 2015 Results/ESRA 2015 Results.pdf.

Tourangeau, R., and Smith, T. W. (1996). Asking sensitive questions: The impact of data collection mode, question format, and question context. *Public Opinion Quarterly*, 60(2), 275–304. https://www.jstor.org/stable/2749691.

United Nations Statistics Division. (2023). *UNData. Population by age, sex and urban/rural residence.* http://data.un.org/Data.aspx?d=POP&f=tableCode%3A22.

Van den Berghe, W. (2020). Factors determining public support for policy measures in road safety. An international survey and its findings for China. Road & Rail Transport Safety: Examples from the International Experience and Implications for China. *Technical Communication*, 1(1), 68–89.

Van den Berghe, W. (2022). *The influence of fairness and ethical trade-offs on public support for road safety measures. An international and intercultural exploration.* [University College London]. https://doi.org/10.13140/RG.2.2.36252.56962.

Van den Berghe, W. (2023). Intercultural differences in road safety performance and support for road safety policy measures. In M. Mohiuddin, S. Jayashree, and T. Aziz (eds.), *Multiculturalism and Interculturalism.* London: IntechOpen. https://www.oxfordlearnersdictionaries.com/.

Van den Berghe, W., Schachner, M., Sgarra, V., and Christie, N. (2020). The association between national culture, road safety performance and support for policy measures. *IATSS Research*, 44(3), 197–211. https://doi.org/10.1016/j.iatssr.2020.09.002.

Van den Berghe, W., Sgarra, V., Usami, D. S., González-Hernández, B., and Meesmann, U. (2022). Public support for policy measures. ESRA2 thematic report (updated version). *ESRA Project (E-Survey of Road Users' Attitudes).* Brussels: Vias Institute and Rome: Research Centre for Transport and Logistics. https://esranet.eu/en/publications/esra2-publications/#thematic_reports.

Vias Institute. (2024). ESRA3 dashboard. https://www.esranet.eu/en/dashboard/.

Vollrath, M., Schumacher, M., Boets, S., and Meesmann, U. (2019). *Guidelines for Assessing the Prevalence of Mobile Phone Use in Traffic* (Issue November). FERSI. https://fersi.org/.

Ward, M. K., and Meade, A. W. (2018). Applying social psychology to prevent careless responding during online surveys. *Applied Psychology*, 67(2), 231–263. https://doi.org/10.1111/apps.12118.

Ward, M. K., and Meade, A. W. (2023). Dealing with careless responding in survey data: Prevention, identification, and recommended best practices. *Annual Review of Psychology*, 74, 577–596. https://doi.org/10.1146/annurev-psych-040422-045007.

Ward, N. J., Watson, B., and Fleming-Vogl, K. (2019). *Traffic Safety Culture: Definition, Foundation, and Application*. Leeds: Emerald Publishing Limited.

Winter, J. C. F. De, and Dodou, D. (2016). National correlates of self-reported traffic violations across 41 countries. *Personality and Individual Differences*, 98, 145–152. https://doi.org/10.1016/j.paid.2016.03.091.

Wolf, C., Joye, D., Smith, T., and Fu, Y. (2016). *The SAGE Handbook of Survey Methodology*. Thousand Oaks, CA: SAGE Publications. https://doi.org/10.4135/9781473957893.

World Health Organization. (2023). *Global Status Report on Road Safety 2023*. https://www.who.int/teams/social-determinants-of-health/safety-and-mobility/global-status-report-on-road-safety-2023.

Yannis, G., and Folla, K. (2023). *Baseline Report on the KPI Driving under the Influence of Alcohol. Baseline Project*. https://www.baseline.vias.be/en/publications/kpi-reports/.

Chapter 7

Automated driving and safety data

Andrew Tarko[1] *and Vamsi Bandaru*[1]

Although automated driving is in its early stage of development, fast progress is expected in the foreseeable future, as well as the growing presence of automated vehicles (AVs) on roads. A central premise for the ultimate AV success is its safety. Fulfilling the safety expectation has become crucial, especially after the involvement of early AVs in collisions with other road users. Timely thorough research and proper standards are key factors in controlling the AV collision risk. Although the US government took a stance in 2016 and declared that machine-driven cars would make the US highways safer [1], AV safety assurance is the primary motivation of the early governmental requirements of AVs testing and evaluation before their approval for public roads. One of the most promising expansions and enhancements of automated driving is communication between road users (V2V) and between road users and road infrastructure (V2I). This enhancement was enabled by modern technology that has led to a new paradigm – connected and automated vehicles (CAVs). The AV acronym used here includes AVs with added communication ability.

This chapter discusses both the data expected to be collected by AV units and the potential use of these data in future safety management and other related areas. Driving, regardless of whether performed by humans or machines, is facilitated via navigation – a sequence of frequently repeated tasks. Section 7.1 explains these tasks briefly in the context of both human driving and automated driving. Section 7.2 discusses the type of data generated by each of the tasks. This data is likely preserved for later inspection and analysis. Although the basics of AV technology and its implementation discussed in this chapter are based on well-established elements at the time of writing, fast progress is expected.

A failure in any of the elements of a driving task – whether during the perception of the surroundings, planning navigation, or its execution – may lead to a safety-related event such as a crash with considerable outcomes, a traffic conflict with a looming crash prevented at the last moment with an evasive action, or a violation some of the operational thresholds with no outcome obvious to vehicle

[1]Lyles School of Civil and Construction Engineering, Purdue University, IN, USA

occupants and external observers. All these various events are considered incidents in this chapter. Some of them are expected to be detected and corrected in real time; other incidents may lead to traffic conflicts and even collisions. All of them should be documented with saved sensors' outputs, results of data processing (perception), navigation decisions, navigation execution, and vehicle motion. Some of these incidents may only be detected after the trip by inspecting data saved by the AV unit. More serious incidents, such as traffic conflicts and collisions, could be detected during the trip by the AV unit and by other AVs and roadside infrastructure-based units.

Traffic conflicts are considered surrogates of crashes due to the same etiology – a navigation failure. The mentioned task execution failures and resulting traffic conflicts and collisions are incidents that may be identified afterward to evaluate the AV safety performance, reveal the causes of incidents, and identify the needed countermeasures to improve navigation. These events are discussed in detail in one of the following chapters.

The importance of collected data for understanding the origination of these incidents is emphasized. Various stakeholders, such as AV users, government, insurance companies, and vehicle fleet owners, should be aware of the data potential applications useful to them and the general public to encourage data accessibility and utilization. The necessary safeguards of privacy, ownership, and business interest are also mentioned to increase the prospect of a full and just use of the data with a focus here on road safety improvements.

7.1 Components of driving task

Automated driving refers to vehicles that are capable of driving by themselves without human involvement. The Society of Automotive Engineers (SAE) has established six levels of automated driving with its J3016 standard [2]. Level 0 represents manual driving, level 5 represents complete automated driving, and intermediate levels represent growing extents of driving autonomy. Most of today's vehicles have some version of adaptive cruise control and thus fall under level 2. A human driver is not required at level 5 under all conditions. Instead, a vehicle must autonomously execute the following components (or subtasks):

1. Perceives the surrounding environment.
2. Communicates with nearby road users and/or infrastructure.
3. Predicts the near-future positions of nearby road users.
4. Plans own motion to minimize the risk of collision.
5. Executes the planned actions subject to frequent updating.

Motion execution in step 5 takes several seconds, while perception of surroundings and planning own motion in steps 1–4 are updated much more frequently. Thus, only the initial part of the planned motion is executed when an updated plan is ready. Frequent updating of the motion delivers a frequent and smooth transition between motion plans.

A failure or imperfection in executing any of these subtasks of the driving task may constitute an incident that is reflected in the saved AV data record. Data and incidents are discussed in the subsequent sections of the chapter.

7.1.1 Perceiving the surrounding environment

AVs use sensors to perceive their own position and motion as well as the relative position and motion of the surrounding stationary and moving objects. To achieve a robust perception of the surrounding environment, the AV industry has been gravitating towards using multiple sensors of various types that complement each other and their multiplicity ensures the information redundancy to increase reliability. Four types of sensors are available: (1) cameras, (2) radars, (3) ultrasonic sensors, and (4) LiDARs. To take the full benefit of the sensors' architecture, their signals are processed simultaneously and in combination.

AV companies have their own ways of arranging these sensors around their vehicle. Notable examples of companies and their sensor configurations include

- Uber's system uses all the four types of sensors [3],
- Tesla's system does not use LiDAR [4],
- Waymo's system does not use ultrasonic sensors [5].

Each of the four sensor types has its own characteristics that complement other sensor types. For instance, a visible-spectrum camera works for various ranges and provides good resolution, but it cannot work in the dark and in adverse weather conditions such as heavy fog and rain. A radar, conversely, operates in a long range and in the dark, fog, snow, and rain but has poor resolution and does not provide color information. All the sensors scan the environment at high frequency (from 10 Hz for LiDAR to 40 Hz for ultrasonic sensors) to quickly refresh the data about the environment. Table 7.1 provides a comparison of the strengths and weaknesses of the four types of sensors. The shaded cells of the table highlight the top performance of individual sensors in individual performance categories. Since none of

Table 7.1 Perception sensors' performance – individually and in combination

Performance dimension	**Sensor type**				**All combined**
	Ultrasonic	**Cameras**	**Radar**	**LiDAR**	
Proximity	5	2	3	2	5
Range	1	5	4	4	5
Resolution	2	5	2	4	5
Color/contrast	0	5	0	1	5
Speed	0	2	5	4	5
Distance measurement	5	2	5	5	5
Works in dark	5	0	5	5	5
Works in bright	5	5	5	5	5
Works in snow/fog/rain	5	2	5	3	5

the sensors can deliver the required performance in all categories by itself, fusing signals from various sensors helps the system derive a cohesive and comprehensive view and interpretation of the environment.

Advancements in machine learning and artificial intelligence gradually improve the accuracy of environment detection and interpretation. Yet, no system is perfect, and there are situations when either the individual sensors, the sensor fusion, or the signal interpretation may fail.

7.1.2 Communicating with nearby road users and/or infrastructure

Perception of the surroundings by an AV is limited to the field of view of its sensors. A long-range LiDAR can provide a signal useful to detect another vehicle at a distance of 200 m or even longer. In good conditions, a long-range camera enables the detection and interpretation of a traffic signal by an AV at a distance of 300 m from a signalized intersection. Thus, such long-distance sensors are effective in avoiding collisions with fixed obstructions or moving vehicles. Nevertheless, these technologies are ineffective in preventing some collisions in certain circumstances. For example, angle or left-turn collisions are difficult to avoid if the other object is outside of the detection area – sometimes narrow. This is one of the cases where vehicle-to-vehicle (V2V) and vehicle-to-infrastructure (V2I) communication (together referred to as V2X) becomes useful [6]. Simulation-based research showed that V2X communication not only helps in improving the efficiency of roads (in terms of capacity) but also in improving safety [7,8].

IEEE 802.11p was an amendment to the broader 802.11 standard (commonly referred to as the Wi-Fi standard) to enable support for Intelligent Transportation System (ITS) applications [9]. The standard forms the basis for European standard ETSI ITS-G5 [10] and the US's dedicated short-range communications technology [11], which are intended for use for V2X communication. Another emerging communication technology using cellular signals, the same one that powers mobile phone connectivity, called cellular vehicle to everything (C-V2X), aims to compete with IEEE 802.11p-based systems. A comparison of the two can be found in [12].

7.1.3 Predicting near-future positions of nearby road users (risk-free situation)

After perceiving the positions of fixed and moving objects, a human driver predicts the relative motion of the nearby road users and fixed obstructions. This capability is learned when walking or running and further improved when driving. AVs, in other words, need specialized algorithms to predict the relative motion of other objects and to select a proper response. Multiple methods are used to accomplish this result.

Conventional methods of motion prediction (including relative motion), such as Kalman filter, rely on physics-informed mathematical models and statistical techniques to make predictions based on previous observations and with a proper treatment of involved observation uncertainty. Kalman filter applied in a two-dimensional space predicts the relative motion of other objects in a short time horizon. Predicting for a sufficiently long horizon of several seconds is needed. It must utilize not only the dynamics and geometrics of the recent and current motion (as the Kalman filter does) but also the situational context where multiple road users perceive and predict others' motion to plan own motion to preserve traffic safety and comfort. Thus, there is a push towards machine learning-based algorithms that are capable of simultaneously predicting the future position of multiple road users while taking their interactions with nearby road users into account [13].

For instance, in [14], a spatio-temporal long-short-term memory (ST-LSTM) neural network model is used where the interactions of multiple vehicles and their effects on the vehicle of interest are modeled. Similarly, the authors in [15] use an LSTM neural network with convolutional social pooling for robustly learning interdependencies in vehicle motions. This work was presented in [16], where the authors also encoded scene context in the neural network. A survey of motion prediction methods for pedestrians and vehicles in the context of autonomous driving was provided in [13].

7.1.4 Planning own motion to eliminate the risk of collision

There are at least two horizons for trip planning: one-time route planning between origin and destination and continuous trajectory planning for the next number of seconds along the planned route. The second case is particularly safety-relevant and is presented below.

AV plans its own motion based on the outcome of the previous step, i.e., predicting the near-future positions of the nearby road users. The motion is planned to keep the vehicle within its traveled way and along its own route and to avoid any collision with other road users and obstructions. When the just-planned motion is being executed, the surrounding area is being perceived again and the current path updated. Thus, only a small portion of the designed path is implemented when a new updated path is ready for execution from AV's new position. Thus, the safety and comfort of AV occupants are achieved with small and frequent path adjustments. Although failure of AV navigation is likely when sensors malfunction, Mobileye, for example,– a leading provider of advanced driver-assistance solutions, posits that it is possible to navigate without crashes [17]. A survey of motion planning methods for autonomous vehicles can be found in [18] and [19].

7.1.5 Execution of the planned trajectory

Like planning, execution of the planned path happens at two levels. At the first level, the overall route planned is followed towards the destination. When the destination is known, a high-level route planning algorithm decides the set of roads that need to be taken to reach the destination. A comprehensive study on this topic can be found in [20]. This route is broken down into a sequence of waypoints that help inform the short-term planning step [20].

The planned short-term path is then executed at the second level, by executing various navigation actions locally by keeping lane, safely changing lanes, maintaining proper distance to other vehicles, etc.

Replacing mechanical linkages and hydraulic systems with drive-by-wire (or "x-by-wire") systems to transfer steering decisions to corresponding actuators via electrical impulses marked a significant step towards automated driving in CAVs [21]. For example, in the electronic throttle control, the sensor-measured position of the accelerator pedal is coded via an electrical signal and received by an engine control module (ECM) [22]. The ECM compares it with the required throttle position (output of calculations performed on the data measured by other sensors such as vehicle speed sensor, engine speed sensor, cruise control settings). The calculated throttle position, if different from the measured current one, triggers sending an appropriate signal to electric motor (actuator) which adjusts the throttle valve. The on-board computer responsible for the above-mentioned driving subtasks computes the necessary throttle position and sends an appropriate signal to the ECM. In the instances where the ECM function is integrated into the on-board computer, the appropriate signals are directly sent to the electric motor controlling the throttle valve. All the generated data can be saved for future inspections.

To ensure high reliability and minimize the impact of possible steering incidents (failures of any of the components in the execution chain), several additional safety measures such as computational redundancy, functional redundancy, fail-operation systems, and other safety measures are required [23].

7.2 AV collected data

An AV generates large amounts of data during its operation. When a potentially unsafe situation occurs, data generated during the several seconds leading up to the event may be stored to diagnose, understand, and improve system responses. This data is crucial for post-incident analysis, system refinement and enhancing overall safety. The types of data stored or available can be detailed with respect to the five driving tasks mentioned earlier.

7.2.1 Own perception sensors data

Data involved in the perception subtask includes sensor data and perception algorithm output. *Sensor data* such as high-resolution images from various cameras installed around the vehicle, LiDAR point clouds, and radar reflections and ultrasonic sensor readings may help reconstruct the vehicle's surroundings leading up to the time of some incident.

Perception algorithm output may include data on detected objects (vehicles, pedestrians, obstacles), their classification, position, and movement. Generated data logs are crucial for detecting sensing incidents when a sensor misses an object while other sensors indicate the object's presence. Such an omission may lead to a violation of certain motion thresholds (lateral acceleration, jerk, etc.).

7.2.2 Data received from other vehicles and roadside units

Records of all exchanges between the vehicle and other entities (vehicles, infrastructure) including but not limited to received warnings, status of traffic lights, approaching emergency vehicle, positions of nearby road users can be used to ascertain if incorrect information was received and acted upon which led to an unsafe event.

7.2.3 Prediction and planning results

Logs of the predictive models' outputs, including the predicted trajectories of other road users relative to the AV, conditions under which those predictions were made such as the speed of the vehicle, actions by the road users and any assumption made by the prediction algorithms. This information helps assess the accuracy of the predictions leading up to the incident.

Also of importance is the trajectory that the AV planned to follow, along with any alternative paths it considered. This includes the rationale behind choosing a specific path over others. Documentation of the decision-making process, including risk assessments, safety margins considered, and any constraints that influenced the path planning, will be of utmost importance to AV manufacturers.

An example of a perception incident is when the AV updated trajectory differs more than allowed for a smooth transition from one trajectory to the other of this AV.

7.2.4 Motion and vehicle control data

Logs of the control commands issued (steering, acceleration, braking) and the vehicle's responses to these commands. Data on the operational status of critical systems (e.g., braking system, steering mechanism) and any faults or warnings that were active at the time of the incident. These pieces of data help identify any discrepancy between the planned and actual vehicle behavior. Such a case may be caused by the mentioned in the previous step too large difference between two consecutively planned trajectories of the same vehicle.

Finally, own trajectories executed by the vehicle are recorded using GPS coordinates, readings from the speedometer, accelerometer, etc.

7.3 Safety-relevant incidents and their causes

Any event caused by an error in perception or navigation by any of the involved vehicles, including AV, is considered a traffic incident. Not all incidents have a risk of collision associated with them. For instance, a momentary interference to a LiDAR sensor caused by direct sunlight or another LiDAR pointed directly at it may cause issues that result in an error in the perception step. If this happens when the vehicle is stopped, for example, when waiting for the green at a traffic signal, there is no risk of a crash. Whereas, if it happens during merging into an adjacent lane occupied by other vehicles, the risk of collision is considerable. This example shows that the perception issue detected in a safe scenario must be considered seriously for its safety effect under other potential scenarios. Documenting all such incidents helps improve the robustness of various algorithms in an AV.

There are two safety-relevant driving incidents: traffic conflicts and crashes. Traffic conflicts are incidents that involve perception or navigation errors and bring a considerable probability of crash. Traffic conflicts may result from an incident in any of the five tasks of automated driving. Traffic conflicts involve delayed response that leads, in most cases, to an evasive maneuver too strong to be acceptable by vehicle occupants. If the evasive maneuver is not successful, it leads to a crash – a collision with considerable outcome that calls for reporting to the police. Such events are reported to law enforcement and/or insurance agencies.

Traditionally crash data is the primary source of information used to estimate safety. This situation is changing due to the increasing proliferation of traffic monitoring devices such as cameras and the number of sensors used in vehicles. A detailed discussion of the data available from AVs is provided in the previous section (Section 7.1.2).

Another category of incidents of interest to AV manufacturers is called disengagements. These are incidents where the human operator had to take over, either on prompts by the vehicle or the operator monitoring the vehicle, sensing that the vehicle is executing a potentially dangerous maneuver. These events provide insights as to scenarios where the AV's algorithm(s) failed and thus are used to improve and evaluate future versions of the AV algorithms.

7.3.1 Causes of conflicts and crashes

As mentioned previously, errors in any of the diving tasks lead to an incident that has a non-zero risk of crash. In the case of human drivers, human errors such as recognition error or decision error are often considered to be the critical reason and often the last failure in the causal chain that puts the human-driven vehicle on a collision path (NMVCCS report).

Conversely, AVs rely on complex systems and algorithms to navigate safely. However, each stage in their operation (outlined in Section 7.1.2) can encounter challenges that may compromise safety. A brief overview of some of the potential challenges in each step is provided below.

Perception

Sensors may have limited range or resolution due to adverse weather conditions (e.g., fog, heavy rain, snow) or get obstructed by dirt and debris. These limitations can lead to incomplete or inaccurate environmental data [24,25]. Even if the sensors are fully functional, the system might misinterpret the data such as failing to recognize a pedestrian due to poor lighting or confusing an overpass with a vehicle. These errors can result in inappropriate responses [26].

Communication (with other vehicles or infrastructure)

Interruptions or delays in communication due to poor network performance, interference or lack of infrastructure can prevent the exchange of critical safety information. Cybersecurity breaches could lead to the interception, manipulation, or blocking of crucial data, potentially causing unsafe situations [27].

Prediction of future positions of road users

Predicting the movements of dynamic and unpredictable road users (like pedestrians and cyclists) can be challenging. Misjudgments or inaccuracies can lead to incorrect anticipatory actions [28]. The system's algorithms might rely on flawed assumptions about typical road user behavior, failing to account for anomalies or unexpected actions [13].

Path planning

The system might plan a path that is theoretically optimal but does not sufficiently account for current road conditions, construction zones, or dynamic obstacles, leading to unsafe maneuvers [18]. Alternatively, the path planning algorithm might not allow sufficient buffer for unexpected events or emergencies, reducing the vehicle's ability to adapt to sudden changes safely [29].

Execution of the planned path

In situations where human intervention is needed (for SAE level 3 and level 4 AVs), the transition from automated to manual control can be problematic if the human driver is unprepared, distracted, or overly reliant on the AV's systems [30]. Hardware malfunctions or software glitches in the vehicle's control system can prevent the execution of the planned path as intended, leading to erratic or unsafe behavior.

Each of these challenges represents a potential risk to the safety and reliability of AV operations. Addressing these issues involves ongoing research, development, and rigorous testing to enhance the robustness and resilience of automated driving technologies.

A notable example of a failure at multiple levels is the incident involving Uber's AV that resulted in the death of a pedestrian [3]. The perception system never correctly classified the pedestrian. Instead classified the female as "other," "unknown," bicycle or vehicle at various points in time in the 5 s leading up to the crash. Further, the system ignored the tracking history when predicting the future path of the pedestrian due to classification changes leading to incorrect predictions [3]. Finally, when the system realized that a crash was imminent 1.2 s before impact, it still did not apply brakes due to a different design decision [3]. This is a failure at the planning stage.

7.3.2 AV data for crash analysis

Currently, crash reports developed and stored by police are used as the primary source of safety information. The crash data collected by police includes the location and time of crash and multiple pieces of information recorded at the crash scene from the officers' own investigation and statements collected from road users involved in a crash and from eyewitnesses. Severe crashes may be followed by forensic investigation to faithfully restore the events leading to and during a crash (refer to Chapter 2). Data acquiring and saving by AVs before and during crashes are expected to be a "game-changer" in documenting crash occurrences.

The most promising are data acquired and recorded by AV units involved in collisions or present in the vicinity of the collision spot, as outlined in Section 7.1.2. The data collected by uninvolved units may be more reliable if the impact at the collision damages the sensors and/or data storage in involved AV units. Thus, the biggest void haunting currently existing crash data – the lack of pre-crash data – will be filled in most cases when AVs are involved in a crash or present nearby.

One most valuable piece of data that cannot be easily replaced with a surrogate measure is the collision outcome. Currently, most of what is known on this matter comes from police-reported crashes and from laboratory experiments with dummies. The first source does not provide reliable information at the moment of collision pertaining to the speeds, vehicle motions, deceleration rates, positions and motions of vehicle occupants, performance of safety devices inside the vehicle, etc. The second source relies on models of human bodies with their simplifications and imperfections. Instrumented AVs may and should include sensors that measure speeds and accelerations of the vehicle, and hopefully, video cameras pointing inside the vehicle. These signals would document the impact phase and provide unprecedented object dynamic tracking and deformation information useful for further improvements of biomechanical models and prediction of bodily injuries and their severity.

Properly acquired and verified crash data collected with AVs may become an important source of safety data for the AV industry and designers to improve their products. Crash data helps traffic engineers and road operators in the identification, analysis and treatment of risks. Safety managers use crash data to identify high-risk locations and allocate a budget for necessary countermeasures after performing in-depth audits of said locations. This allows for effective spending of taxpayers' money. Policymakers and safety managers use crash data to set crash reduction targets, develop road safety action plans and monitor the progress in accomplishing the targets.

Although crashes are the most definite manifestation of unsafety, they are scarce and do not warrant sufficient flow of safety information for timely interventions where the system is deficient, and this fact is yet to be discovered. Thus, the role of crashes in preventing other crashes from occurring is somewhat debatable, and a more attractive alternative will be presented in the next section of this chapter.

7.3.3 AV data for traffic conflicts analysis

This section presents the concept of failure-based traffic conflicts and conditions applicable to AV-collected and other data necessary to detect these types of conflicts. The remaining section briefly presents a procedure for estimating the frequency of crashes based on the observed conflicts and provides initial evidence of the method's validity. More details can be found in [31].

One may consider a traffic encounter as an event that exposes a road user to a crash potential. As such, the number of traffic encounters may serve as an overall exposure of an AV unit to crash. This measure seems to be better than the overall time spent when driving or the overall distance covered by the unit, because the mentioned overall time of travel and distance traveled, include portions of travel when the potential of colliding with another road user is non-existent. This improvement may lead to a more adequate comparison of crash rates estimated based on aggregate data for various AV solutions and under various external conditions. Furthermore, improving existing crash frequency models could be possible by applying more adequate exposure measures together with other safety factors.

A further improvement of the concept of exposure, and consequently, estimation of safety, is incorporating road user's failure, such as a lapse of attention, as a necessary component of a traffic conflict. Such a failure must be manifested with a late response to the hazard. This modification requires telling apart truly dangerous failure-cased traffic conflicts from controlled aggressive behaviors. Tarko [31] discussed this issue and proposed a practical method of setting a critical threshold of crash nearness and companion conditions to eliminate the majority of aggressive behaviors observed by external observers.

The usefulness of traffic conflict and other data is conditioned on the ability to extract crash frequency and severity information. Two alternative methods proposed to accomplish this are as follows: (1) extreme value models [32–35] and (2) counterfactual and causal models [31,36,37].

The recent large-scale naturalistic driving program SHRP2 [38,39] provided an unprecedented opportunity to test both the usefulness of data collected with instrumented vehicles and the analysis methods for traffic safety evaluation and management. Tarko [31,37] applied the failure-based traffic conflicts in a counterfactual model to data collected with instrumented vehicles driven by human drivers. The human component made the task difficult but did not invalidate the conclusions about data usefulness, even in relation to AV units.

Naturalistic driving studies brought a glimpse of the benefits and challenges possible when applying real-time road and traffic sensing along with in-vehicle technologies. Although the purpose of sensing in naturalistic driving is much less ambitious than that of AVs, certain issues arising from using sensors attached to a vehicle could be revealed. The required accuracy of measurements could be met by the SHRP2 sensors only after serious testing, cleaning, and preprocessing of obtained data. Data collected with front-facing radar during severe near-crashes and crash events may be problematic due to missing or incorrect readings. Vehicle "diving" during strong braking pushes a radar beam downward beneath a preceding vehicle. A strongly swerving vehicle directs fixed radar beams in directions not relevant to the intended direction of travel. The imperfect performance of radar raises a concern about the reliability of directional sensors subject to irregular motions of a vehicle's body during abrupt evasion maneuvers. The inherent limitations of sensors could be mitigated with sensor redundancy, diversification, and adding omnidirectional sensors with the 360-degree field of view.

After ensuring the data sufficient coverage, completeness, and accuracy, the next important question about validity should address the data adequacy for analyzing safety. Tarko [31] and Tarko and Lizarazo [40] compared the rear-end crash frequencies estimated from crash data and conflict data. The analysis was applied to three categories of drivers: young male drivers, mature male drivers, and mature female drivers. The results clearly indicated that speed and its higher derivatives supplemented with the distance to another vehicle provide sufficient information to detect the failure-caused traffic conflicts and to estimate the expected frequency of crashes for studied scenarios. In this study, the scenarios were based on the age and gender of a driver following another vehicle during car-following. The estimated rear-end crashes for these driver categories followed closely the actual crash rates recorded during the program over several years and thousands of drivers. The ratio of times required to collect traffic conflicts and crashes was 1:70.

The most challenging were the heterogeneous drivers' behavioral preferences of following another vehicle changing across drivers and across time and conditions. Even more challenging was proper interpretation of the data to eliminate false positives and interactions at speeds that did not pose any real hazard; thus, affecting drivers' behavior towards accepting smaller distances to other vehicles. These challenges should be easier to address with data expected to be collected with AV units.

7.4 Data sharing and users

Although AV-collected data are expected to be of volume and quality higher than the SHRP2 data, the challenges to overcome would be data handling and sharing with proper protection of business ownership and privacy of AV users. These challenges are less related to technology and more to legal regulations, business practices, and privacy concerns. New protocols of data exchange and cooperation are required between automanufacturers, owners and users of AVs, and law enforcement and transportation agencies that protect the interests of all involved groups [41].

One such effort is IEEE 1616.1-2023 [42] that focuses on Data Storage Systems for Automated Driving. It defines the requirements, goals, and metrics for data storage systems specifically designed to support automated driving technologies. The standard outlines functions and common technical requirements for these in-vehicle data storage systems, ensuring that data relevant to automated driving system (ADS). Levels 3, 4, and 5 are adequately captured, stored, and managed. This includes critical data elements necessary for the operation, analysis, and improvement of automated driving systems, addressing aspects such as performance, safety, and incident analysis. By standardizing the approach to data storage for AVs, IEEE 1616.1-2023 aims to enhance the reliability, security, and interoperability of these systems, facilitating advancements in automated driving technologies and their integration into broader transportation systems [42].

The need for a robust privacy framework is underscored by potential vulnerabilities that could be exploited by unauthorized surveillance, identity theft, or manipulation of vehicle functionality. Given the intricacies of AV data ecosystems, privacy regulations by the government need to evolve to address these unique challenges, ensuring data minimization, encryption, and secure data-sharing mechanisms. Establishing comprehensive privacy standards is imperative, not only to protect individuals but also to foster public trust in AV technologies, as emphasized in scholarly discourse and regulatory discussions [41, 43]. These measures are essential in mitigating risks associated with data breaches and unauthorized access, thereby safeguarding the integrity of AV operations and the privacy of road users. Ultimately, privacy regulations are inherently regional and thus must be resolved by individual country and/state governments. The standards, such as the IEEE1616.1-2023 [42], provide a basis to build upon the user of the standard to comply with local regulations.

Apart from the AV manufacturers, other groups may also be interested in the AV data. Some of the stakeholders with their unique needs and potential, sharing issues are discussed in the remainder of this section.

7.4.1 Government

Governments are responsible for setting policies regarding AVs in their respective countries. They are also responsible for mainlining road infrastructure that directly impacts the AVs, especially connected vehicles.

During the transition period from the current fleet of human-operated vehicles to a proliferation of AVs, understanding the safety performance of AVs is key to setting up the right regulations that allow AVs to thrive but also protect road users.

California in the USA is at the forefront of AV testing on public roads. Programs enacted in California grant manufacturers two permissions: first to test AVs with backup drivers and second to test driverless vehicles [44]. The state requires that the testers report not only collisions but also incidents where vehicles disengage from the automated mode to let a human take control [45]. These incident reports not only help the manufacturer understand the shortcomings of their algorithms but also help the government understand the scenarios in which these AVs fail.

One of the concerns of California's approach is that reporting AV disengagements is difficult if not impossible to enforce. Another difficulty is comparing disengagement statistics across manufacturers. Lack of standards for driving scenarios and the shortage of exposure data, one can neither compare the rates of disengagement between manufacturers nor draw meaningful conclusions. For obvious reasons, manufacturers do not support these systems [46]. Manufacturers have their own development programs and schedules. Some manufacturers test AVs for level 5 autonomy in all scenarios, while others may only test AVs along certain routes for taxi services. To make it more difficult, disengagement reports do not include routes. Advanced driver support systems, such as Tesla's Autopilot [4], are exempt from reporting [46]. At the time of this writing, California's program is the only publicly available source of data on AV safety performance on public roads.

It is tempting for AV manufacturers and developers to take advantage of the imperfect disengagement reporting rules by reducing the AV annual exposure (miles driven) to lower the number of reported disengagements. On the other hand, a likely public mistrust towards AVs [47] after possible AV crashes may force AV manufacturers to take all precautions including thorough testing before introducing their product to the market. Governments might have to engage AV manufacturers in collaboration for a better understanding AV requirements in the context of road infrastructure design and planning. In addition to public perception of AV safety, governments have means of enforcing the proper operation of AV units by certifying the AV units only for certain conditions called operational design domain [48]. The failure-related risk analysis based on traffic conflicts may be used to define these domains without waiting for crashes.

Highway administrators with transportation and other engineers involved in safety management form another group of potential users. Roadside instrumentation at road intersections can track all road users within intersection areas, identify traffic conflicts, and estimate the risk of crashes. The location-specific safety analysis supplemented with traffic and weather conditions provides a traffic flow-level perspective that differs from a vehicle-centered analysis. Although the idea that AVs can operate successfully in the current environment designed for human road users is attractive, the truth is that an instrumented transportation infrastructure can help accomplish safe and efficient travel by adjusting the roads and traffic control to the presence of automated driving. Traffic engineers must use proactive approaches in analyzing safety data with attention to updating their knowledge about changes in safety. Factors such as road geometry, signage, traffic signals and interactions between road users, as well as human and machine interaction must be considered. The flow-level perspective on safety and its management requires quick feedback in response to equally fast changes in safety and its relationships to these factors.

7.4.2 Insurance companies

One of the motivations for developing and implementing AVs is the promise of roads safer and more efficient than today. One of the stakeholders that will benefit from the advent of safer vehicles on roads is the insurance industry. For instance, it is well known that a considerable percentage of all auto-related insurance claims are a result of parking lot accidents. The proliferation of AVs with their ability to park themselves is expected to significantly reduce such accidents.

Insurance companies need safety data to estimate the effects of modern safety features, for example, blind spot detection or automatic emergency braking, to determine the proper insurance premiums paid by owners of modern vehicles with such features. It gives the auto-insurance industry an edge over competitors. This sort of data takes time to build up to the levels that they can have a considerable impact and AV data will be helpful [49]. For example, Tesla – an American manufacturer of electric cars – has equipped its vehicles with an advanced driver support system – Autopilot. Tesla cars have driven over billions of miles with or without their Autopilot system engaged [50]. Even when the Autopilot system is not engaged, it passively executes its decision process based on the sensor input and compares its outcome with the decisions made by the user. The system learns from the actions taken by a driver and sends them to Tesla's servers. The knowledge learned is shared with all Tesla vehicles [51] and Tesla's automated platform is improving continuously. Equipped with this tool applied to their vehicles, Tesla has entered the insurance industry itself [52]. Tesla's example indicates that automakers, if they follow Tesla's example, may hold massive data relevant to their vehicles and drivers of these vehicles once they start selling AVs. They may share or sell data to insurance companies or establish their own auto-insurance service.

It is already clear that AV-collected data are beneficial for the insurance industry. It is not obvious yet if combining production of AV units, selling them to end users and providing auto-insurance (even if AV users are occupants and not drivers) will be beneficial to both automakers/insurers and their customers. Otherwise, certain regulations may be needed to encourage AV data sharing among various industries that currently exist separately and complement each other to maintain this model if preferable.

7.4.3 Vehicle owners

Two distinct models of AV ownership are possible: (1) individual ownership for the exclusive use of own unit and (2) fleet ownership by a business that offers its units for rent or use-as-needed to end customers.

7.4.4 Individual AV owners

In the scenario of AV units owned by individuals, the AV data could be utilized by organizations or businesses specialized in processing the safety data to maintain safety ratings for various AV units available on the market, similar to the currently existing star-based rankings by IIHS [53]. These rankings help individual buyers make an informed purchasing or renting decision. Current tests by the IIHS are limited to crash-based test results. Their website gives detailed descriptions of how the tests are performed and how they assign ratings. Only one category, namely, "Front Crash Prevention" test, deals with how the car responds to imminent collision scenarios [53]. The pool of AV data extended to traffic conflicts would allow expanding these rankings to other safety functions of AV units.

With a considerable proliferation of AVs, a completely new set of tests needs to be developed and applied in addition to those being carried out currently. These tests need to be designed to simulate real-world scenarios and test both common scenarios that AVs normally encounter and corner cases where the automated driving systems are likely to struggle. Unlike current crash-based testing done, IIHS may need manufacturer help to get some diagnostic information to exactly ascertain what the car "sees" and what its planned course of action is. Traffic conflict data will play a crucial role in this new development.

7.4.5 AV fleet owners

Fleets owned by businesses may include taxi cabs, rental cars, and commercial trucks. Safety rankings described in [53] for the benefit of individual owners may also be useful for businesses exploring the AV market when investing in a new fleet or modernizing the existing one. Decisions of customers to use an AV taxi service are most likely to be affected by safety records and ratings of offered vehicles. Thus, fleet owners will be forced to maintain safe fleets or will risk losing to competition. Thus, the safety data collected by various AV types will be critical.

There is a strong need for collecting safety data for these early solutions other than crashes. Some uses of traffic incident data for the trucking industry currently would be to help improve driver behavior through targeted training.

In the event of a crash, the wealth of data collected by AVs (as described in Section 7.1.2) allows for accident reconstruction, which removes a lot of ambiguity regarding who (or which vehicle) is at fault. This allows for more streamlined crash investigation, damage claims, insurance claims, etc. In the long term, this helps lower insurance premiums for fleet operators.

7.4.6 AV manufacturers

There are two different approaches taken by manufacturers to develop AVs. The first approach is to build vehicles with as much capability as possible today and then let them operate in a geofenced area that is well-mapped. A human safety driver is always present monitoring the system and intervening when necessary. The second approach is to collect data from existing vehicles that are instrumented with sensors to slowly develop increasing automated functions.

Companies such as Waymo [5], Uber [54], and May Mobility [55] in the US have taken the first approach. Waymo has progressed to providing a true driverless taxi service, thereby becoming a fleet owner as well. The primary and obvious use of incident data for AV manufacturers would be to improve the overall safety of the vehicle by improving algorithms responsible for each of the driving tasks described in Section 7.1.1. All incidents with non-zero probability of crash, namely traffic encounters and disengagements are of vital importance to AV manufacturers. The data recorded by the various perception sensors onboard the vehicle help recreate the scenario leading up to and a few seconds after the incident in a simulation. These simulations serve as test cases against which future versions of the AV algorithms are tested before being deployed in vehicles. The data recorded by the sensors also helps maintain updated maps of the location in which the vehicle operates.

The second approach is adopted by Tesla. A portion of vehicles sold by Tesla have a suite of sensors that Tesla claims is enough to allow for the vehicle to drive by itself. When these vehicles are operated by human drivers, the sensor data is fed to a computer that perceives the environment, predicts the future position of surrounding road users, and finally plans its own motion. It compares this plan with the action taken by the human driver, and if the differences are large, the data leading up to the scenario is recorded and eventually sent to Tesla's servers. Using this approach, Tesla gathers a lot more information than other competitors who operate a limited fleet in a limited geographical area.

This data can support root-cause analysis of AV units' failures to improve their reliability by units' manufacturers. The data may also help identify contributing factors attributable to other vehicles and road users. The countermeasures in the first category include identifying conditions and situations leading to AVs' involvement in crashes, improvements of AVs' design in all the phases of its navigation discussed previously in this chapter, supporting AVs' functions in real time by communicating operational information or more tactical warning messages at intersections.

7.5 Conclusion

This chapter has given only a glimpse of present and anticipated needs, opportunities, and solutions in the area of data collected by current and future AVs. Dynamically developing technology will possibly lead to new ground-breaking solutions in sensing and operations. The future will most likely undermine many of today's expectations and will revise the current expectations about the future techniques and technologies of data collection and data use.

One expectation seems to be indisputable though – the revolutionary changes in surface transportation brought by the emerging AV technology. Data already being collected, and emerging new data will play up most important role in tracking and even predicting the safety trends, detecting new safety issues, and helping point out timely solutions and revisions of the technological development.

One dramatic shift will take place in the safety analysis. The current struggle to understand human behavior and limitations under the omnipresent shortage of data and information will be replaced with the struggle of managing large amounts of various types of data and process them timely to extract useful safety-relevant information needed to evaluate and improve new solutions, and to implement the identified countermeasures timely. The long-foretold transition from too little data to too much data is happening.

Human operators – the weak point of transportation systems safety – are being gradually replaced with automated units and this process is expected only to speed up. The better controlled transportation environment with potentially more effective automata in avoiding crashes, will require a complete revamping of the safety management system. Attempting to better understand the natural components and environment of the transportation system will be supplemented with collecting feedback from the added man-made components and environment. This shift will require engaging in safety management in addition to civil engineering experts. They include system engineers, software engineers, automata experts, mechanical engineers, industrial engineers, and others.

The contribution of the private sector to the transportation system will expand with instrumentation added to vehicles and infrastructure. Data use, sharing, and access control will require novel legal and technological solutions that will address privacy issues and fair sharing of the information. The government's role will be crucial in these regulatory needs.

It seems quite likely that the perceived risk of crashes among travelers may be growing with reducing their involvement in driving. At the same time, the importance of crash data for safety management will be diminishing with the decline in the number of crashes and the increased ability to observe and document traffic encounters, particularly traffic conflicts (near-crashes). This new type of data is critical for developing new theories of traffic safety that connect in a proactive way the non-crash events and their conditions with the risk of crashes and their outcome severity.

References

[1] C. Kang, "Self-driving cars gain powerful ally: The government," *The New York Times*, 20 Sep 2016. [Online]. Available from: https://www.nytimes.com/2016/09/20/technology/self-driving-cars-guidelines.html [accessed: 7 Feb 2024].

[2] Society of Automotive Engineers, "J3016_202104: Taxonomy and definitions for terms related to driving automation systems for on-road motor vehicles – SAE International," [Online]. Available from https://www.sae.org/standards/content/j3016_202104/ [accessed 28 Feb 2024].

[3] National Transportation Safety Board, *"Vehicle automation report,"* National Transportation Safety Board, HWY18MH010, Nov 2019. [Online]. Available from: https://data.ntsb.gov/Docket/Document/docBLOB?ID=40477717&FileExtension=.PDF&FileName=Vehicle%20Automation%20Report-Master.PDF [accessed 28 Feb 2024].

[4] Tesla, "Autopilot," [Online]. Available from https://www.tesla.com/en_IE/autopilot [accessed 28 Feb 2024].

[5] S. Jeyachandran, "Introducing the 5th-generation Waymo driver: Informed by experience, designed for scale, engineered to tackle more environments," Waymo. [Online]. Available from https://blog.waymo.com/2020/03/introducing-5th-generation-waymo-driver.html [accessed 28 Feb 2024].

[6] G. D. Krueger, *VII Michigan Test Bed Program Concept of Operations*, Michigan Department of Transportation, 2005.

[7] H. Chen, V. K. Bandaru, Y. Wang, M. A. Romero, A. Tarko, and Y. Feng, "Cooperative perception system for aiding connected and automated vehicle navigation and improving safety," *Transportation Research Record*, 2024. https://doi.org/10.1177/03611981241252779.

[8] J. Cui, H. Qiu, D. Chen, P. Stone, and Y. Zhu, "Coopernaut: End-to-end driving with cooperative perception for networked vehicles," in *IEEE/CVF Conference on Computer Vision and Pattern Recognition (CVPR)*, 2022.

[9] D. Jiang and L. Delgrossi, "IEEE 802.11 p: Towards an international standard for wireless access in vehicular environments," in *VTC Spring 2008-IEEE Vehicular Technology Conference*, IEEE, Piscataway, NJ, 2008, pp. 2036–2040.

[10] ETSI, "Intelligent Transport Systems (ITS): Access layer specification for Intelligent Transport Systems operating in the 5 GHz frequency band," European Telecommunications Standards Institute (ETSI), EN 302 663 V1.2.1, 2013.

[11] "Dedicated Short Range Communications (DSRC) Service | Federal Communications Commission," [Online]. Available from https://www.fcc.gov/wireless/bureau-divisions/mobility-division/dedicated-short-range-communications-dsrc-service [accessed 25 Mar 2024].

[12] K. Kiela, V. Barzdenas, M. Jurgo, *et al.*, "Review of V2X–IoT standards and frameworks for ITS applications," *Applied Sciences*, vol. 10, no. 12, p. 4314, 2020, doi:10.3390/app10124314.

[13] M. Gulzar, Y. Muhammad, and N. Muhammad, "A survey on motion prediction of pedestrians and vehicles for autonomous driving," *IEEE Access*, vol. 9, pp. 137957–137969, 2021, doi:10.1109/ACCESS.2021.3118224.

[14] S. Dai, L. Li, and Z. Li, "Modeling vehicle interactions via modified LSTM models for trajectory prediction," *IEEE Access*, vol. 7, pp. 38287–38296, 2019, doi:10.1109/ACCESS.2019.2907000.

[15] N. Deo and M. M. Trivedi, "Convolutional social pooling for vehicle trajectory prediction," in *2018 IEEE/CVF Conference on Computer Vision and Pattern Recognition Workshops (CVPRW)*, 2018, pp. 1549–15498. doi:10.1109/CVPRW.2018.00196.

[16] T. Zhao, Y. Xu, M. Monfort, *et al.*, "Multi-agent tensor fusion for contextual trajectory prediction," in *2019 IEEE/CVF Conference on Computer Vision and Pattern Recognition (CVPR)*, 2019, pp. 12118–12126. doi:10.1109/CVPR.2019.01240.

[17] S. Shalev-Shwartz, S. Shammah, and A. Shashua, "Vision zero: Can roadway accidents be eliminated without compromising traffic throughput?" [Online]. Available from https://static.mobileye.com/website/corporate/rss/vision_zero_with_map.pdf [accessed 29 Feb 2024].

[18] B. Paden, M. Čáp, S. Z. Yong, D. Yershov, and E. Frazzoli, "A survey of motion planning and control techniques for self-driving urban vehicles," *IEEE Transactions on Intelligent Vehicles*, vol. 1, no. 1, pp. 33–55, 2016, doi:10.1109/TIV.2016.2578706.

[19] S. Aradi, "Survey of deep reinforcement learning for motion planning of autonomous vehicles," *IEEE Transactions on Intelligent Transportation Systems*, vol. 23, no. 2, pp. 740–759, 2022, doi:10.1109/TITS.2020.3024655.

[20] H. Bast, D. Delling, A. Goldberg, *et al.*, "Route planning in transportation networks," *Algorithm Engineering: Selected Results and Surveys*, Springer, Berlin, 2016, pp. 19–80.

[21] R. Isermann, R. Schwarz, and S. Stolzl, "Fault-tolerant drive-by-wire systems," *IEEE Control Systems Magazine*, vol. 22, no. 5, pp. 64–81, 2002, doi:10.1109/MCS.2002.1035218.

[22] R. D. Garrick, "Sensitivity of contact electronic throttle control sensor to control system variation," *SAE Technical Paper*, 2006.

[23] "ISO 26262-1:2018(en), Road vehicles — Functional safety — Part 1: Vocabulary," [Online]. Available from https://www.iso.org/obp/ui/#iso:std:iso:26262:-1:ed-2:v1:en [accessed 6 Mar 2024].

[24] J. Janai, F. Güney, A. Behl, and A. Geiger, "Computer vision for autonomous vehicles: problems, datasets and state of the art," *Foundations and Trends® in Computer Graphics and Vision*, vol. 12, no. 1–3, pp. 1–308, 2020.

[25] D. Feng, C. Haase-Schütz, L. Rosenbaum, *et al.*, "Deep multi-modal object detection and semantic segmentation for autonomous driving: Datasets, methods, and challenges," *IEEE Transactions on Intelligent Transportation Systems*, vol. 22, no. 3, pp. 1341–1360, 2020.

[26] R. McAllister, Y. Gal, A. Kendall, *et al.*, "Concrete problems for autonomous vehicle safety: Advantages of Bayesian deep learning," *International Joint Conferences on Artificial Intelligence*, pp. 4745–4753, 2017.

[27] J. Petit and S. E. Shladover, "Potential cyberattacks on automated vehicles," *IEEE Transactions on Intelligent Transportation Systems*, vol. 16, no. 2, pp. 546–556, 2014.

[28] S. Lefèvre, D. Vasquez, and C. Laugier, "A survey on motion prediction and risk assessment for intelligent vehicles," *ROBOMECH Journal*, vol. 1, no. 1, pp. 1–14, 2014.

[29] W. Schwarting, J. Alonso-Mora, and D. Rus, "Planning and decision-making for autonomous vehicles," *Annual Review of Control, Robotics, and Autonomous Systems*, vol. 1, pp. 187–210, 2018.

[30] S. M. Casner, E. L. Hutchins, and D. Norman, "The challenges of partially automated driving," *Communications of the ACM*, vol. 59, no. 5, pp. 70–77, 2016.

[31] A. Tarko, *Measuring Road Safety With Surrogate Events*. Amsterdam: Elsevier, 2020.

[32] K. L. Campbell, H. C. Joksch, and P. E. Green, "A bridging analysis for estimating the benefits of active safety technologies," University of Michigan, Transportation Research Institute, Ann Arbor, MI, Final Report UMTRI-96-18, Apr 1996. [Online]. Available from: https://deepblue.lib.umich.edu/bitstream/handle/2027.42/1165/89736.0001.001.pdf?sequence=2 [accessed 29 Feb 2024].

[33] S. Coles, J. Bawa, L. Trenner, and P. Dorazio, *An Introduction to Statistical Modeling of Extreme Values*, vol. 208, Springer, Berlin, 2001.

[34] P. Songchitruksa and A. P. Tarko, "The extreme value theory approach to safety estimation," *Accident Analysis & Prevention*, vol. 38, no. 4, pp. 811–822, 2006.

[35] A. P. Tarko, "Use of crash surrogates and exceedance statistics to estimate road safety," *Accident Analysis & Prevention*, vol. 45, pp. 230–240, 2012.

[36] G. A. Davis, J. Hourdos, H. Xiong, and I. Chatterjee, "Outline for a causal model of traffic conflicts and crashes," *Accident Analysis & Prevention*, vol. 43, no. 6, pp. 1907–1919, 2011.

[37] A. Tarko, "Estimating the expected number of crashes with traffic conflicts and the Lomax Distribution–A theoretical and numerical exploration," *Accident Analysis & Prevention*, vol. 113, pp. 63–73, 2018.

[38] A. Blatt, J. Pierowicz, M. Flanigan, *et al.*, *Naturalistic Driving Study: Field Data Collection*, Transportation Research Board, Washington, DC, 2014, p. 22367. doi:10.17226/22367.

[39] J. M. Hankey, M. A. Perez, and J. A. McClafferty, "Description of the SHRP 2 naturalistic database and the crash, near-crash, and baseline data sets," Virginia Tech Transportation Institute, 2016.

[40] A. Tarko and C. Lizarazo, "Validity of failure-caused traffic conflicts as surrogates of rear-end collisions in naturalistic driving studies," *Accident Analysis & Prevention*, vol. 149, p. 105863, 2021.

[41] N. H. T. S. Administration and others, "Automated driving systems 2.0: A vision for safety," Washington, DC: US Department of Transportation, DOT HS, vol. 812, p. 442, 2017, [Online]. Available from: https://www.nhtsa.gov/sites/nhtsa.gov/files/documents/13069a-ads2.0_090617_v9a_tag.pdf?xid=PS_smithsonian [accessed: 29 Feb 2024].

[42] IEEE, *IEEE 1616.1-2023: IEEE Standard for Data Storage Systems for Automated Driving*. Institute of Electrical and Electronics Engineers. [Online]. Available from: https://standards.ieee.org/ieee/1616.1/10939/

[43] D. J. Fagnant and K. Kockelman, "Preparing a nation for autonomous vehicles: Opportunities, barriers and policy recommendations," *Transportation Research Part A: Policy and Practice*, vol. 77, no. C, pp. 167–181,2015, [Online]. Available from: https://EconPapers.repec.org/RePEc:eee:transa:v:77:y:2015:i:c:p:167-181

[44] "Autonomous vehicles," California DMV. [Online]. Available from https://www.dmv.ca.gov/portal/vehicle-industry-services/autonomous-vehicles/ [accessed 29 Feb 2024].

[45] California DMV, "Disengagement reports," California DMV. [Online]. Available from https://www.dmv.ca.gov/portal/vehicle-industry-services/autonomous-vehicles/disengagement-reports/ [accessed 29 Feb 2024].

[46] A. J. Hawkins, "Everyone hates California's self-driving car reports," The Verge. [Online]. Available from https://www.theverge.com/2020/2/26/21142685/california-dmv-self-driving-car-disengagement-report-data [accessed 29 Feb 2024].

[47] Partners for Automated Vehicle Education, "PAVE poll: Americans wary of AVs but say education and experience with technology can build trust | PAVE campaign," [Online]. Available from https://pavecampaign.org/pave-poll-americans-wary-of-avs-but-say-education-and-experience-with-technology-can-build-trust/ [accessed 29 Feb 2024].

[48] E. Thorn, S. C. Kimmel, M. Chaka, and B. A. Hamilton, "A framework for automated driving system testable cases and scenarios," United States. Department of Transportation. National Highway Traffic Safety Administration, Final Report, 2018.

[49] A. J. Hawkins, "We're never going to get meaningful data on self-driving car testing," *The Verge*. [Online]. Available from https://www.theverge.com/2020/6/15/21292014/dot-nhtsa-self-driving-car-test-data-voluntary. [accessed 29 Feb 2024].

[50] J. Steve, "CONFIRMED: Tesla autopilot projected to hit 2 billion miles engaged on Monday (or sooner)," Whole Mars Catalog. [Online]. Available from https://wholemars.net/2019/11/08/tesla-autopilot-projected-to-hit-2-billion-miles-engaged-on-monday-or-sooner/ [accessed 29 Feb 2024].

[51] M. Harris, "Tesla's autopilot depends on a deluge of data," *IEEE Spectrum*. [Online]. Available from https://spectrum.ieee.org/tesla-autopilot-data-deluge [accessed 29 Feb 2024].

[52] Tesla, "Introducing Tesla insurance," *Tesla*. [Online]. Available from https://www.tesla.com/insurance [accessed 29 Feb 2024].

[53] IIHS, "About our tests," *IIHS-HLDI Crash Testing and Highway Safety*. [Online]. Available from https://www.iihs.org/ratings/about-our-tests [accessed 29 Feb 2024].

[54] Uber Technologies Inc, "Explore the Uber platform | Uber United States," *Uber*. [Online]. Available from https://www.uber.com/ [accessed 6 Mar 2024].

[55] May Mobility, "Home | May Mobility." [Online]. Available from https://maymobility.com/ [accessed 6 Mar 2024].

Chapter 8

Naturalistic (driving/cycling/walking) studies

Niels Agerholm[1]

8.1 Theoretical background

Dark figures refer to the under-reporting or non-reporting of traffic accidents [1]. A fundamental problem in traffic safety work is that there is no comprehensive registration of all traffic accidents [1]. The official accident statistics based on police-registered accidents are, despite variation between countries, in general, burdened by a very large dark figure [1].

The advent of Big Data, characterized by large data volumes generated and made available on the Internet and through digital media ecosystems, has significant implications for traffic safety [2]. The global digitization of products, processes, and business models is reshaping the opportunity with data. With advancing technological infrastructure and the advent of the "Internet of Things," companies continue to innovate, finding new ways to capture and leverage ever-expanding amounts of data [2].

Traditional techniques of collecting and storing data have become insufficient, with an overwhelming amount of data being generated [3]. New technologies have appeared, each one targeting specific aspects of large-scale distributed data processing [3]. However, in the Big Data context, data characteristics, such as volume, multi-heterogeneous data sources, and fast data generation, increase the risk of quality degradation and require efficient mechanisms to check data worthiness [4].

In conclusion, the increasing volume of data collected can play a crucial role in enhancing traffic safety knowledge. It provides a wealth of information that can be analyzed and utilized to make informed decisions and strategies for improving road safety further than traditional traffic safety data allows.

8.2 What are naturalistic driving studies?

Naturalistic driving study (NDS) and, in a broader perspective, naturalistic transportation study is a method to collect data from road users' behavior under real driving/transportation conditions. Data are continuously recorded and reflect

[1]The Danish Road Directorate, Aalborg, Denmark

selected aspects of the driving behavior. The basic philosophy is that one can get thorough knowledge about driving behavior and any incidents during transportation activities as a tool to measure safety/accessibility/comfort, etc., in a transport system. The data volume is very big, and analyses of data are often difficult, because more than scratching the surface of data requires data management skills as well as the appurtenant hard- and software. NDS's contain manifold data types such as Global Navigation Satellite System (GNSS) (GPS) data, acceleration data, video recording, thermal data, and temperature, among others. While historically most data for research were collected with onboard units/wearables in a broad perspective, more and more data collection tasks are carried out with smartphones [5].

NDS capture and analyze driving behavior and can be seen as a significant supplement to traditional accident data. Partly, it creates new perspectives on driving behavior in different driving environments, and partly, it mitigates some part of the knowledge we miss due to the large dark figure in accident data [5]. To collect data, such studies require that the road user is equipped with sensors of various types of cameras as the most used ones. The studies monitor various driving activities such as speed, acceleration, braking, lane changes, and turning.

By these data collections and especially the appurtenant analyses, researchers can gain insights into driver decision-making, risk factors, and hopefully areas for improvement in road safety via a change in infrastructure or driving habits.

Likewise, naturalistic cycling studies capture and analyze driving behavior among cyclists. A special perspective here is any detailed acceleration data collected via wearable sensors, which indicate near accidents, conflicts or more towards a maintenance perspective as uneven surfaces. This information can then be used to develop a more cycling-friendly environment.

Naturalistic walking studies involve the observation and analysis of pedestrians in their natural environments to understand behavior and interactions with the surroundings. Like cycling studies, these studies use various methods such as video surveillance, wearable sensors, and other data collection technologies. This information can then support the development of interventions, policies, and urban planning strategies to enhance pedestrian safety and create more walkable communities. In a marked perspective, naturalistic walking studies are mainly used to understand pedestrian flows and where bottlenecks occur in airport or station environments.

8.3 Methods

To measure any effect of any traffic safety intervention/development, it is paramount that we are able to know what would have been the case if the intervention was not implemented. For this purpose, before–after studies and with–without studies are used.

A before-after study in the transportation field is a research method used to evaluate the impact of a specific intervention by comparing data collected before and after the intervention. This method is particularly useful for assessing the effectiveness of changes in road safety measures, public transportation systems, or traffic management strategies.

For example, a before–after study might be conducted to evaluate the impact of installing speed cameras on a particular road. Before the intervention, data on traffic speeds, accident rates, and traffic volume would be collected [6]. After the intervention, the same data would be gathered to determine any changes [7]. The study would analyze whether the speed cameras led to a reduction in speeding incidents and accidents, thereby improving road safety [8].

Such studies are valuable because they provide empirical evidence on the effectiveness of transportation interventions, helping policymakers and planners make informed decisions [6]. They can also identify any unintended consequences of the intervention, such as changes in traffic patterns or driver behavior [7].

A with–without study in the transportation field is a research method used to evaluate the impact of a specific intervention by comparing data from scenarios where the intervention is present (with) and absent (without). This method helps in understanding the effectiveness of transportation policies, technologies, or infrastructure changes.

Such studies are valuable because they provide empirical evidence on the benefits and drawbacks of transportation interventions, helping policymakers and planners make informed decisions [6]. They can also highlight areas where the intervention may need adjustments to achieve the desired outcomes [7].

8.4 Data for NDS

NDS provide valuable insights into driver behavior and performance in real-world conditions. They utilize various data collection methods, including GNSS data, video recording, and jerk/unevenness measurements [9].

GNSS data is used to track the vehicle's position, speed, and direction. This data helps researchers understand the driving environment, such as urban or rural areas, and analyze driving patterns like speed variations and route choices [10].

Video Recording: Cameras installed inside and outside the vehicle capture video footage of the driver, passengers, vehicle controls, and the surrounding environment. This allows researchers to observe driver behavior (like eye movement and hand position), interactions with vehicle controls, and responses to external events [9].

Jerks/Unevenness: Sensors installed in the vehicle measure jerks and unevenness. Jerk is the rate of change of acceleration, i.e., how quickly an acceleration or deceleration occurs. It is a measure of the smoothness of the ride. Unevenness, like potholes or speed bumps, affects jerk and can indicate road conditions. These measurements help researchers understand how drivers respond to different road conditions and how these conditions affect driving behavior and vehicle control.

Deceleration vs Jerk: Deceleration is a decrease in speed over time, while jerk is the rate at which this deceleration changes. For example, if a car slows down at a constant rate, that is deceleration. If the rate of slowing down changes abruptly, that is a jerk. In NDS, both deceleration and jerk are important as they can indicate driver reactions to events (like sudden braking) and the comfort of the ride.

By combining these data sources, NDS can provide a comprehensive picture of driving behavior, vehicle performance, and roadway conditions. This information is crucial for developing strategies to improve road safety and driver performance.

8.4.1 GNSS data in NDS

GNSS data is a central data source in NDS. It is used to track the vehicle's position, speed, and direction. They allow for a detailed analysis of driving behavior in relation to the geographical context and road conditions and hence give an understanding of the driving environment. Also, they describe the overall driving pattern like speed variations and route choices [11].

GNSS data can be combined with Geographic Information Systems (GIS) for more comprehensive analyses. It gives a deeper understanding of our driving behavior, and in combination it can clarify the spatial aspects of driving behavior, such as route choices, driving speed in different road types, and the impact of geographical features on driving such as roadside conditions, urbanization level, any road signs to mention a few [12].

Also, the impact of exogenous and endogenous factors on driver safety can be measured [13]. For example, it can help to understand how road users behave in different conditions; how and when normal traffic conditions or ordinary behavior evolve into critical events or (near-)crashes [11]. More knowledge on how weather affects driving behavior can be gained from combining GNSS from NDS and Meteorological Data. They provide valuable insights into driver behavior and the impact of environmental conditions on driving.

Route analysis based on GNSS data provides precise location information and is used to analyze the routes taken by drivers and any impact from the weather conditions. This can help identify patterns, such as preferred routes or risky behavior, if drivers change their driving behavior too little to mitigate any increased risk due to weather conditions, such as high winds, slippery surface, or similar [14]. For example, researchers can study how rain, snow, or fog affect driving speed, route choice, and accident risk [12]. Furthermore, real-time GNSS data combined with meteorology has recently enabled real-time meteorological observations. This allows for immediate analysis of weather conditions and their impact on driving behavior as well as real-time communication to the road users in order to suit the driving behavior with the driving conditions [12].

8.4.2 Video recording in NDS

Video recording is a central data type in NDS. It is a rich source of data that allows researchers to observe and analyze driver behavior, vehicle movement, and the surrounding environment in real-world driving conditions [15,16].

Video data can be used to study driver behavior in detail. For instance, researchers can observe where drivers are looking, their reactions to different situations, and their interactions with vehicle controls. Video recordings capture the behavior of the vehicle (e.g., speed, lane position) and the behavior of other road users with whom the driver interacts (e.g., other drivers, motorcyclists, cyclists, and pedestrians). It can also be used for identifying critical incidents or near-crash events. It can reveal events that may not be detected using numerical parameters alone [15].

Video data can be used in conjunction with computer vision methods to automatically annotate behaviors, making the analysis of large datasets more manageable [17], and can be combined with other data types, such as GNSS data, to provide a comprehensive picture of driving behavior [18].

NDS is an effective tool for the design, testing, and evaluation of driver assistance systems. They can support various stages of a user-center systems design process [15]. The rich data from video recordings allow for the detection of sequences of events and conditions and may potentially help gain insights into their causal relationships, for example, between driver distraction and safety-critical events [18].

8.5 Examples of projects based on GNSS data

The ITS Platform North Denmark was a project used as a test platform for Intelligent Transportation System (ITS) solutions carried out 2009–2014. The platform consisted of a GNSS/GPRS On Board Unit (OBU), which was installed in nearly 500 cars, a backend server, and a specially designed digital road map for ITS applications [19].

The platform was freely accessible, which meant that third-party applications could be run on the platform, and a series of test applications were developed on the platform [19]. The ITS Platform had a total budget of 5 million €. The project was funded with DKK 50% from the EU Regional Development Fund, the North Jutland Development Fund, as well as from Aalborg University, GateHouse, and Inntrasys. In conclusion, the ITS Platform North Denmark served as a significant initiative in the field of Intelligent Transportation Systems, providing an open platform where a range of ITS applications were developed and tested.

The Pay As You Speed (PAYS) project was a Danish Intelligent Speed Assistance (ISA) initiative that was based on the Pay As You Drive principles. The ISA equipment used in this project served two main functions: It notified the driver when they were speeding. It applied penalty points for speeding, which decreased the driver's part of a 30% discount on their automobile insurance.

The preliminary results from the first 38 of 180 participants showed that the combination of information and incentive almost eliminated speeding on rural roads. Significant reductions in speeding were also found for urban roads and, to some extent, motorways. On roads with speed limits of 50, 80, and 110 km/h, the proportion of distance traveled while speeding was significantly reduced. However, no significant results were found for motorways with speed limits of 130 km/h [20,21].

Also, prominent Swedish and English studies have been conducted, and these and many other studies have significantly contributed to the understanding of real-world driving behaviors, pedestrian and cyclist safety, and the development of interventions to improve road safety [22–24].

Despite fading interest in new ISA trials in the last decade, their outcome has affected the traffic safety policy markedly recently. Hence, ISA is mandatory for all new cars sold in the EU from July 2024. This regulation aims to enhance road safety by using cameras and GPS to recognize speed limits and provide feedback to drivers. The ISA systems offer four feedback options: acoustic warnings, vibrating warnings, haptic feedback through the accelerator pedal, and speed control functions [25]. The introduction of ISA is expected to reduce speeding by 30% and road deaths by 20%. Car manufacturers must implement these systems in all new models introduced after July 2022, with full compliance by July 2024. This move aligns with the EU's goals of improving road safety and reducing traffic accidents [25].

8.6 Examples of NDS based on video recording

The Strategic Highway Research Program 2 (SHRP 2): This large-scale NDS conducted in the United States collected data from over 3,000 participants, capturing real-world driving behaviors to gain insights into various aspects of driver performance, safety, and behavior [9].

The Transportation Research Center for Livable Communities' NDS (NRAD): This study aimed to understand the interactions between drivers, pedestrians, and cyclists in urban environments. It collected data from over 200 drivers and analyzed their behavior in different traffic scenarios.

The Naturalistic Teenage Driving Study (NTDS): This study focused on teenage drivers and aimed to understand the factors contributing to their higher crash risk. It collected data from over 100 newly licensed teenage drivers and analyzed their driving behaviors and habits [26].

The Naturalistic Cycling Study by the University of California, Berkeley: This study aimed to understand the factors influencing the safety of cyclists on urban roads. It collected data from a diverse group of cyclists in different cities to analyze their interactions with traffic, infrastructure, and other environmental factors [27].

The Virginia Tech Transportation Institute's Naturalistic Driving Study: This ongoing study involves continuously monitoring a large sample of drivers over an extended period of time. It collects data on various aspects of driving behaviors, distractions, and safety-critical events to gain insights into driver performance and safety.

8.7 Type of analyses

NDS is open for a variety of analyses. Only the creativity sets the borders. However, while some are straightforward and easy to conduct, others are more difficult due to, among other data quality, data volume, and costs. Some samples of analyses are these:

- Driver Behavior Analysis: Understanding how drivers behave in real-world conditions, including reactions to various traffic situations [5].
- Crash and Near-Crash Analysis: Identifying factors leading to crashes or near-crashes to improve safety measures [29].
- Distraction and Inattention Studies: Examining the impact of distractions (e.g., mobile phone use) on driving performance [29].
- Fatigue and Drowsiness Detection: Analyzing signs of driver fatigue and its effects on driving safety [29].
- Traffic Flow and Congestion Analysis: Studying how different driving behaviors affect traffic flow and congestion levels [5].
- Impact of Advanced Driver Assistance Systems (ADAS): Evaluating the effectiveness of ADAS technologies in preventing accidents [29].
- Environmental Impact Studies: Assessing how driving behaviors influence fuel consumption and emissions [29].
- Road Infrastructure Evaluation: Analyzing how road design and conditions affect driving behavior and safety [30].
- Speed and Aggressive Driving Analysis: Investigating the prevalence and impact of speeding and aggressive driving behaviors [5].
- Policy and Regulation Impact: Evaluating the effectiveness of traffic laws and regulations on improving road safety [29].

8.8 Pitfalls regarding data analyses

NDS open for a variety of analyses and data collection approaches. Many projects have been conducted and not all have been fully successful. Therefore, we will highlight some pitfalls which are worth considering when analyzing data from NDS:

- Data Complexity: NDS data is often vast and complex, making it challenging to manage and analyze effectively [28].
- Privacy Concerns: Ensuring the privacy and confidentiality of participants can be difficult, especially with video and GPS data [31].
- Bias in Data Collection: The data may be biased if the sample is not representative of the general drivers [32].
- Data Quality Issues: Inconsistent or incomplete data can lead to inaccurate analyses and conclusions [28].
- Observer Effect: Drivers may alter their behavior because they know they are being observed, which can bias results [33].
- Technical Limitations: The technology used to collect data may have limitations, such as sensor inaccuracies or data loss [28].
- Ethical Considerations: Ethical issues may arise regarding the use of personal data and the potential impact on participants [31].

- Data Integration Challenges: Combining data from different sources or studies can be problematic due to varying formats and standards [32].
- Interpretation Difficulties: Analyzing and interpreting a vast amount of data requires sophisticated statistical methods and expertise [28].
- Cost and Resource Intensive: Conducting and analyzing NDS can be expensive and resource-intensive [33].

References

[1] Lahrmann H. Researchers shed light on traffic dark figures and find a tripling in the number of accidents, Aalborg University, Denmark, 2024.

[2] Mazzei M.J. and Noble D. Big data and strategy: theoretical foundations and new opportunities. In *Strategy and Behaviors in the Digital Economy*. London: Intech Open; 2020.

[3] Suthakar U., Magnoni L., Smith D.R. *et al.* An efficient strategy for the collection and storage of large volumes of data for computation. *Journal of Big Data*. 2016, 3(21), pp. 1–17.

[4] Taleb I., Serhani M.A., Bouhaddioui C. *et al.* Big data quality framework: a holistic approach to continuous quality management. *Journal of Big Data*. 2021, 8(76), pp. 1–41.

[5] Barnard Y., Utesch F., van Nes N., Eenink R., and Baumann M. The study design of UDRIVE: the naturalistic driving study across Europe for cars, trucks and scooters. *European Transport Research Review*. 2016, 8(14), pp. 1–10.

[6] Choi Y.Y., Kho S.-E., Jang K., and Kim D-K. Evaluating time trend correction approaches associated with empirical Bayes before-after study for road safety. *KSCE Journal of Civil Engineering*. 2018, 22, pp. 4593–4601.

[7] Brenac T. Common before-after accident study on a road site: a low-informative Bayesian method. *European Transport Research Review*. 2009, 1, pp. 125–134.

[8] Federal Transit Administration. 2021. *Before and After Studies of New Starts Projects*. Available from: https://www.transit.dot.gov/funding/grant-programs/capital-investments/and-after-studies-new-starts-projects [accessed 31 Aug 2024].

[9] José Balsa-Barreiro J., Valero-Mora P.M., Berné-Valero J.L., and Varela-García F.-A. GIS mapping of driving behavior based on naturalistic driving data. *International Journal of Geo-Information*. 2019, 8(5), pp. 1–12.

[10] Australian Academy of Science. *What Is the Global Navigation Satellite System?* Available from: https://www.science.org.au/curious/technology-future/what-global-navigation-satellite-system [accessed 31 Aug 2024].

[11] van Malderen R., Santos M., and Zhang K. Special issue title: climate modelling and monitoring using GNSS. *Remote Sensing*. 2022, 14(17), pp. 1–2.

[12] Hadas T., Hobiger T., and Hordyniec P. Considering different recent advancements in GNSS on real-time zenith troposphere estimates. *GPS Solutions*. 2020, 24(99), pp. 1–14.

[13] Varela A. GIS mapping of driving behavior based on naturalistic driving data. *ISPRS International Journal of Geo-Information*. 2019, 8(5), p. 226.

[14] Van Malderen R., Santos M., and Zhang K. Special issue: Climate modelling and monitoring using GNSS. *Remote Sensing*. 2022, 14(17), p. 4371.

[15] Hanwen Miao H., Zhang S., and Flannagan C. Driver behavior extraction from videos in naturalistic driving datasets with 3D ConvNets. *Journal of Big Data Analytics in Transportation*. 2022, 4, pp. 41–55.

[16] Fitch G.M. and Hanowski R.J. Using naturalistic driving research to design, test and evaluate driver assistance systems. In: Eskandarian, A. (ed), *Handbook of Intelligent Vehicles*. London: Springer, 2012, pp. 559–580.

[17] VicRoads. n.d. *Australian Naturalistic Driving Study*. Available from: https://www.vicroads.vic.gov.au/safety-and-road-rules/road-safety-programs/australian-naturalistic-driving-study [accessed 31 Aug 2024].

[18] Lin Q., Feng R., Cheng B., *et al.* Analysis of causes of rear-end conflicts using naturalistic driving data collected by video drive recorders. *SAE Technical Paper*. 2008.

[19] Lahrmann H., Agerholm N., Juhl J., *et al.* ITS Platform North Denmark: idea, content, and status. *Proceedings of the 19th ITS World Congress*. Vienna, Austria. 2012.

[20] Agerholm N., Waagepetersen R. Tradisauskas N., Harms L., and Lahrmann H. Preliminary results from the Danish Intelligent Speed Adaptation Project Pay As You Speed. *IET Intelligent Transport Systems*. 2008, 2(2), pp. 143–153.

[21] Lahrmann H., Agerholm N., Tradisauskas N., Næss T., Juhl J., and Harms L. Pay as You Speed, ISA with incentives for not speeding: a case of test driver recruitment. *Accident Analysis & Prevention*. 2012, 48(9), pp. 10–16.

[22] Biding T. and Lind G. 2002. *Intelligent Speed Adaptation (ISA), Results of Large-scale Trials in Borlänge, Lidköping, Lund and Umeå during the Period 1999–2002*. Swedish National Road Administration. Borlänge, Sweden.

[23] Carsten O.M.J. and Tate F.N. Intelligent speed adaptation: accident savings and cost-benefit analysis. *Accident Analysis and Prevention*, 2005, 37(3), pp. 407–416.

[24] Carsten O., Fowkes M., Lai F., *et al. ISA – UK – Intelligent Speed Adaptation – Final report*. Department for Transport, United Kingdom, 2008.

[25] European Road Safety Chartern.d. Intelligent speed assistance (ISA) set to become mandatory across Europe. Available from: https://road-safety-charter.ec.europa.eu/resources-knowledge/media-and-press/intelligent-speed-assistance-isa-set-become-mandatory-across [accessed 31 Aug 2024].

[26] Simons-Morton B.G., Ehsani J.P., Gershon P., Klauer S.G., and Dingus T.A. Teen driving risk and prevention: naturalistic driving research contributions and challenges. *Safety*. 2017, 3(4), pp. 1–16.

[27] Leyva H., Dozza M., and Baldanzini N. E-bikers' braking behavior: results from a naturalistic cycling study. *Traffic Injury Prevention*, 2019, 20(Suppl. 3), pp. 62–67.
[28] Bärgman J. 2015. *On the Analysis of Naturalistic Driving Data – Development and Evaluation of Methods for Analysis of Naturalistic – Driving Data from a Variety of Data Sources*. Department of Applied Mechanics. Chalmers, Gothenburg, Sweden.
[29] Ehsani J.P., Harbluk J.L., Bärgman J., *et al.* Naturalistic driving studies: an overview and international perspective. *International Encyclopedia of Transportation*. 2021, 7, pp. 20–38.
[30] Singh H. and Kathuria A. Analyzing driver behavior under naturalistic driving conditions: A review. *Accident Analysis and Prevention*. 2020, 150, pp. 1–21.
[31] Zong R., Wang Y., Ding J., and Deng W. Statistical risk and performance analyses on naturalistic driving trajectory datasets for traffic modeling. *World Electric Vehicle Journal*. 2024, 15(77), pp. 2–15.
[32] Venkatachalapathy A., Sharma A., and Vanajakshi L.D. A naturalistic data-driven study of Indian traffic conditions and driver's physiological stress. *Transportation in Developing Economies*. 2022, 8, p.23.
[33] Victor T., Dozza M., Bärgman J., *et al. Analysis of Naturalistic Driving Study Data: Safer Glances, Driver Inattention, and Crash Risk*. Transportation Research Board, 2014.

Chapter 9

Surrogate measures of safety: site-based observations

Aliaksei Laureshyn[1] *and Nicolas Saunier*[2]

Traditionally, traffic safety is expressed through the number of accidents (crashes) or people injured, since these are the most direct manifestations of an unsafe traffic environment. As we discuss in this chapter, such an approach has its limitations due to the low frequency of accidents and some other issues. Neither is it unusual that accident data is not available, not reliable or lacks important details, particularly in the context of low- and middle-income countries. Surrogate measures of safety (SMoS), based on the observation of traffic conflicts or near-misses, may serve as a viable alternative, or a complement, to accident/injury counts.

9.1 Why surrogate measures?

9.1.1 Limitations of accident data

Even though accidents and injuries represent the most direct consequences of an unsafe traffic environment, accident counts as such are not 'direct' measures of (un)safety. Due to the random nature of accident occurrence, their number during equally long time periods will vary, even if there are no real changes in any safety-related properties taking place. The true safety is rather reflected by the 'expected number of accidents', the underlying property of the system that cannot be observed but only estimated using statistical methods (Hauer 1997).

Traffic accidents are rare events (Elvik *et al.* 2009b). Particularly if we 'zoom in' to smaller units of the traffic system, such as individual road links or intersections, it may take months and sometimes years between accident registrations. As a result, it takes time to accumulate enough data for any meaningful analysis. As traffic safety gradually improves, the datasets tend to contain less and less of actual data, mostly dominated by counts of 'zeros' or 'ones'.

Accidents are underreported (Janstrup *et al.* 2016; Amoros *et al.* 2011; Elvik *et al.* 2009b; Amoros *et al.* 2006; Alsop and Langley 2001). It would not be a big

[1]Department of Technology and Society, LTH, Lund University, Sweden
[2]Civil, Geological and Mining Engineering Department, Polytechnique Montreal, Canada

issue if the under-reporting was relatively 'stable', since statistics has tools for working with samples rather than the entire population. However, the under-reporting rate varies depending on the context, accident type, location, road users involved, outcome severity, etc. Underreporting reduces even further the datasets that are already quite sparce with accidents.

The ultimate goal of accident analysis is to understand what factors contribute to accident occurrence and come up with suggestions on how similar accidents can be prevented in the future. The standard police report is a relatively poor source of information for such a task. Being filled in without witnessing the accident, under time pressure, and often with focus on assigning guilt rather than trying to understand the complex causal mechanisms, such reports tend to be very sparse on details, leaving a researcher wondering what actually happened and why.

Indeed, accident analysis is always retro-active – accidents happen first, and analysis is done later. Waiting a long time for accidents to accumulate is not acceptable. For example, if a new road design feature is evaluated, we want quick feedback on its safety effects, especially on whether the safety improves (and we should build more of such roads) or deteriorates (and we should remove the feature immediately and halt any other similar ongoing constructions). From the ethical perspective, it is hard to motivate that we simply wait for accidents to happen and people to suffer before any conclusions can be reached. This is akin to building a bridge without being sure it will not collapse in normal operating conditions.

Surrogate measures of safety (SMoS) based on the observation of traffic conflicts offer solutions to some of these issues. Traffic conflicts:

- occur more frequently, thus shortening the data collection period (days and weeks instead of years) and allowing to accumulate larger conflict counts;
- can be directly observed, thus allowing more nuanced understanding of their development process and contributing factors;
- are pro-active, thus allowing conclusions and interventions can be made before real accidents happen.

The term 'surrogate' implies that traffic conflicts are used instead of accidents, thus being an 'indirect' and inferior measure of safety. However, as we wrote in the beginning of this section, accident counts are quite 'indirect', too. With the proper statistical techniques, SMoS may outperform accident counts in estimating the 'true' (unobservable directly, yet present) safety level at a given site (Svensson 1998).

9.1.2 Historical note

The idea of collecting reports on critical situations in traffic was tested as early as in the 1950s (Forbes 1957). However, the real start of traffic conflicts era is usually attributed to the study by General Motors (Perkins and Harris 1968), in which actual observations of conflicts (loosely defined as any situation requiring an intervention, such as braking or lane changing) were performed in traffic. The approach rapidly gained popularity, leading to development of numerous conflict observation techniques in countries like Austria, Canada, Finland, France, Germany, Israel, the Netherlands, Sweden, the United Kingdom and the United States (Asmussen 1984). The techniques used different protocols to define traffic conflicts, and their comparability and validity remained unclear.

Starting from the late 1970s, the International Committee on Traffic Conflict Techniques (ICTCT) became a major forum for discussion on these issues. ICTCT organized a series of workshops on traffic conflict techniques (TCTs) (ICTCT 1986; Asmussen 1984; Kraay 1982; Older and Shippey 1980; Amundsen and Hyden 1977) as well as international calibration studies (Kraay and van der Horst 1985; Grayson 1984), in which various teams could apply their tools simultaneously and at the same site and compare and discuss the results afterwards. The committee developed a solid theoretical foundation for SMoS, and the proceedings of the early ICTCT workshops remain an absolute must-read for anyone seriously interested in the subject[*].

By the mid-1990s, TCTs somewhat declined in popularity. While practitioners liked the idea behind the method, they were quite hesitant to adopt it due to its labour intensity and subjectivity risks – as data was still collected manually by human observers. A real break-through came in 2000s with the rapid improvements in video analysis techniques (Laureshyn 2010; Saunier and Sayed 2007; Songchitruksa and Tarko 2004) and other sensor technologies (Fu *et al.* 2017; Tarko *et al.* 2017; Kirchhof and Stilla 2006). By now, there is a wide plethora of both free and commercial tools available[†], though, again, their comparability and cross-validation deserve further studying.

9.2 Theoretical background

9.2.1 Hierarchy of events

The idea behind SMoS relies on the assumption that all traffic events (within defined spatial and temporal limits) form a continuum, in which relevant information about the events of interests, accidents, can be obtained through studying the remaining non-accident events. This concept is frequently illustrated by a safety pyramid (Hydén 1987), as shown in Figure 9.1.

The most severe and most rare events are accidents (which in their turn can be divided into fatal, injury and property damage accidents) at the top of the pyramid. The massive base of the pyramid is formed by events, encounters or interactions, that are 'normal' in all aspects, i.e. in which the road users interact smoothly, adjust their travel accordingly to ensure sufficient safety margins, etc. In-between these two there is a layer of events in which the actual collision was avoided (no accident happened); however, the safety margins were compromised to such degree that people involved in or observing the situations would perceive them as 'dangerous'. These events are called traffic conflicts (other frequently used names are

[*]The proceedings can be accessed the ICTCT library at www.ictct.net/smos/library

[†]Examples can be found at www.ictct.net/smos/software and www.ictct.net/smos/related

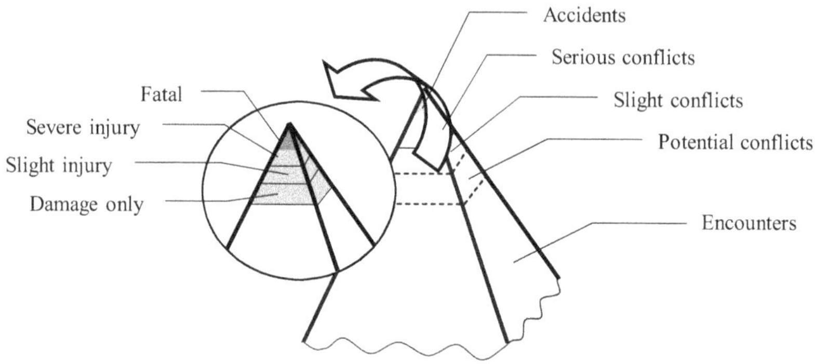

Figure 9.1 Safety pyramid [adopted from Hydén (1987)]

almost-accidents and near-misses). The nearness of traffic conflicts to the actual accidents within the safety hierarchy generally makes them quite similar when it comes to the scenarios, contributing factors, and behaviour and reactions. It is also believed that there is a quantitative association that allows estimation of the expected accidents frequency from the frequency of the observed conflicts, which become therefore surrogate measure of safety.

Originally, the development of traffic conflict theory was entirely focused on the accident scenario of two moving road users colliding with each other (usually, collision of two *motor vehicles*, which was indeed the operational definition of the core subject of traffic safety at that era). However, there is nothing limiting application of the same reasoning to other accident types – for example, run-off-road or bicycle fall accidents – as a similar continuity in severity of such events may be constructed.

Hydén's pyramid was intended to merely illustrate the concept and it should not be interpreted as if the relations between the events belonging to different severity layers are actually linear. Migletz *et al.* (1985) presented the same idea through mathematical language, replacing the abstract pyramid with a frequency distribution function that could take any shape (Figure 9.2).

Svensson (1998) specifically studied how the shapes of the event distribution change depending on the traffic environment and interaction types. The uncovered variability (see Figure 9.3) confirmed that the relation between conflicts and accidents is indeed quite complex, but also allowed for hypothesising on how a more holistic understanding of safety could be developed from examining the distributions. For example, argued Svensson, frequent presence of 'moderate' conflicts could be seen as positive for safety since they keep the road users alert and provide them with safety-relevant learning experiences (roundabouts are a good illustration of such environment). Conversely, an environment that has almost no conflicts, but occasionally 'peaks' with very severe ones (a signalised intersection) may give a false impression of safety and control, while any eventual mistake by a road user would catch everyone off guard, potentially leading to much more severe outcomes.

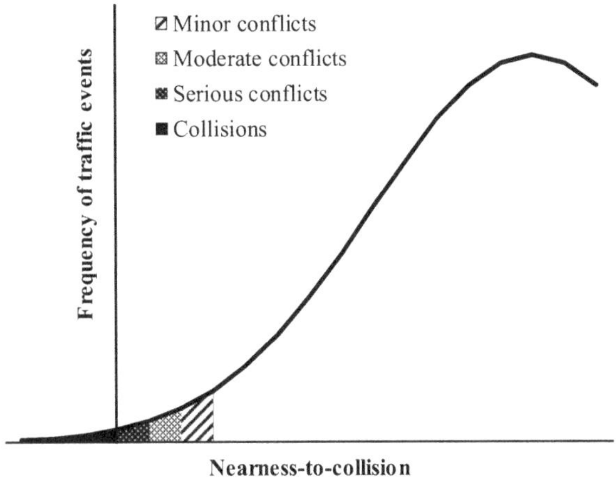

Figure 9.2 *Frequency density function illustrating the continuity between accidents, traffic conflicts and normal traffic events [adopted from Migletz et al. (1985)]*

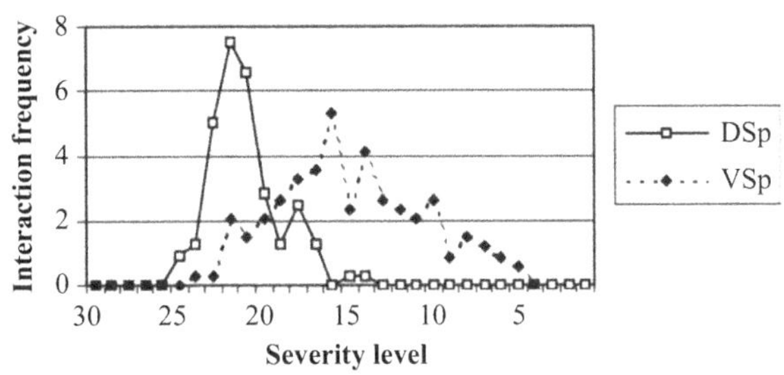

Figure 9.3 *Distribution of traffic conflicts between motor vehicles and pedestrians at signalized (VSp) versus non-signalised (DSp) crossings (Svensson 1998); severity levels are defined according to the Swedish Traffic Conflicts technique (Laureshyn and Varhelyi 2020)*

9.2.2 Concept of severity

Intuitively, we can imagine that two traffic situations, none resulting in an actual collision, could be perceived quite different in their dangerousness. 'Severity' of a traffic situation is an abstract concept reflecting just that – how close a situation is to an accident or, being more in line with the Safe System principles, how close it is to a *personal injury*.

Two dimensions contributing to the event severity can be distinguished (see Figure 9.4). One is the *proximity* to a collision, i.e. the size of the safety margin measured either in terms of distance, time or some more elaborated measure. However, events with high collision proximity might be of low relevance for traffic

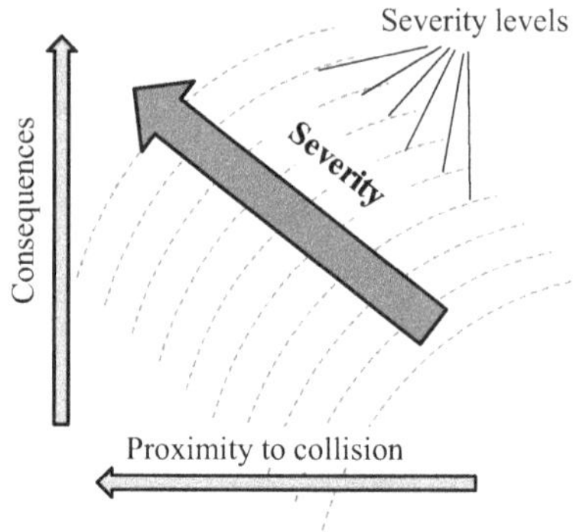

Figure 9.4 Two dimensions of traffic event severity [adopted from Laureshyn et al. (2017)]

safety if the energy involved is not likely to generate any severe outcomes in terms of personal injuries. Thus, the second dimension is the *consequences* in case a collision taken place.

To illustrate the point, consider interactions (or even collisions) taking place at a parking lot. While the margins between manoeuvring cars could be very small, it is very seldom that someone gets seriously injured due to very low speeds. Similarly, two cars crossing the same spot at high speeds, but with 20 s gap between them, could hardly be seen as a severe situation either. However, when both factors coincide (a 'close shave' taking place at a high speed), one would definitely feel that it was a dangerous situation.

Interestingly, humans seem to possess some built-in 'instrument' that allows us to judge the danger level in traffic. Numerous studies on traffic conflict observers' reliability showed a very high degree of agreement between individuals, both when it comes to detecting potentially dangerous situations and in ranking them by degree of severity once detected (Madsen 2018; Kruysse and Wijlhuizen 1992; Kruysse 1991; Hydén 1987; Grayson 1984; Lightburn and Howarth 1979). A recent study by Yastremska-Kravchenko *et al.* (2024) confirmed that human perception of danger in traffic is quite in line with the two-dimensional severity model described above – both the nearness and the energy involved were found to be very important variables explaining the severity ranking decisions taken by humans. Another interesting finding in an earlier study by Kruysse (1991) is that humans seem to realise the severity level at the very early stage of the situation development, i.e. it is rather the *potential* for a severe outcome – not the *actual* outcome – that makes us feel the danger.

9.2.3 Traffic conflicts and exposure

In general terms, exposure is a measure of activity that generates accidents (Chapman 1973). In the traffic safety context, exposure is usually defined by the number of vehicles passed, kilometres driven or other similar and relatively simple measures.

Such simplicity is an advantage when it comes to data collection, but the interpretation of such measures as 'opportunities' for accident is often problematic. For example, a passage of a vehicle can only become a collision with another vehicle if such another vehicle is present, yet this is not captured if traffic flow (a sum of individual passages) is used as the exposure measure. This issue is partly addressed in accident prediction models by introducing a product of the two conflicting traffic flows (each raised to the power of some constants); while providing the model fit to the data, the actual meaning of such products is still hard to interpret in causal terms.

Elvik *et al.* (2009a) suggested to be more thoughtful in defining exposure and start indeed with the question of what a meaningful 'opportunity' for the accident type in question could be. For example, in case of vehicle-to-vehicle collisions, a suitable event could be a simultaneous arrival of two vehicles on a crossing course. In case of run-off-road accidents, a meaningful 'opportunity' could be, for example, a change in the road direction (e.g. a start of a curve), assuming that the accident scenario is that the driver fails – due to fatigue, distraction, or other reasons – to adjust the course accordingly. Operational definition of such event-based exposure measures, as well as the data collection procedures, undeniably become more complex; however, several studies have confirmed both the theoretical and practical advantages of this approach to exposure definition (Bulla-Cruz *et al.* 2020; Johnsson *et al.* 2020; Elvik 2010, 2015).

Within the SMoS discourse, a clear distinction must be drawn between the event-based exposure in general and traffic conflicts (and accidents) that are just a small subset of the exposure events. Already in early 1980s, the principle difference was explained by Ezra Hauer: while traffic conflicts are intended to measure safety in absolute terms (expected number of accidents), the exposure is used to calculate the accident rates (probability of an exposure unit to result in an accident) (Hauer 1982).

Johnsson *et al.* (2021a) illustrate how these two distinct functions can be mixed up (see Figure 9.5). The operational definition of a threshold for classifying an event as a traffic conflict may vary anywhere between 'including all events' to

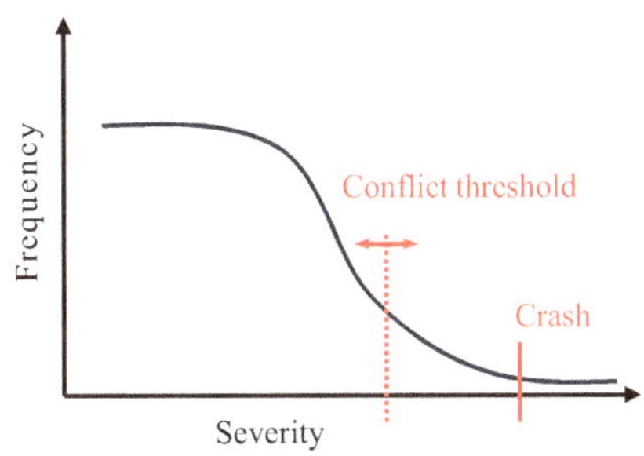

Figure 9.5 Moving the conflict threshold away from accidents yields more events classified as conflicts. However, the conflicts become less similar to accidents and more similar to regular exposure units [adopted from Johnsson et al. (2021a)]

'including only accidents'. For practical reasons, it is always preferable to lower the threshold, which will increase the conflict counts with all the benefits a larger dataset brings to a researcher. On the other hand, the further from accidents an event is on the severity scale (i.e. the more *unlike* an accident it is), the less relevant and valid it becomes as a surrogate for accidents.

When the conflict definition is too inclusive, the conflicts become virtually a 'noisy' reflection of the exposure. Since exposure is a relevant factor explaining the accident frequency, some association between conflicts and accidents is always found even if the conflict threshold is set ridiculously low. A relevant question, however, is why one should bother collecting traffic conflicts instead of collecting the exposure directly at much lower costs (Johnsson *et al.* 2021b). To motivate the additional costs and efforts, traffic conflicts should be a better predictor of accident frequency compared to mere exposure at a particular site (Güttinger 1982). For example, when the exposure is the same, traffic conflicts still should be able to detect differences in safety related to the presence or absence of other risk factors or traffic safety measures.

9.2.4 Traffic conflicts and accidents

Nearly all traffic conflict validation studies do report some association between conflicts and accidents (Johnsson *et al.* 2018). However, as it was argued in the previous section, some of the associations might be spurious, e.g. they occur due to the fact that conflicts are – just as the accidents – a product of the exposure. To get more trust in our traffic conflict measure, a persuasive theory relating conflicts to accidents is necessary.

Güttinger (1982) presented two alternative models of this relation (see Figure 9.6). In Figure 9.6(a), the conflict is defined as a set of *initial conditions* in which an accident becomes possible (given the range of future scenarios, at least some of them have non-zero probability of a collision). Depending on the success of the evasive action intended to mitigate the situation, it may develop into a collision (unsuccessful) or be resolved peacefully with the collision avoided (successful). The conflict thus precedes the accident on the causal chain, taking the meaning of a statistical trial that may – or may not – result in an accident. In order to estimate the expected number of accidents based on the conflict observations, it

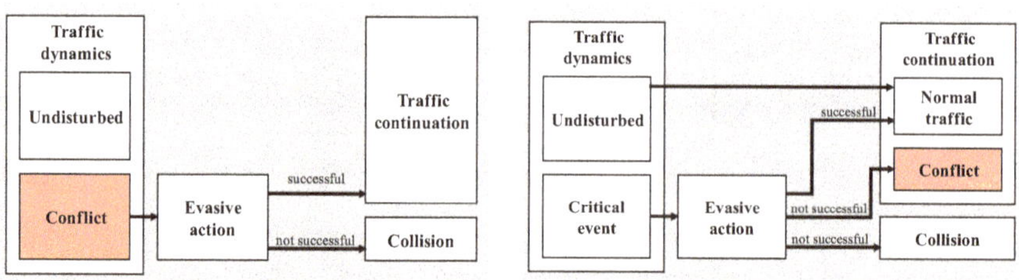

Figure 9.6 (a) and (b) Two models illustrating the relation between accidents and conflicts (Güttinger 1982)

is necessary: (i) to establish the accident probability as a function of the initial conditions of the traffic conflict, and (ii) to describe the distribution of the initial conditions over a longer time period (e.g. a year). This approach has been formalized in the causal model described by Davis *et al.* (2011).

The alternative model presented in Figure 9.6(b) defines the conflict as the *outcome* of the evasive action. Such an approach is inherently adopted when conflict definition utilizes indicators such as braking intensity or the time gap by which road users missed each other, i.e. it is based on the revealed behaviour or situation outcome rather than hypothetical scenarios as the model in Figure 9.6(a) would require. Though from the practical perspective such indicators are easier to observe and measure, this model has a serious flaw – traffic conflicts and accidents appear on two parallel causal paths, undermining the entire idea of unbiased accident frequency estimation by measuring the conflicts. In fact, the association between the two can go in any direction (e.g. improved safety could 'convert' accidents into conflicts that would increase their number).

There are different strategies for approaching the practical task of 'converting' observed traffic conflicts into the expected accident number. The causal approach, as described by Davis *et al.* (2011), has the most solid theoretical foundation. However, its practical implementation requires numerous assumptions about the properties of the studied system that cannot be directly observed. For example, even for the simplest case of collisions between vehicles travelling in the same lane, the information about the speed and position of road users (observable properties) must be complemented with assumed distributions of the drivers' reaction time, applied braking force, friction between the tires and asphalt, etc. In complex contexts, such as urban traffic with mixed directions and road user types, the amount of 'unknowns' makes the practical implementation of the approach virtually impossible.

Regression modelling is a common tool to reveal the relation between two variables without any requirement to actually understand the causal mechanism behind it. It became apparent very early that the simple linear relation (conflicts multiplied by a conversions factor equals accidents) is too crude and should be abandoned. More complex functional relationships perform better, yet they lack 'universality' (there is a great variation between best-fitting functions and their parameter values coming from different studies, i.e. it is hard to know beforehand what model to choose to convert conflict counts into expected accident number – see Tarko (2018) for an overview).

A breakthrough seems to be coming with applications of the extreme value theory (EVT) to handle the conflict-accidents relationship (Chen *et al.* 2024; Niu *et al.* 2024; Songchitruksa and Tarko 2006). Unlike many other areas of statistics that focus on central tendencies and variation of random variables, the EVT is primarily interested in the 'distribution tails', i.e. the frequency of very infrequent and thus 'extreme' observations. Technically, EVT uses the most extreme observations in the dataset to estimate the frequency of even more extreme events that have not yet been observed (but can realistically be observed). This problem formulation fits well with the problem of conflicts (rare events) to be used to estimate the frequency of accidents or injuries (even more rare events).

The EVT toolbox allows working with a single variable (univariate model) or two and more variables (bi- and multivariate models). In the univariate case, the variable of interest is usually some measure of temporal or spatial proximity, which allows to estimate the probability of a collision ('what is the probability of proximity reaching zero?'). The multivariate models allow combined usage of several criteria, such as 'what is the probability of the proximity reaching zero while kinetic energy being high enough to cause a severe injury?'.

9.2.5 Reliability and validity

Reliability refers to the ability of a tool used for collecting SMoS to keep approximately the same level of accuracy regardless of the study location, type of situations, traffic characteristics, environmental conditions, etc. Basically, it ensures that the measurement error is always within certain (known) limits. Importantly, reliability is not concerned with whether the collected data have any meaning, i.e. it is possible to accurately (reliably) measure indicators that fail to reflect safety.

Whether an indicator is a good representation of the property that we want to measure is reflected by its validity. Within the traffic safety context, *absolute validity* of a surrogate measure of safety refers to its ability to estimate the expected number of accidents (inherent yet unobservable safety property of a traffic system). While this can be seen as the ultimate goal for the SMoS methodology development, in many cases obtaining a *relative validity* is quite satisfactory. A surrogate measure of safety possesses relative validity when it can sense whether safety improves or deteriorates (i.e. a direction of change in safety) yet may not be able to express it in absolute terms (e.g. lives saved or injuries prevented). Finally, *process validity* refers to the degree of similarity between traffic conflicts and real accidents in their development process and contributing factors. If a strong similarity exists, the observation and detailed analysis of traffic conflicts can be used to study accident causation. It is likely that safety interventions that can prevent the occurrence or mitigate the severity of the conflicts would have a similar effect on the accidents, too.

As argued by Hauer and Gårder (1986), validity is not a strict 'yes'-or-'no' property but rather a continuum, with some indicators (or their combinations) being more valid than others. While striving to improve validity is a worthy pursuit, an SMoS methodology does not need to be 'perfect' to be useful.

9.3 Indicators, measures and techniques

The first ICTCT workshop on Traffic Conflicts held in 1977 in Oslo, Norway, resulted in the first and up till now, probably, the most frequently cited definition of a traffic conflict (Amundsen and Hyden 1977):

> *'A traffic conflict is an observable situation in which two or more road users approach each other in space and time to such an extent that there is a risk of collision if their movements remain unchanged.'*

Capturing the general idea quite well, it is too unspecific for practical use. Statements like 'there is a risk of collision' or 'movements remain unchanged' require operationalization by objective criteria expressed through indicators that can be observed and measured without any ambiguity. Even though at its early stages, the traffic conflict method heavily relied on human observers who were quite good at applying unprecise definitions yet still getting meaningful results (see a discussion on that in Yastremska-Kravchenko *et al.* (2022)), automation of conflict detection tasks through the analysis of video or other types of sensor data requires very specific calculation instructions and threshold values that can be implemented in a programming code.

Over the years, many indicators to measure the severity of conflicts have been suggested. A review by Mahmud *et al.* (2017) listed over 40 different indicators, and the list has probably grown since that even further. What follows describes the conceptual 'families' of indicators without any ambition to provide a comprehensive list.

9.3.1 TTC family

According to a systematic review on the actual usage of SMoS found in the scientific literature, *Time-to-Collision* (*TTC*), is by all means the most frequently used indicator (Laureshyn *et al.* 2016). It was first introduced by Hayward (1972) who defined it as 'the time until a collision between the vehicles would occur if they continued on their present course at their present rates'.

TTC has some distinct properties:

- It cannot be directly observed or measured but is instead 'predicted' using some assumptions about the intended future motion of the involved road users; depending on the assumptions, different TTC values might be obtained for the same situation.
- TTC is defined and can be calculated only when the road users are on a *collision course*, i.e. in a situation where they would eventually collide if they continue travelling as originally intended. Again, the assumptions about the intended travelling are crucial for whether a collision course is identified or not. Another practical problem is that even a very minor deviation from the collision course (the road users miss each other by 0.01 s – a situation with clearly compromised safety) leaves the researcher with no meaningful measure at all.
- TTC is a continuous indicator, and it may be calculated for any time instance as long as the collision course is present. This raises a question about which moment is the most representative or relevant, and whether the change of TTC over time carries some relevant information, too (Saunier and Mohamed 2014).

Figure 9.7 illustrates the theoretical TTC curve. According to Hayward (1972), the TTC will decrease as the two road users approach each other. The rate of decrease may change if, for example, they realize the danger and start braking or taking other evasive action. The TTC value at the moment of evasive action onset

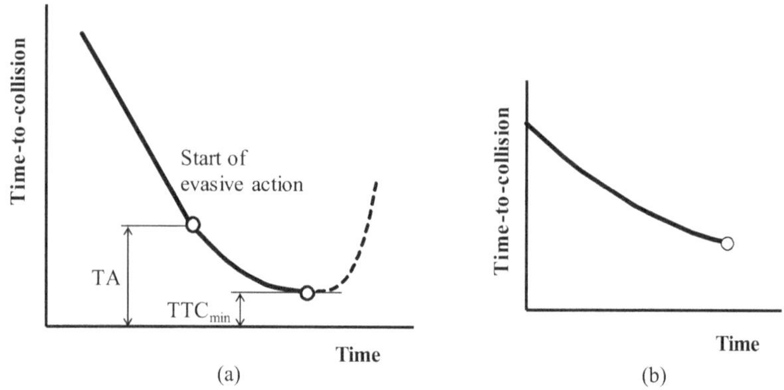

Figure 9.7 Theoretical TTC curve: (a) original curve discussed by Hayward (1972); (b) a special case when the collision course ceases without TTC increasing after a minimum (Laureshyn 2010)

was named Time-to-Accident (TA) by Hydén (1987). If the evasion is successful, the TTC reaches its lowest value (TTC_{min}) and then increases before becoming unavailable when the collision course ceases. Otherwise, if no evasive action took place, or it was not efficient enough, the TTC would reach zero, indicating a collision. It was also shown that in some scenarios, the TTC can cease to exist without increasing after a minimum [e.g. if one of the road users 'accelerate out of collision course' or swerves (Laureshyn 2010)].

TA and TTC_{min} are a good illustration of how the choice of an indicator determines, whether the researcher is aware of it or not, the choice of the underlying theoretical model as discussed by Güttinger (1982) (see above). While at the moment of TA registration, it is still unknown whether the evasive action will succeed or not, a non-zero TTC_{min} value indicates that the collision was avoided. TA can be applied both in near-miss situations and collisions, revealing a more versatile picture of the relation and continuity between the two situation types; moreover, it can even be used to grade the severity of the collisions, while TTC_{min} will always yield zero in case of a collision. Despite these strong arguments in favour of TA, TTC_{min} is used overwhelmingly more often. The reasons for that are purely practical – identifying a minimum value in a sequence of measurements is straightforward, while detecting an evasive action requires an elaborate understanding of how the situation is developing. This is not impossible, though – see, for example, a method suggested by Johnsson and Laureshyn (2022).

Future motion prediction required for TTC calculation is another problematic area frequently avoided in the reporting of studies using traffic conflict analysis. There are many scenarios in which the original Hayward's definition of 'present course' and 'present rates' does not hold. In case a road user intends to make a turn, it does not make much sense to predict that he or she will travel along a straight line. Similarly, starting from a full stop at a traffic light or negotiating a 90-degree turn at an intersection involves a deliberately intended change of speed.

The motion prediction methods can be roughly divided into kinematic and context-based methods (Mohamed and Saunier 2013). Kinematic methods usually employ simplistic assumptions about the dynamics of the involved road users (the original Hayward's definition belongs to this category). More complex assumptions have been suggested, such as constant angular speed and acceleration (the driver 'freezes' with the steering wheel and gas pedal); however, the predictions are often even more unrealistic compared to constant speed and heading, with trajectories leaving the road or speeds 'shooting' up to unreasonable values (van der Horst 1990).

Context-based methods utilize the knowledge about what could be the reasonable ways to travel through the studied area (Johnsson and Laureshyn 2022; Mohamed and Saunier 2013). This is usually done by analysing the historical trajectories of other road users who passed the same area from similar initial conditions (position, speed, intended manoeuvre, etc.). It seems reasonable to assume that when travelling freely (i.e. not being involved in any interactions) the current road user would intend to travel approximately in the same way that many others did before him/her. However, the predictions become less reliable (and meaningful) after the road user starts to interact or initiate an evasive action. For example, the logic of a particular interaction may require a full stop at an 'inappropriate' location; however, such behaviour would be difficult to predict from the historical data since the specific conditions and spatial-temporal relations between the road users in this particular interaction would be very unique and hardly present in the historical data. Theoretically, the issue could be addressed by introducing so-called interaction-aware models of road user behaviour (Lefèvre *et al.* 2014). However, we are not aware of any practical implementation of such models for SMoS.

Besides the 'classic' TTC, several modified indicators utilizing the same basic concept have been suggested. For example, the inversed TTC (i.e. 1/TTC) is claimed to be more appropriate to reflect the 'dangerousness', both since it increases with the TTC getting smaller, and due to the property of the inverted function yielding much higher values when TTC gets small (Chin *et al.* 1992). Minderhoud and Bovy (2001) suggested measuring the duration of the time spent with TTC being below a certain level, as well as measuring the area of the TTC curve below this threshold (naming the measures Time Exposed TTC, TET, and Time Integrated TTC, TIT). To avoid the problem of having no value in the case of the absence of the collision course, the time of arrival could be measured until a certain spatial position with a special meaning for safety. For example, Várhelyi (1998) introduced Time-to-Zebra indicator, namely the time remaining until a vehicle enters the zebra crossing markings. This indicator can be calculated in all vehicle-pedestrian interactions, regardless of the evasive actions taken by the pedestrians (which in many cases lead to termination of the collision course).

In a similar vein, Laureshyn *et al.* (2010) suggested using the expected time of arrival to the conflict point of the second – the latest – road user. This indicator, named T_2, has no difference from TTC in case of a collision course. However, it continues to deliver values even when the collision course has terminated, providing a smooth transition between the two states with no peculiar jumps or missing values at the moment of the transition. Moreover, it provides a connection to another family of indicators based on Post-Encroachment Time (PET, discussed in the following section) since T_2 becomes equal to the predicted PET at the moment the first road user leaves the conflict area.

One more concept related to TTC deserves to be mentioned here. Since the future motion of two road users involved in a conflict has not just one, but many possible options, various combinations of the future trajectories will not or will have a collision course, and if so, will yield different TTC values. Thus, the initial set of possible trajectory alternatives will be characterised by (i) the proportion of the trajectory combinations that result in a collision course [Probability of a Collision Course (PCC) (Johnsson and Laureshyn 2022, Mohamed and Saunier 2013)]; and (ii) the distribution of TTC values that can (but not necessarily have to) be aggregated to some 'probabilistic TTC' value (Saunier *et al.* 2010). Both measures convey distinct yet relevant meaning for safety analysis. The former reflects how often road users end up on a collision course and need to adjust for that (may be relevant for flow efficiency and subjective safety aspects), while the latter reflects, in a more sophisticated manner yet in a spirit similar to 'normal TTC', how close they come to a collision (i.e. the objective collision probability).

9.3.2 PET family

The *PET* was first introduced by Allen *et al.* (1978) to be used in situations when a collision was avoided but with a relatively small margin. PET equals the time between the first road-user leaving the area in which a collision was possible ('conflict area') and the second road user arriving there (see Figure 9.8). The actual meaning of PET is thus by how much time the road users 'missed' each other.

PET is the second most used indicator following the TTC (Laureshyn *et al.* 2016). To a high degree, this popularity can be attributed to its practical usability. PET can be directly observed and measured without any prior assumptions (such as the various methods for motion predictions in the case of TTC), in the simplest case just using a stopwatch. It also serves as a natural connection between safety and efficiency in traffic – while low PET values can be seen as compromised safety margins, the values in the 'moderate' range are rather an indication of efficiency, that is how well the available gaps between the vehicles are being used. As such, there is no real need to increase PET indefinitely to achieve safety, but, on the

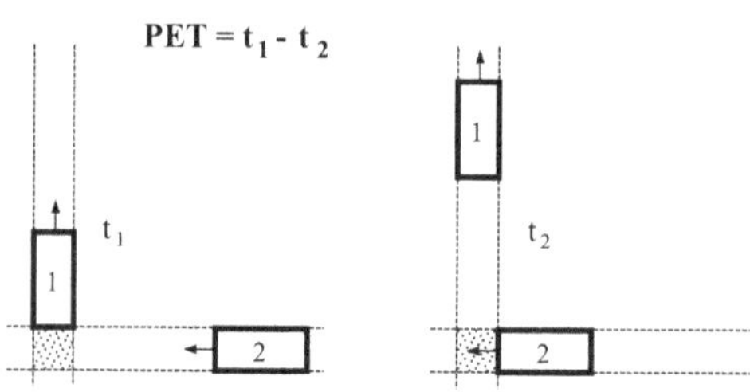

Figure 9.8 Definition of PET

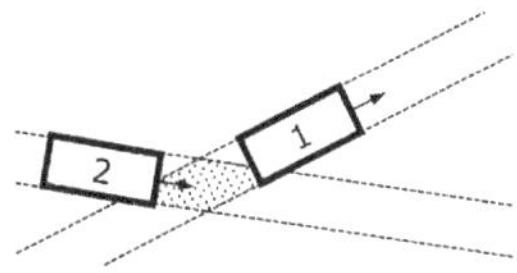

Figure 9.9 Problems with PET definition at angles different from 90° (Laureshyn et al. 2010)

contrary, as long as safety is not compromised, lower values are beneficial for the capacity and efficiency of traffic flow. Time gap and time headway are two indicators of a similar 'spirit' to PET, only defined for vehicles travelling in the same direction (in which case Time Gap is equal to PET and can be measured continuously). While well-established within the traffic flow theory context, their relation to safety remains questionable (Vogel 2003).

However, there are also drawbacks. First, PET becomes ill-defined when the road users interact at angles different from the classical 90 degrees shown in Figure 9.8. Figure 9.9 illustrates such a case in which both road users are present in the geometrically defined conflict area, yet the actual collision is avoided. Similarly, if the road users avoid collision by changing the trajectory, the location of the conflict area changes, affecting the arrival time values. This is particularly a problem in the case of interactions with cyclists since 'slaloming' is a very common strategy used by cyclists in case of perceived danger (Laureshyn *et al.* 2015).

Another problematic scenario is when the second road user applies hard braking that leads to a complete stop before the conflict area; it may take quite long time to start moving again, resulting in a very large PET, even though the original potential for collision was high.

Finally, there is no obvious transition point from PET being a safety indicator to becoming an indicator of traffic flow efficiency. It is apparent, however, that if the threshold for PET-defined traffic conflicts is set too inclusive, it just becomes an elaborate measure of exposure (any interaction, aka elementary exposure unit, would be counted as a conflict). For example, Johnsson *et al.* (2021b) report nearly perfect linear relations between the total number of interactions and 'traffic conflicts' defined with PET thresholds chosen at 2 seconds and higher.

Attempts to modify PET to address some of the mentioned issues have been made. For example, a predicted PET, pPET [also called Time Advancement, TAdv, by Hansson (1975)] estimates what PET will be if the road users continue travelling as initially intended (Saunier *et al.* 2010). But mitigating the problem of PET being affected by the evasive actions, it however becomes subject to the same motion prediction issues as in the case of TTC.

9.3.3 Measures related to evasive actions

The indicators of this category are based on the idea that more dangerous traffic situations are reflected by more intense evasive actions, either revealed by road users or calculated hypothetically. Since braking is by all means the most frequent evasive action observed (Hydén 1987), the largest share belongs to the deceleration-based indicators. For example, (*Initial*) *Deceleration Rate, DR*, is the actual deceleration value at the moment of the evasive action onset. *Maximum Deceleration, MD*, is the highest deceleration during the entire evasive action (Gettman and Head 2003). Some other related measures include *Deceleration to Avoid Crash, DRAC*, minimal deceleration necessary to avoid a collision in car-following scenarios (Mahmud *et al.* 2017), and *Deceleration-to-Safety Time, DST*[‡], applicable in crossing scenarios and equal to the minimal deceleration necessary to leave the collision course (i.e. to convert a 'TTC-situation' into 'PET-situation') (Hupfer 1997).

It has been shown that it is hard to distinguish intensive deceleration in safety-critical situations from 'normal' braking exercised by drivers with a bit more aggressive driving style (Bagdadi and Várhelyi 2011; af Wåhlberg 2004). The derivative of the acceleration, *jerk*, seems to be more sensitive to differences between intended and emergency-induced intensive braking (Bagdadi and Várhelyi 2011; Nygård 1999).

Swerving is commonly performed in conflict situations by two-wheeled vehicles, such as motorcycles and bicycles (Laureshyn *et al.* 2015). Indicators related to rotational rather than longitudinal movement might be better candidates to describe such manoeuvres [see, e.g., *Yaw rate* in Guo *et al.* (2018) and Tageldin *et al.* (2015)]. For pedestrians, rapid change of speed and direction ('jumping out of the situation') are even more common. Possible indicators to address such behaviour are the gait descriptors (e.g. step length and frequency) as suggested by Tageldin and Sayed (2016).

9.3.4 Outcome-based indicators

Indicators of this family reflect the severity of potential outcomes in case of an accident. Not surprisingly, most of them are related to the speed, the main determinant of the energy being released during the collision.

Conflicting Speed represents the speed at the moment of initiation of an evasive action (Hydén 1987). While originally Hydén (1987) used the speed of the road user who initiated the evasive action (so-called *relevant road user*), Shbeeb (2000) argued for a more flexible approach – for example, in conflicts between motor vehicles and pedestrians, it is nearly always the speed of the motor vehicle that has the largest effect on the severity in case of an accident, whether the driver is first to react to the dangerous situation or not.

Obviously, when both road users have non-zero speeds, it is rather their relative motion that affects the severity of a collision (consider head-on vs. rear-end collisions of two road users travelling at the same speeds). This aspect is better reflected by the *Relative speed*, which is the geometric difference between the two road users' speed vectors (Laureshyn *et al.* 2010; Saunier *et al.* 2010).

[‡]It remains unclear why the indicator was called 'time' while it is measured in m/s^2 units.

Potential Collision Speed (PCS) is the expected impact speed, given certain assumptions about the reaction time and intensity of the evasive action (MacCarley *et al.* 2007). *Delta-V* is defined as the change in speed as a result of a collision (Shelby 2011). The advantage of Delta-V is that it takes into account both the actual directions of the speed vectors of the two road users, and the difference in their masses. The latter can serve as a proxy for road user's vulnerability – for example, an unprotected pedestrian or cyclist weighs 10+ times less than a car, which will also generate a much higher Delta-V value for them compared to a motor vehicle.

9.3.5 Combined indexes

The rationale behind combining several indicators into a single value (an index) is that individual indicators might not be sufficiently universal, each representing only 'part of the truth'; put together, however, they can present a more integral view (Ismail *et al.* 2011). For example, Lu *et al.* (2012) combined non-complete braking time and TTC to calculate *conflict severity index*; Wang and Stamatiadis (2014) used required braking rate, maximum available braking rate and TTC to create an *Aggregate Crash Propensity Metric*; Ismail *et al.* (2010) combined TTC, PET, DST and Gap Time.

While the approach has its merits, a general observation could be that such indexes are seldom used repeatedly (one index will be found in one single publication), the decisions on how to combine individual indicators are taken quite arbitrary, and validation is mostly missing.

9.3.6 TCTs

TCTs are presented here as a separate category since, unlike individual indicators, they offer holistic methodologies for safety analysis based on traffic conflicts, including recommendations for observation duration, positioning of the observer, conflict recording sheets, criteria for conflict classification, etc. The 'golden age' for TCTs was in the 1980s when many countries developed their own TCTs as listed below:

- Austria (Risser *et al.* 1991; Risser and Schutzenhofer 1984)
- Belgium (Mortelmans *et al.* 1986)
- Canada (Sayed and Zein 1999; Brown 1994)
- Czech Republic (Kocárková 2012)
- Finland (Kulmala 1984)
- France (Muhlrad and Dupre 1984)
- Germany (Erke and Gstalter 1985; Erke 1984)
- the Netherlands (DOCTOR) (Kraay *et al.* 2013)
- Sweden (Laureshyn and Varhelyi 2020; Hydén 1987)
- UK (Baguley 1984)
- USA (Parker and Zegeer 1989).

There is a high degree of variation in the formal definitions and procedures used in these conflict techniques, while many rely on the previously mentioned indicators. However, the joint ICTCT calibration studies, in which several international teams applying different techniques observed the same traffic site simultaneously, showed that there was a strong agreement between the teams in rating the conflicts by their severity; the differences came mostly in the range of minor conflicts which were recorded by some techniques but disregarded by others (Kraay and van der Horst 1985; Grayson 1984). Significant efforts were also made to validate the techniques, in some cases using very large datasets with hundreds of observed locations (Shbeeb 2000; Svensson 1998; 1992; Hydén 1987).

Reflecting the traffic safety understanding of their time, all techniques were primarily designed to study vehicle-to-vehicle conflicts. Their application on traffic situations involving vulnerable road users is questionable or at least not as straightforward since many of the assumptions, such as braking being the most expected evasive manoeuvre, are not particularly relevant for cyclists (who often swerve rather than brake) or pedestrians (who can change the speed and direction of movement virtually instantly).

To our knowledge, it is only the Swedish TCT and the Dutch technique DOCTOR that still remain in active use (van der Horst *et al.* 2017; Laureshyn *et al.* 2015). Particularly in the context of low- and middle-income countries, these tools provide reliable and cost-effective alternatives to accident-based analysis, which often cannot be performed due to the poor quality of historical accident data.

9.4 Data collection tools

9.4.1 Human observations

The original TCTs from 1970s to 1980s relied almost entirely on human observers as the conflict detection and analysis tool. This has shaped the techniques towards more simplistic conflict definitions (to limit the cognitive load on the observer who performed the task in real time and often over extended periods), but also opened up for subjective elements used in rating the conflict severity. Frequently, a conflict would be based on criteria like 'fully controlled manoeuvre' or 'rapid evasion', which are intuitively understandable, but lack a strict definition in kinematic terms. Naturally, this raised concerns about the reliability of the observer's judgements and whether different results would be obtained if someone else performed the observations.

A series of validation and calibration studies persuasively showed that there is a high degree of agreement between trained observers in what constitutes a conflict and also in rating the conflicts by their severity once they have been identified (Kruysse 1991; Hydén 1987; Grayson 1984; Lightburn and Howarth 1979). It has also been shown that it is possible to train observers to estimate speeds and distances with relatively high precision (Hydén 1987).

Generally, it appears that humans possess a 'built-in' instrument for judging the severity of traffic situations, well aligned between individuals (Yastremska-Kravchenko *et al.* 2022; Kruysse and Wijlhuizen 1992; Kruysse 1991). This instrument, however, seems to be based on a holistic perception of the situation rather than on some simple kinematic measure. Experienced conflict observers usually admit that even if the guidelines explicitly require them to rely on objective measures only, it is not unusual to 'stretch' the formal definitions to better fit with their own gut feeling. For that likely reason, it turned out to be difficult to replicate the traditional TCTs in computer-aided conflict analysis based on objective measurements since this subjective component was missing.

A better understanding of which additional aspects contribute to 'severity' and how they can be expressed in objective terms is a research area deserving further attention. However, one should not forget that the objective is to measure safety, not to replicate human observer judgment as such.

9.4.2 Video recording

Field observations of traffic conflicts are quite demanding. Therefore, the observers were eventually replaced by video cameras installed at the locations of interests. Video recordings as such do not fundamentally change the task of the observer who still must perform the detection and classification tasks, though in a more comfortable office environment, and with the possibility to re-watch the situations of interest or take a break when the attention level declines.

Using video recordings introduces new challenges, both of a technical and legal nature. To get a good perspective, a camera should be installed at a sufficient height, but for short-term recording, the researcher can rely mostly on already existing infrastructure like lamp posts or balconies of nearby buildings. While it is possible to get some usable view, it is rarely the best one. Other practical problems include access to the power supply (e.g. lamp posts usually do not have power during daytime, which means some batteries are necessary, too), protection of the equipment from the weather, potential theft or vandalism, getting permissions and technical assistance from the authorities responsible for the infrastructure being used, etc.

From the legal perspective, recording and storage of video data are regulated in most countries by personal data protection laws. Since it is theoretically possible to recognise individual persons (by seeing the face or through other features like unusual clothing, gait features, etc.) or vehicles (by reading the number plates), the recording becomes sensitive data that can be misused. Despite attempts to unify the requirements [the most ambitious is probably the GPDR law in the European Union – GDPR (2016)], there is still a large variety in how individual states solve the trade-off between protecting the individuals and providing some other values for the society. Checking the local regulations is always a good idea prior to planning any recordings.

There are some alternatives that allow making video data less sensitive. For example, the usage of thermal rather than regular RGB cameras makes it quite impossible to distinguish the face features or read number plates, while still containing enough information to see that it is a person or a vehicle moving (see Figure 9.10). Flying drones equipped with a camera provides the best possible view (no occlusions by trees or large vehicles), but at the cost of limited flying time and some legal limitations related to drone piloting (pilot licencing, maximum flight height, equipment weight, proximity to special areas like airports, etc.). If the

Figure 9.10 Same situation recorded by an RGB and thermal cameras, installed next to each other [courtesy of Horizon 2020 project InDeV (2015–2018)]

recorded video is processed in real time (see next section), without storing the actual footage, no records of sensitive data are created. It is however often useful to have some video stored and not bluntly rely on trajectory analysis, both for getting a more holistic view of the detected situations and for communicating the results at a later stage.

9.4.3 *Video analysis and other emerging technologies*

Video analysis basically means interpreting the information contained in the video stream (a sequence of still images taken with high frequency) in terms relevant to the purpose of the analysis, for example, detecting road users' presence, their location, direction and velocity of movement, preferably expressed in the real-world (rather than image-based) coordinates. The technical aspects of video analysis constitute a very broad and complex subject, so what follows is rather a very simplified overview meant for a non-expert to be able to understand the practical implications of using video analysis tools for traffic conflict studies.

The simplest applications of video analysis in traffic are so-called 'virtual loops' (Yung and Lai 2001). A selected area of the camera view is monitored for rapid changes in colour and intensity – for example, if the pixels that are normally grey (asphalt) become red, it may indicate a presence of a road user (red car). A bit more sophisticated 'loops' may identify the direction of motion or long presence (idling); what is common, however, is that all analyses are done within the image reference system, i.e. the real-world distances or speeds are not available.

Several 'loops' placed strategically can act as a watchdog, detecting situations that should be further investigated. Despite the utter simplicity of the approach, significant savings in time can be reached, as up to 85%–90% of the video footage can be excluded from further analysis as not relevant. It is particularly efficient for simple and well-defined scenarios (e.g., a simultaneous presence of a pedestrian at a crossing and an approaching car), and when the frequency of the events of interest is low (Madsen and Lahrmann 2017; Laureshyn *et al.* 2009).

To be able to transform image pixel positions into real-world coordinates, a camera calibration procedure needs to be performed. Special models exist for this task. For example, the frequently used Tsai model (Tsai 1987) requires calibration of 17 parameters that both describe the orientation of the camera in relation to the road plane and the optical properties of the camera itself to be able to handle the radial distortions introduced by its lenses. Usually, camera calibration requires a set of points with known positions in the real world that can also be easily identified in the camera image. A well-calibrated camera becomes an accurate instrument for measuring both positions and speeds of road users (Laureshyn and Nilsson 2018).

With a calibration model in place, road users can be either marked manually in the video frames, or fully automated algorithms can be employed to detect road users and connect their positions into continuous trajectories. The manual processing is labour intensive and can only be used for very short video sequences (e.g. to process a traffic conflict that has already been detected); its advantage is a very high precision of the trajectories and also the general quality check performed by the operator. To process large volumes of video data (days, possibly weeks long footage), automated tracking is the only feasible option. It is important to realise, though, that detection and tracking performance is not perfect, in particular in complex urban traffic and high volumes with occlusions (in case of a relatively low camera angle). Users may be missed, or others falsely detected, or wrongly connected, and the automated trajectories contain some amount of noise, for example, the positions oscillating around the 'true' values.

Other technologies for tracking road users involve special types of cameras (thermal, airborne), LiDAR, and combinations of sensors (sensor fusion) (Fu *et al.* 2017; Tarko *et al.* 2017; Alldieck *et al.* 2016; Knake-Langhorst *et al.* 2016). Fortunately, the field of computer vision and machine learning evolves very rapidly, with significant improvement being made every year. It is reasonable to expect that in the relatively near future, the technical problem of road user detection and tracking will be resolved. It is important to note, however, that the trajectories provide only raw data for the calculation of the SMoS indicators. The problem of choosing the right indicators, specifying their exact definitions and calculation procedures, and, most importantly, confirming their validity will not go away regardless of how accurate trajectory data might be.

9.5 Future directions

Starting from the 1970s, the number of traffic fatalities has been decreasing in most high-income countries. Today, many of these countries have adopted Vision Zero (no one killed or severely injured in traffic) as the long-term goal, which will most probably lead to the further reduction of fatalities, injuries and accidents in general. On the contrary, in the low- and middle-income countries (LMICs), the burden of traffic accidents seems to remain unchanged or even grow in size (WHO 2023), while these countries still struggle with poor quality and limited access to the accident data. Despite the dramatic differences between the two country categories, the ability to rely on alternative data sources that are not accident-based grows in importance, and SMoS are most likely to play an even greater role in the future traffic safety work.

The development of technical tools for automated SMoS data collection is likely to continue at a fast pace, decreasing the costs both for the equipment and for the data processing itself. Yet, it is not yet time to discard human observers completely. In the context of low- and middle-income countries, using trained observers is still the most cost-effective option compared to relying on advanced (and costly) hard- and software, which might not perform as reliably in the messy traffic conditions often found in these countries. Moreover, general caution should be taken about the aspects of SMoS analysis that benefit from automation. A strong advantage of traffic observations as a study method (compared to post hoc accident analysis) is that it provides a better understanding of 'how traffic really works', often revealing unconventional yet quite common behaviours (and related safety problems). Usually, these are hard to anticipate, which means that a fully automated traffic conflict detector machine might be 'blind' to certain traffic safety problems that an open-minded observer easily spots out. The great advantage of being able to actually see the safety-critical situations and reflect on what might have contributed to their occurrence gets wasted if automation goes too far; counting conflicts, without their further examination, is not a practice to be encouraged.

Further development of the SMoS theory, its validation, and particularly developing methods for estimating the expected number of accidents/injuries based on SMoS input are still highly relevant pursuits. While the EVT might be the solution, the approach is very 'heavy' theoretically, and it is difficult both to apply and explain for non-statisticians. From this perspective, one still can argue for the existence of a 'light version' of SMoS tools, potentially less precise, yet easier to apply and interpret.

We have already stressed that the adoption of the Safe System mindset shifts the focus of SMoS applications from predicting accident frequency (expectedly dominated by property damage only accidents) to dealing with injury and fatality risks (which becomes a more complex task). Fundamentally different from the traditional view of safety as 'optimization' of risks, the Safe System advocates a rather radical approach of removing any opportunities for severe injuries to occur, regardless of their probability. From this perspective, SMoS could be used to reveal whether unsafe interactions are possible given the infrastructure design and road user behaviour (e.g., despite the presence of a speed-secured pedestrian crossing, many of them may take shortcuts outside the crossing where vehicle speeds are high). If the *possibility*, rather than frequency of such conflicts is of interest, relatively simple and straightforward SMoS measures could be used to perform the task (Johnsson and Laureshyn 2024).

Historically, SMoS focused much on vehicle-to-vehicle interactions, ignoring or down-prioritizing the unprotected road users (cyclists, pedestrians, etc.). While the SMoS concepts are general and valid even for these road user groups, indicators and techniques may not always be adjusted or sensitive enough to handle their vulnerability and higher injury proneness. These groups are still greatly over-represented in injury statistics (particularly in urban areas), and further development of SMoS to accommodate them is necessary.

During the last two decades, a great bulk of literature has been generated exploring whether SMoS analysis can be based on trajectories generated by microscopic simulations, as implemented in the Surrogate Safety Assessment Model (Gettman *et al.* 2008). In their current shape, microscopic simulation programs are mostly developed for other purposes than safety analysis, and the models representing the 'behaviour' of road users are generally collision-free and calibrated for normal traffic rather than extreme safety-compromised conditions as found in traffic conflicts. While correlations have been found between simulated conflicts and recorded accidents, it is unclear whether it is just a mere result of both accidents and the modelled interactions being functions of the traffic flows (exposure). More confidence in the approach can be gained after developing and validating models representing the entire spectrum of human interactions (including 'emergencies'), occurrence of errors leading to conflicts and accidents, as well as the behaviour in critical situations (e.g. braking intensity might be higher than comfortable level, spontaneous reactions might be less rational, etc.).

Another promising data source for SMoS analysis is the data logged by vehicles that are increasingly equipped with sensors and processing capabilities as driving becomes more and more automated. To be able to navigate in a complex traffic environment, a self-driving vehicle must be 'aware' of the surrounding environment and other road users, including their positions, headings, speeds, etc. Moreover, since the 'behaviour' of automated vehicles is pre-programmed, prediction of future motion (e.g. to calculate TTC) may be done without complex assumptions or even be directly obtained. As the presence of automated vehicles in traffic increases, and as their manufacturers and owners may become more open to the idea of data sharing (or be forced by regulation), analysing data logs of vehicles that pass a certain geographical area might be more efficient compared to trying to collect the same data from dedicated sensors installed in the infrastructure. This evolution of traffic is also a challenge as traffic becomes more heterogeneous with varying levels of automation and vehicles of various forms and weights (e.g., micromobility), changing the relationship between conflicts and safety. In such quickly evolving conditions, SMoS may be the only way to diagnose safety without having to wait years for accident data.

References

Af Wåhlberg, A. E. (2004), 'The stability of driver acceleration behavior, and a replication of its relation to bus accidents', *Accident Analysis & Prevention*, 36 (1), 83–92, https://doi.org/10.1016/S0001-4575(02)00130-6.

Alldieck, T., C. H. Bahnsen, and T. B. Moeslund (2016), 'Context-aware fusion of RGB and thermal imagery for traffic monitoring', *Sensors*, 16 (11), 1947, https://doi.org/10.3390/s16111947.

Allen, B. L., B. T. Shin, and P. J. Cooper (1978), 'Analysis of traffic conflicts and collisions', *Transportation Research Record*, 667, 67–74, https://onlinepubs.trb.org/Onlinepubs/trr/1978/667/667-009.pdf.

Alsop, J., and J. Langley (2001), 'Under-reporting of motor vehicle traffic crash victims in New Zealand', *Accident Analysis & Prevention*, 33 (3), 353–359, https://doi.org/10.1016/S0001-4575(00)00049-X.

Amoros, E., J.-L. Martin, and B. Laumon (2006), 'Under-reporting of road crash casualties in France', *Accident Analysis & Prevention*, 38 (4), 627–635, https://doi.org/10.1016/j.aap.2005.11.006.

Amoros, E., M. C. Brosnan, F. Wegman, *et al.* (2011), *Reporting on Serious Road Traffic Casualties: Combining and Using Different Data Sources to Improve Understanding of Non-fatal Road Traffic Crashes* (Paris, France: International Transport Forum).

Amundsen, F. H., and C. Hyden (eds.) (1977), *Proceedings from the First Workshop on Traffic Conflicts, 26–27 September 1977*, Oslo, Norway, https://www.ictct.net/wp-content/uploads/XX-Oslo-1977/1977-Proceedings.pdf.

Asmussen, E.(ed.) (1984), *Proceedings of the NATO Advanced Research Workshop on International Calibration Study of Traffic Conflict Techniques*. 25–27 May 1983, Copenhagen, Denmark (NATO ASI Series F: Computer and System Science vol. 5), https://www.ictct.net/wp-content/uploads/XX-Copenhagen-1983/1983-Proceedings.pdf.

Bagdadi, O., and A. Várhelyi (2011), 'Jerky driving – An indicator for accident proneness', *Accident Analysis & Prevention*, 43, 1359–1363, https://doi.org/10.1016/j.aap.2011.02.009.

Baguley, C. J. (1984), 'The British traffic conflicts technique', presented at *NATO Advanced Research Workshop on International Calibration Study of Traffic Conflict Techniques* (Copenhagen, Denmark), https://www.ictct.net/wp-content/uploads/SMoS_Library/LIB_Baguley-1984.pdf.

Brown, G. R. (1994), 'Traffic conflicts for road user safety studies', *Canadian Journal of Civil Engineering*, 21, 1–15, https://www.ictct.net/wp-content/uploads/SMoS_Library/LIB_Brown_1994.pdf.

Bulla-Cruz, L. A., A. Laureshyn, and L. Lyons (2020), 'Event-based road safety assessment: A novel approach towards risk microsimulation in roundabouts', *Measurement: Journal of the International Measurement Confederation*, 165, 108192, https://doi.org/10.1016/j.measurement.2020.108192.

Chapman, R. (1973), 'The concept of exposure', *Accident Analysis & Prevention*, 5, 95–110, https://doi.org/10.1016/0001-4575(73)90018-3.

Chen, Z., O. Yastremska-Kravchenko, A. Laureshyn, C. Johnsson, and C. D'Agostino (2024), 'Stochastic method based on copulas for predicting severe road traffic interactions', *Analytic Methods in Accident Research*, 44, 100347, https://doi.org/10.1016/j.amar.2024.100347.

Chin, H. C., S. T. Quek, and R. L. Cheu (1992), 'Quantitative examination of traffic conflicts', *Transportation Research Record*, 1376, 86–91, https://onlinepubs.trb.org/Onlinepubs/trr/1992/1376/1376-012.pdf.

Davis, G. A., J. Hourdos, H. Xiong, and I. Chatterjee (2011), 'Outline for a causal model of traffic conflicts and crashes', *Accident Analysis & Prevention*, 43 (6), 1907–1919, https://doi.org/10.1016/j.aap.2011.05.001.

Elvik, R., A. Erke, and P. Christensen (2009a), 'Elementary units of exposure', *Transportation Research Record*, 2103 (1), 25–31, https://doi.org/10.3141/2103-04.

Elvik, R., A. Høye, T. Vaa, and M. Sørensen (2009b), *The Handbook of Road Safety Measures* (Bingley: Emerald Group Publishing Limited), https://doi.org/10.1108/9781848552517.

Elvik, R. (2010), 'An exploratory study of mechanisms by which exposure influences accident occurrence', presented at *89th TRB Annual Meeting*, 10–14 January 2010 (Washington, DC, USA).

Elvik, R. (2015), 'Some implications of an event-based definition of exposure to the risk of road accident', *Accident Analysis & Prevention*, 76 (Supplement C), 15–24, https://doi.org/10.1016/j.aap.2014.12.011.

Erke, H. (1984), 'The traffic conflict technique of the Federal Republic of Germany', presented at *NATO Advanced Research Workshop on International Calibration Study of Traffic Conflict Techniques* (Copenhagen, Denmark), https://www.ictct.net/wp-content/uploads/SMoS_Library/LIB_Erke_1984.pdf.

Erke, H., and H. Gstalter (1985), 'Verkehrskonflikttechnik – Handbuch für die Durchführung und Auswertung von Erhebungen [Traffic conflict technique – manual]' (Unfall- und Sicherheitsforschung Straßenverkeh), *Heft*, 52, https://www.ictct.net/wp-content/uploads/SMoS_Library/LIB_Erke_Gstalter_1985.pdf.

Forbes, T. W. (1957), 'Analysis of "near accident" reports', *Highway Research Bulletin*, 152.

Fu, T., J. Stipancic, S. Zangenehpour, L. F. Miranda-Moreno, and N. Saunier (2017), 'Automatic traffic data collection under varying lighting and temperature conditions in multimodal environments: Thermal versus visible spectrum video-based systems', *Journal of Advanced Transportation*, 2017, 5142732, https://doi.org/10.1155/2017/5142732.

GDPR (2016), '*General Data Protection Regulation 2016/679 of the European Parlament and the Council*', https://eur-lex.europa.eu/legal-content/en/TXT/?uri=CELEX:32016R0679.

Gettman, D., and L. Head (2003), 'Surrogate safety measures from traffic simulation models', *Transportation Research Record*, 1840, 104–115, https://doi.org/10.3141/1840-12.

Grayson, G. B. (1984), 'The Malmö study: A calibration of traffic conflict techniques' (SWOV), *R-84-12*, https://www.ictct.net/wp-content/uploads/SMoS_Library/LIB_Grayson_1984.pdf.

Guo, Y., T. Sayed, and M. H. Zaki (2018), 'Exploring evasive action-based indicators for PTW conflicts in shared traffic facility environments', *Journal of Transportation Engineering, Part A: Systems*, 144 (11), 04018065, https://doi.org/10.1061/JTEPBS.0000190.

Güttinger, V. A. (1982), 'From accidents to conflicts: Alternative safety measurement', presented at *Third International Workshop on Traffic Conflicts Techniques* (Leidschendam, The Netherlands), https://www.ictct.net/wp-content/uploads/XX-Leidschendam-1982/Guttinger_1982.pdf.

Hansson, A. (1975), 'Studies in driver behaviour, with applications in traffic design and planning. *Two examples' (Lund Institute of Technology, Department of Traffic Planning and Engineering, University of Lund), Bulletin* 11.

Hauer, E. (1982), 'Traffic conflicts and exposure', *Accident Analysis & Prevention*, 14 (5), 359–364, https://doi.org/10.1016/0001-4575(82)90014-8.

Hauer, E., and P. Gårder (1986), 'Research into the validity of the traffic conflict technique', *Accident Analysis & Prevention*, 18 (6), 471–481, https://doi.org/10.1016/0001-4575(86)90020-5.

Hauer, E. (1997), *Observational Before–After Studies in Road Safety: Estimating the Effects of Highway and Traffic Engineering Measures on Road Safety* (London: Emerald Group Publishing Limited).

Hayward, J. C. (1972), 'Near-miss determination through use of a scale of danger', *Highway Research Record*, 384, 22–34, https://onlinepubs.trb.org/Onlinepubs/hrr/1972/384/384-004.pdf.

Hupfer, C. (1997), 'Deceleration to Safety Time (DST) – A useful figure to evaluate traffic safety?', presented at *10th ICTCT Workshop*, 5–7 November 1997 (Lund, Sweden), https://www.ictct.net/wp-content/uploads/10-Lund-1997/10-Hupfer.pdf.

Hydén, C. (1987), 'The development of a method for traffic safety evaluation: The Swedish traffic conflict technique', *Doctoral thesis*, Department of Traffic Planning and Engineering, Lund University (Lund, Sweden), Bulletin 70, https://www.ictct.net/wp-content/uploads/SMoS_Library/LIB_Hyden_1987.pdf.

ICTCT (1986), *Proceedings of the Workshop on Traffic Conflicts and Other Intermediate Measures*. 8–10 September 1986, Budapest, Hungary (Institute of Transport Sciences), https://www.ictct.net/wp-content/uploads/XX-Budapest-1986/1986-Proceedings.pdf.

InDeV (2015–2018), 'InDeV: In-depth understanding of accident causation for vulnerable road users' (EU Horizon 2020, Framework Programme for Research and Innovation), project No. 635895, https://doi.org/10.3030/635895.

Ismail, K., T. Sayed, and N. Saunier (2010), 'Automated safety analysis using video sensors: Technology and case studies', presented at *Canadian Multidisciplinary Road Safety Conference* (Ontario, Canada: 6–9 June 2010), http://n.saunier.free.fr/saunier/stock/ismail10cmrsc.pdf.

Ismail, K., T. Sayed, and N. Saunier (2011), 'Methodologies for aggregating indicators of traffic conflict', *Transportation Research Record*, 2237, 10–19, https://doi.org/10.3141/2237-02.

Janstrup, K. H., S. Kaplan, T. Hels, J. Lauritsen, and C. G. Prato (2016), 'Understanding traffic crash under-reporting: Linking police and medical records to individual and crash characteristics', *Traffic Injury Prevention*, 17 (6), 580–584, https://doi.org/10.1080/15389588.2015.1128533.

Johnsson, C., A. Laureshyn, and T. De Ceunynck (2018), 'In search of surrogate safety indicators for vulnerable road users: A review of surrogate safety indicators', *Transport Reviews*, 38 (6), 765–785, https://doi.org/10.1080/01441647.2018.1442888.

Johnsson, C., A. Laureshyn, C. D'Agostino, and T. De Ceunynck (2020), 'The "safety in density" effect for cyclists and motor vehicles in Scandinavia: An observational study', *IATSS Research*, 45 (2), https://doi.org/10.1016/j.iatssr.2020.08.003.

Johnsson, C., A. Laureshyn, and C. Dágostino (2021a), 'A relative approach to the validation of surrogate measures of safety', *Accident Analysis & Prevention*, 161, 106350, https://doi.org/10.1016/j.aap.2021.106350.

Johnsson, C., A. Laureshyn, and C. Dágostino (2021b), 'Validation of surrogate measures of safety with a focus on bicyclist–motor vehicle interactions', *Accident Analysis & Prevention*, 153, 106037, https://doi.org/10.1016/j.aap.2021.106037.

Johnsson, C., and A. Laureshyn (2022), 'Identification of evasive manoeuvres in traffic interactions and conflicts', *Traffic Safety Research*, 3, 000012, https://doi.org/10.55329/erqd8683.

Johnsson, C., and A. Laureshyn (2024), 'Worst case analysis, a video analysis approach to injury potential in traffic', *Journal of Road Safety*, 35(4), 10–18, https://doi.org/10.33492/JRS-D-24-4-2450246.

Kirchhof, M., and U. Stilla (2006), 'Detection of moving objects in airborne thermal videos', *ISPRS Journal of Photogrammetry and Remote Sensing*, 61 (3–4), 187–196, https://doi.org/10.1016/j.isprsjprs.2006.09.003.

Knake-Langhorst, S., K. Gimm, T. Frankiewicz, and F. Köster (2016), 'Test site AIM – Toolbox and enabler for applied research and development in traffic and mobility', *Transportation Research Procedia*, 14, 2197–2206, https://doi.org/10.1016/j.trpro.2016.05.235.

Kocárková, D. (2012), 'Traffic conflict techniques in Czech Republic', presented at *SIIV – 5th International Congress – Sustainability of Road Infrastructures* (Rome, Italy: 29–31 October 2012), https://doi.org/10.1016/j.sbspro.2012.09.952.

Kraay, J. H. (ed.) (1982), *Proceedings of the Third International Workshop on Traffic Conflicts Techniques. April 1982*, Leidschendam, The Netherlands (SWOV, R-82-27), https://www.ictct.net/wp-content/uploads/XX-Leidschendam-1982/1982-Proceedings.pdf.

Kraay, J. H., and R. van der Horst (1985), *The Trautenfels Study: A Diagnosis of Road Safety Using the Dutch Conflict Observation Technique DOCTOR* (SWOV), R-85–53.

Kraay, J. H., A. R. A. van der Horst, and S. Oppe (2013), *Manual Conflict Observation Technique DOCTOR* (Voorburg, the Netherlands: Foundation Road safety for all), https://www.ictct.net/wp-content/uploads/SMoS_Library/LIB_Kraay_et_al_2013.pdf.

Kruysse, H. W. (1991), 'The subjective evaluation of traffic conflicts based on an internal concept of dangerousness', *Accident Analysis & Prevention*, 23 (1), 53–65, https://doi.org/10.1016/0001-4575(91)90035-4.

Kruysse, H. W., and G. J. Wijlhuizen (1992), 'Why are experts not better in judging the danger of filmed traffic conflicts?', *Accident Analysis & Prevention*, 26 (3), 227–235, https://doi.org/10.1016/0001-4575(92)90002-Z.

Kulmala, R. (1984), 'The Finnish traffic conflict technique', presented at *NATO Advanced Research Workshop on International Calibration Study of Traffic Conflict Techniques* (Copenhagen, Denmark), https://www.ictct.net/wp-content/uploads/SMoS_Library/LIB_Kulmala-1984.pdf.

Laureshyn, A., H. Ardö, T. Jonsson, and Å. Svensson (2009), 'Application of automated video analysis for behavioural studies: Concept and experience', *IET Intelligent Transport Systems*, 3 (3), 345–357, https://doi.org/10.1049/iet-its.2008.0077.

Laureshyn, A. (2010), 'Application of automated video analysis to road user behaviour', *Transport & Roads*, Department of Technology & Society, Faculty of Engineering, LTH, Lund University, Bulletin 253, https://portal.research.lu.se/en/publications/0d1697f5-8e26-4f16-b933-679295264d75.

Laureshyn, A., Å. Svensson, and C. Hydén (2010), 'Evaluation of traffic safety, based on micro-level behavioural data: Theoretical framework and first implementation', *Accident Analysis & Prevention*, 42, 1637–1646, https://doi.org/10.1016/j.aap.2010.03.021.

Laureshyn, A., M. D. Goede, N. Saunier, and A. Fyhri (2015), 'Cross-comparison of three surrogate safety methods to diagnose cyclist safety problems at intersections in Norway', presented at *International Cycling Safety Conference*, 15–16 September (Hannover, Germany).

Laureshyn, A., C. Johnsson, T. De Ceunynck, *et al.* (2016), 'Review of current study methods for VRU safety. Appendix 6 – Scoping review: Surrogate measures of safety in site-based road traffic observations' (*InDeV, Horizon 2020 project), Deliverable 2.1 – part 5*, https://www.ictct.net/wp-content/uploads/SMoS_Library/LIB_Laureshyn_et_al_2016.pdf.

Laureshyn, A., T. De Ceunynck, C. Karlsson, Å. Svensson, and S. Daniels (2017), 'In search of the severity dimension of traffic events: Extended Delta-V as a traffic conflict indicator', *Accident Analysis & Prevention*, 98, 46–56, https://doi.org/10.1016/j.aap.2016.09.026.

Laureshyn, A., and M. Nilsson (2018), 'How accurately can we measure from video? Practical considerations and enhancement of the camera calibration procedure', *Transportation Research Record*, 2672 (43), 24–33, https://doi.org/10.1177/0361198118774194.

Laureshyn, A., and A. Varhelyi (2020), *The Swedish Traffic Conflict Technique: Observer's Manual* (Lund, Sweden: Lund University), https://www.ictct.net/wp-content/uploads/SMoS_Library/LIB_Laureshyn_Varhelyi_2020.pdf.

Lefèvre, S., D. Vasquez, and C. Laugier (2014), 'A survey on motion prediction and risk assessment for intelligent vehicles', *ROBOMECH Journal*, 1 (1), https://doi.org/10.1186/s40648-014-0001-z.

Lightburn, A., and C. I. Howarth (1979), 'A study of observer variability and reliability in the detection and grading of traffic conflicts', *presented at Second International Traffic Conflicts Technique Workshop*, 10–12 May 1979 (Paris, France), https://www.ictct.net/wp-content/uploads/XX-Paris-1979/1979-Lightburn-Howarth.pdf.

Lu, G., M. Liu, Y. Wang, and G. Yu (2012), 'Quantifying the severity of traffic conflict by assuming moving elements as rectangles at intersection', *Procedia – Social and Behavioral Sciences*, 43, 255–264, https://doi.org/10.1016/j.sbspro.2012.04.098.

MacCarley, C. A., C. Ackles, and T. Watts (2007), 'A study of the response of highway traffic to dynamic fog warning and speed advisory messages', *presented at 86th TRB Annual Meeting*, 21–25 January 2007 (Washington, DC, USA).

Madsen, T., and H. Lahrmann (2017), 'Comparison of five bicycle facility designs in signalized intersections using traffic conflict studies', *Transportation Research Part F: Traffic Psychology and Behaviour*, 46 (Part B), 438–450, https://doi.org/10.1016/j.trf.2016.05.008.

Madsen, T. K. O. (2018), 'Video analysis and mapping of vulnerable road users' safety', *PhD Thesis*, Aalborg University, Denmark, https://doi.org/10.5278/vbn.phd.eng.00057.

Mahmud, S. M. S., L. Ferreira, M. S. Hoque, and A. Tavassoli (2017), 'Application of proximal surrogate indicators for safety evaluation: A review of recent developments and research needs', *IATSS Research*, 41 (4), 153–163, https://doi.org/10.1016/j.iatssr.2017.02.001.

Migletz, D. J., W. D. Glauz, and K. M. Bauer (1985), 'Relationships between traffic conflicts and accidents' (U.S. Department of Transportation, Federal Highway Administration), FHWA/RD-84/042.

Minderhoud, M. M., and P. H. L. Bovy (2001), 'Extended time-to-collision measures for road traffic safety assessment', *Accident Analysis & Prevention*, 33 (1), 89–97, https://doi.org/10.1016/S0001-4575(00)00019-1.

Mohamed, M. G., and N. Saunier (2013), 'Motion prediction methods for surrogate safety analysis', *Transportation Research Record*, 2386, 168–178, https://doi.org/10.3141/2386-19.

Mortelmans, J., A. Vits, L. Venstermans, M. van Essche, M. Boogaerts, and E. Vergisson (1986), 'Analyse van de verkeersveiligheid met behulp van de Bijna-Ongevallen methode [Analysis of road safety by means of the near-accident method]' (Katholieke Universiteit Leuven, Departement

Bouwkunde), https://www.ictct.net/wp-content/uploads/SMoS_Library/LIB_Mortelmans_1986.pdf.

Muhlrad, N., and G. Dupre (1984), 'The French conflict technique', presented at *NATO Advanced Research Workshop on International Calibration Study of Traffic Conflict Techniques* (Copenhagen, Denmark).

Niu, D., T. Sayed, C. Fu, and F. Mannering (2024), 'A cross-comparison of different extreme value modeling techniques for traffic conflict-based crash risk estimation', *Analytic Methods in Accident Research*, 100352, https://doi.org/10.1016/j.amar.2024.100352.

Nygård, M. (1999), *A Method for Analysing Traffic Safety with Help of Speed Profiles', Master's thesis, Department of Civil Engineering, Tampere University of Technology* (Espoo, Finland).

Older, S. J., and J. Shippey (eds.) (1980), *Proceedings of the Second International Traffic Conflicts Technique Workshop*, Paris, France, 10–12 May 1979 (Transport and Road Research Laboratory), https://www.ictct.net/wp-content/uploads/XX-Paris-1979/1979-Proceedings.pdf.

Parker, M. R., and C. V. Zegeer (1989), 'Traffic conflict techniques for safety and operation' (US Department of Transportation, Federal Highway Administration), FHWA-IP-88-027, https://www.fhwa.dot.gov/publications/research/safety/88027/88027.pdf.

Perkins, S. R., and J. L. Harris (1968), 'Traffic conflict characteristics – Accident potential at intersections', *Highway Research Record*, 225, 35–43, https://onlinepubs.trb.org/Onlinepubs/hrr/1968/225/225.pdf.

Risser, R., and A. Schutzenhofer (1984), 'Application of traffic conflict technique in Austria', presented at *NATO Advanced Research Workshop on International Calibration Study of Traffic Conflict Techniques* (Copenhagen, Denmark), https://www.ictct.net/wp-content/uploads/XX-Copenhagen-1983/1983-Proceedings.pdf.

Risser, R., W. D. Zuzan, W. Tamme, J. Steinbauer, and A. Kaba (1991), *Handbuch zur Erhebung von Verkehrskonflikten mit Anleitungen zur Beobachterschulung [Handbook for the collection of traffic conflicts with instructions for observers' training]* (Vienna, Austria: Literas Universitätsverlag), https://www.ictct.net/wp-content/uploads/SMoS_Library/LIB_Risser-1991.pdf.

Saunier, N., and T. Sayed (2007), 'Automated analysis of road safety with video data', *Transportation Research Record*, 2019, 57–64, https://doi.org/10.3141/2019-08.

Saunier, N., T. Sayed, and K. Ismail (2010), 'Large scale automated analysis of vehicle interactions and collisions', *Transportation Research Record*, 2147, 42–50, https://doi.org/10.3141/2147-06.

Saunier, N., and M. G. Mohamed (2014), 'Clustering surrogate safety indicators to understand collision processes', presented at *93rd TRB Annual Meeting*, 12–16 January 2014 (Washington DC, USA).

Sayed, T., and S. Zein (1999), 'Traffic conflict standards for intersections', *Transportation Planning & Technology*, 22, 309–323, https://doi.org/10.1016/j.ssci.2012.04.015.

Shbeeb, L. (2000), 'Development of a traffic conflicts technique for different environments – A comparative study of pedestrian conflicts in Sweden and Jordan', Department of Technology and Society, Traffic Engineering, University of Lund, Lund Institute of Technology.

Shelby, S. G. (2011), 'Delta-V as a measure of traffic conflict severity', presented at *90th TRB Annual Meeting*, 23–27 January 2011 (Washington DC, USA).

Songchitruksa, P., and A. P. Tarko (2004), 'Using imaging technology to evaluate highway safety' (Indiana Department of Transportation and Purdue University), FHWA/IN/JTRP-2004/27, https://doi.org/10.5703/1288284313343.

Songchitruksa, P., and A. P. Tarko (2006), 'The extreme value theory approach to safety estimation', *Accident Analysis and Prevention*, 38, 811–822, https://doi.org/10.1016/j.aap.2006.02.003.

Svensson, Å. (1992), 'Vidareutveckling och validering av den svenska konfliktekniken [Further development and validation of the Swedish traffic conflict technique]' (Lund University, Institute of Technology, Department of Traffic Planning & Engineering), https://www.ictct.net/wp-content/uploads/SMoS_Library/LIB_Svensson_1992.pdf.

Svensson, Å. (1998), 'A method for analysing the traffic process in a safety perspective', Doctoral thesis, Department of Traffic Planning and Engineering, University of Lund, Lund Institute of Technology, https://www.ictct.net/wp-content/uploads/SMoS_Library/LIB_Svensson_1998.pdf.

Tageldin, A., T. Sayed, and X. Wang (2015), 'Can time proximity measures be used as safety indicators in all driving cultures? A case study of motorcycle safety in China', *presented at 94th TRB Annual Meeting*, 11–15 January 2015 (Washington DC, USA).

Tageldin, A., and T. Sayed (2016), 'Developing evasive action-based indicators for identifying pedestrian conflicts in less organized traffic environments', *Journal of Advanced Transportation*, 50, 1193–1208, https://doi.org/10.1002/atr.1397.

Tarko, A. P., M. A. Romero, V. K. Bandaru, K. B. Ariyur, and C. G. Lizarazo (2017), 'Feasibility of tracking vehicles at intersections with a low-end LiDAR', *presented at 96th TRB Annual Meeting*, 8–12 January 2017.

Tarko, A. P. (2018), 'Surrogate measures of safety', in Lord, D. and S. Washington (eds.) *Transport and Sustainability, Volume 11 – Safe Mobility: Challenges, Methodology and Solutions* (London: Emerald Publishing Limited), pp 383–405, https://doi.org/10.1108/S2044-994120180000011019.

Tsai, R. Y. (1987), 'A versatile camera calibration technique for high-accuracy 3D machine vision metrology using off-the-shelf TV cameras and lenses', *IEEE Journal of Robotics and Automation*, RA-3 (4), 323–344, https://doi.org/10.1109/JRA.1987.1087109.

van der Horst, A. R. A. (1990), 'A time-based analysis of road user behaviour in normal and critical encounters', *PhD Thesis*, Delft University of Technology.

van der Horst, A. R. A., M. C. Thierry, J. M. Vet, and A. F. Rahman (2017), 'An evaluation of speed management measures in Bangladesh based upon alternative accident recording, speed measurements, and DOCTOR traffic conflict observations', *Transportation Research Part F*, 46, 390–403, https://doi.org/10.1016/j.trf.2016.05.006.

Várhelyi, A. (1998), 'Driver's speed behaviour at a zebra crossing: A case study', *Accident Analysis & Prevention*, 30 (6), 731–743, https://doi.org/10.1016/S0001-4575(98)00026-8.

Vogel, K. (2003), 'A comparison of headway and time to collision as safety indicators', *Accident Analysis & Prevention*, 38, 427–433, https://doi.org/10.1016/S0001-4575(02)00022-2.

Wang, C., and N. Stamatiadis (2014), 'Evaluation of a simulation-based surrogate safety metric', *Accident Analysis & Prevention*, 71, 82–92, https://doi.org/10.1016/j.aap.2014.05.004.

WHO (World Health Organization) (2023), 'Global status report on road safety', https://iris.who.int/bitstream/handle/10665/375016/9789240086517-eng.pdf.

Yastremska-Kravchenko, O., A. Laureshyn, C. D'Agostino, and A. Varhelyi (2022), 'What constitutes traffic event severity in terms of human danger perception?', *Transportation Research Part F: Traffic Psychology and Behaviour*, 90, 22–34, https://doi.org/10.1016/j.trf.2022.08.001.

Yastremska-Kravchenko, O., A. Laureshyn, J. Rahm, *et al.* (2024), 'Video analysis of bicyclist and pedestrian movement on shared-use paths under daylight and electric lighting conditions – Method exploration', *Journal of Cycling and Micromobility Research*, 2, 100032, https://doi.org/10.1016/j.jcmr.2024.100032.

Yung, N. H. C., and A. H. S. Lai (2001), 'An effective video analysis method for detection red light runners', *IEEE Transactions on Vehicular Technology*, 50 (4), 1074–1084, https://doi.org/10.1109/25.938581.

Chapter 10
Behavioural observations

Matúš Šucha[1] and Ralf Risser[1]

10.1 Theoretical background

Traffic safety means the absence of unintended harm to living creatures or inanimate objects. Most studies and reports focusing on road safety consider only fatal and (severe) injury crashes when discussing traffic safety, thereby ignoring crashes in which only minor or no injuries at all occur. These studies use **crash data** retrieved from police and hospital departments. The most important limitation of crash data regards underreporting or incorrect reporting. Generally, accidents involving cyclists are highly underreported, while underreporting of injuries of car occupants is very low when hospital data and police reports are compared. However, the precise underreporting rates remain unclear since it is very likely that accidents without physical or property damage are not reported to either source. Accident data only provides information on the outcome of an unsafe traffic event, but valuable information regarding situational and behavioural processes leading to the crash occurrence is missing. Thus, research investigating road safety has also focused on the traffic events that precede accidents, for instance, traffic conflicts[*].

Apart from the use of crash data, road user behaviour can be described by using various other types of data, collected under different conditions, which can be categorised as follows:

1. Controlled data collection:
   - stated behaviour data (surveys, questionnaires, interviews, focus groups)
   - data collected from simulations using driving simulators or virtual reality
2. Naturalistic data collection:
   - behaviour observation
   - traffic conflict observation
   - naturalistic driving (using instrumented vehicles to collect behaviour data)

[1]Palacky University Olomouc, Department of Psychology, Czech Republic

[*]"A traffic conflict is an event where there is a collision course with a time to collision shorter than ~1 second, and where an accident is avoided thanks to the evasive action of at least one of the involved road users" [1] (see also Section 10.3.2.)

Naturalistic data collection methodologies focus on the observation of behaviour that is shown in the real traffic environment. A distinction can be made between studies in which road users are aware of being observed (e.g. ride-along observations, naturalistic driving studies) and studies in which unobtrusive observations of road user behaviour are made (e.g. behavioural observation in the field, traffic conflict studies). The main difference between behavioural observation studies and traffic conflict studies is that traffic conflict studies try to measure traffic safety in terms of the expected number of accidents (theoretically reflected by the number of conflicts), while road user behavioural observation studies focus on observing and understanding behaviour and interaction phenomena, rather than on quantifying traffic safety levels.

Controlled data collection techniques can be further divided into stated and simulated research methodologies. Stated research designs investigate attitudes and behaviours based on questioning participants, usually with the use of questionnaires, interviews or focus groups. These methodologies are limited in the sense that collected data is dependent on behaviour that is stated by respondents (e.g. attitudes, opinions and beliefs), rather than behaviour actually shown in a natural setting. The advantage of simulated studies is that different scenarios or new types of infrastructure can be tested without the necessity of implementing any measures in reality.

Behaviour observation of road users is defined as 'the discipline of psychology which studies the relationship between psychological processes and behaviour of road users' [2]. Behavioural observation approaches might be considered proxies for traffic safety since they do not analyse or evaluate crashes themselves but focus on processes and events that precede them. Observation can be direct (obtrusive) or unobtrusive. Obtrusive observation means that the road users will see the researcher or research teams and may alter the desired behaviour, for instance, in the sense of social desirability. Unobtrusive observation means that the observed drivers – or road users – do not know that they are being observed. For this purpose, researchers choose places where they are hard to recognize, and if the video camera is used to observe behaviour, then it is concealed in such a way that the road users are not aware of being observed.

Unobtrusive behaviour observation studies in road safety research collect naturalistic data of road users who are not informed (beforehand) of their participation in a research project. It enables the observation of behavioural and situational processes that contribute to unsafe traffic events, while possible behavioural adaptations due to the road users' recognition of being observed are minimized.

Behavioural observation or registration involves the systematic observation and recording of specific domains of behaviour in typical environments, e.g. at pedestrian crossings or along defined routes, such that the resulting descriptions of behaviour are replicable. For this, up to now, mostly video recordings and/or human observers (observations on the spot, in-car observations) or different types of sensors (e.g. infrared cameras) are used. The use of nomadic devices allows to follow virtually every step of observed persons, of course, with the consent of the observed persons. The disadvantage of this method is that behaviour of, and communication among, road users cannot be observed nor registered with nomadic devices.

From a social-scientific perspective, behaviour observations are **ethnographic methods.** Ethnography is defined as an illuminative account of social life and culture in a particular social system based on multiple detailed observations of what people actually do in the social setting being observed [3]. Ethnographic methods can be described as both qualitative and quantitative research procedures that are used when studying specific groups, communities or institutions as part of a larger complex society. They provide information about people or groups by registering the behaviour of people and groups in a systematic manner. Answers to many questions in the realm of traffic and mobility require the collection of behaviour data, as in many study problems in traffic, not least safety problems, are attributed to human behaviour (e.g. [2, 4–7]).

Van Haperen *et al.* [8] summarized the research efforts in the domain of road user behavioural observation research using a scoping review. They collected 600 journal articles and found out that the number of publications regarding road user behaviour observation studies has increased rapidly during recent years, indicating the importance of such studies for analysing traffic safety. On-site human observers have been most often used for more than 60 years until the mid-2000s. The use of video cameras as a primary data source surpassed the use of direct human observations during the mid-2000s, the same period in which the number of publications started to increase rapidly.

Most studies collected data on car drivers (81%), while vulnerable road users have been observed in 32% of all studies, with pedestrians, cyclists and motor cyclists as the most common road user types. From the summed percentage being higher than 100%, one can understand that in a couple of studies, both car drivers and other road users, and also their interactions, were observed.

The results showed that the main goal of behavioural observation is to

- Monitor (51%): one location or multiple similar locations are observed to monitor road user behaviour. The goal of such studies is to 'look what happens' and not to examine a certain intervention or safety-improving measure.
- Evaluate certain safety-improving measures (38%): using a before-after, with-without or cross-sectional research design, the effect of an infrastructural intervention or any other safety-improving measure is evaluated.
- Development of behavioural models (10%): real-world data is used to develop or validate behavioural models for simulation tools.

The same distribution of study purposes was found for both VRU[†] and driver studies, indicating that the approaches towards examining VRU and driver safety are similar.

Concerning studies that focused on topics that relate to interactions between road users, three categories were identified:

[†]VRU = Vulnerable Road Users, encompassing pedestrian and cyclists, including bicycles with motors of all types and scooters etc. Some authors also include powered two-wheelers.

- The first category, direct interaction, is composed of situations in which two or more road users need to negotiate with each other and anticipate each other's actions to pass each other safely (e.g. crossing and yielding behaviour). This type of behaviour has been researched most often; in the case of VRU studies, it even comprises 66% of all retrieved publications.
- Second, an indirect interaction is defined as an event where a road user's behaviour is influenced by the presence of another road user but where direct explicit communication is not necessary (e.g. an overtaking manoeuvre).
- Third, a possible interaction is defined as an event where a road user's behaviour is not necessarily influenced by the presence and may be more influenced by the absence of other road users (e.g. speeding on an empty road).

The findings indicate that **direct interactions** between car drivers have received the most research attention, but also those between car drivers and VRUs. This may be because during these traffic events VRUs are most endangered. When looking at car driver studies, apart from direct interactions (30%), the topic of 'speeding' has received most attention (16%).

As is very well known, the recording of **single-vehicle crashes** is subject to underreporting, especially with regard to cyclists, where reporting levels are estimated to be only around 10% [9]. Van Haperen *et al.* [10] show that most publications have focused on traffic events where interaction between road users is required (either directly or indirectly); behaviours that are related to single-vehicle crashes and that are not influenced by the presence of other road users have received less attention. A wide variety of **behavioural indicators** was found and showed an important challenge of using direct human observations. Over time, many binary measurements (e.g. yes/no answers) have been replaced with continuous variables (e.g. Likert scales), providing more depth in the data analysis process and data interpretation.

10.2 Methods/types of observations

10.2.1 Obtrusive vs. unobtrusive observation

In his master's thesis, where he, in the frame of a literature study, summarises information about behaviour observation, Riaz [11] points out the difference between obtrusive and non-obtrusive observation of subjects. He then focusses on unobtrusive observation, following the assumption that people change their behaviour when they know that they are observed. However, Höfner [12] did not find any differences between the results of obtrusive and unobtrusive observation when he observed the same subjects (moped riders) in the same situation, first without them knowing and then after having informed them of the observation. To him, this did not come unexpectedly as his theory was that driving behaviour is highly automated behaviour, and according to social psychology (taking the starting point in the law of [13]) such behaviour would even improve or become more accentuated, under social pressure represented by the knowledge of being observed.

Kiegeland [14] achieved similar results. With a little bit of salt, Rathmayer *et al.* [15], when analysing the Wiener Fahrprobe‡ [16], an observation method while sitting in the car together with the observed subject (= ride-along observation), came to the same conclusion. They found that the dynamicity of driving (speed, acceleration and deceleration) can be controlled to a certain degree by the observed persons so that they present themselves in a somewhat better light.

Behaviour observation along routes can also be applied in an unobtrusive way by following objects without them knowing it [17,18]. As well, the behaviour of pedestrians and cyclists can be observed by following them [19]. The disadvantage of car-following observation compared to ride-along observation is that handling aspects, concentration, preparedness to take action and verbal and non-verbal expressions – both informational and emotional ones – can hardly be registered from outside the car.

Ride-along observation, usually along a standardised route, can be applied to many different ends. For example, it may be used to assess individual drivers and their aptitude to drive a car safely. However, in such cases, one can only observe a small section of their 'driver's life'. What the person does in the evening when going to the pub, how he or she behaves on long-distance trips, or in hundreds of other situations that cannot be observed within a 1-h observation remains unknown, though. Such observation can also be used to learn about the behaviour of groups, for example, concerning professional drivers [20] or how older drivers tackle driving a car [21,22]. For instance, one can analyse communication of older drivers with pedestrians by leading the route through areas with many pedestrian crossings. Another application is connected with the question of how certain vehicle characteristics or equipment influence the behaviour of drivers in before-after or without-with settings (e.g. [23]). Not least, infrastructure elements can be analysed concerning how they are used by road users or by certain road user groups.

10.2.2 Stationary observation

So far, we have spoken of observing persons along certain trajectories. Observing individuals along a route with the help of human observers may provide a rather holistic information concerning the subjects' behaviour as road users. However, the procedure is time consuming and still only gives information about a short period of time (max 1 h). If one wants to know more about the behaviour of larger samples of persons in certain situations, one may make use of stationary observations, at crossings, at pedestrian crossings or at any other selected spot where things of interest to researchers are expected to happen. In this respect, traffic-conflict registrations are well known (e.g. [24]). Traffic conflict registration theoretically allows to assess the risk of crashes happening at selected spots in the road network. This type of observation is very much focussing on narrowly defined events (traffic conflicts), while psychologists usually choose a broader approach. For them,

‡In the following, the labels "Wiener Fahrprobe" and the English translation "Vienna driving test" will be used as synonyms.

shorter and longer sequences of behaviour that are either interesting with respect to selected criteria (safety, comfort, easiness, etc.), typical for special groups or situations, or deviating from certain norms, etc., are of interest, depending on the defined focus of research (see, for instance [25], also Chapter 3.3.2). Observations on site are usually unobtrusive, although it might be the case that road users realise that there are observers or video-cameras at a certain place, and they might adapt their behaviour, then. Paradoxically, the short-coming of stationary observations is that they are limited to specific sites. A more holistic behavioural frame is usually lacking, there. For instance, the question of how subjects prepare themselves when they approach certain sites, to which degree and in which way their behaviour is anticipatory, etc. Conversely, what happens at specific sites themselves can be found out efficiently with the help of stationary behaviour observation. For instance, when residents complain to the authorities that a special site seems dangerous to them and they would not let their children walk alone, there, the representative of the authority might answer that there never happened an accident at that site, so far. However, behaviour observation could show that there were frictions between road users at that site and that one had to conclude that an accident could happen sooner or later. Indeed, the stochastic character of accidents tells us that a place might be hazardous in spite of the fact that no accident was registered there, so far[§].

10.2.3 Observation with technical help or solely by technical equipment

Both, when following subjects along routes or in the frame of stationary observations, human observers are most often supported by video registrations. Videos have the shortcoming that one optical dimension is missing. However, when trying to remember what exactly happened at a certain point of time or site, or when different observers come to different conclusions concerning observed events, video is a welcome back-up [26].

In connection with the sole use of technical equipment without involving observers, as happens in the frame of naturalistic driving studies, the word observation has to be put in quotation marks. No observation in the field takes place there, but observations are made from the video screen, and other technical registrations have to be interpreted: of sensors, of data-logs, etc. (see [27]). This means that it is not obvious whether a complete picture of the subjects' interaction with the physical and/or the social environment is produced. To our knowledge, only very few studies exist that compare behaviour observations by human observers (and the descriptions of behaviour based on these observations) and the registrations in the frame of naturalistic driving (see [28]). Thus, it is not fully clear in

[§]The Poisson-distribution describes discrete occurrences over an interval. However, the occurrences in each interval can range from zero to infinity. Theoretically, there can be time interval after time interval at a dangerous intersection where no accidents happen; https://www.onlinemathlearning.com/poisson-distribution.html.

which way data produced by human observers and technical registrations lead to qualitatively different conclusions.

Of course, not only vehicle drivers can be 'observed' with the help of technical equipment, including video and eye-movement registrations (e.g. see [29]), but also other road users.

Registration of behaviour data with the help of (solely) technical equipment takes place mainly in the frame of so-called field-operational tests FOT and will be taken up shortly further below.

10.2.4 Behaviour observation on the traffic simulator

The use of simulators allows observing subjects' behaviour under fully controlled experimental conditions. There is always doubt about the validity of such observations: Can behaviour while sitting in the simulator and driving in an artificially produced environment be compared with behaviour in real life? There is much discussion about this. Meuleners and Fraser [30] compared behaviour during rides in real life and on the simulator under, as they say, comparable conditions. They concluded that 'there was no difference between on-road and simulator performance for many outcomes and results provide ... support for the relative validity of the driving simulator'. While one might doubt whether such conclusions can be generalised, the driving simulator without doubt provides good opportunities to test subjects under experimental conditions (as said above), but it also allows to test aspects that cannot be tested in real life, like the behaviour in high-risk situations, which would be prohibitive for observations in the field.

In the following sections, we will not deal with behaviour observation on the simulator any further. We focus on behaviour observation in the real environment.

10.3 Description of procedures and data

In the following, we will first refer to the Wiener Fahrprobe (Vienna driving test), a ride-along behaviour observation method. Then, we will shortly discuss behaviour observation at selected sites, with a special focus on traffic conflicts. The rather technical aspects connected to traffic-conflict observations and the recent state of the art will be discussed. After that, while still dealing with observations at selected spots, we will have a closer look at the interaction between car drivers and pedestrians. Finally, we will give some examples of naturalistic driving observation studies, especially field operational tests (FOT).

10.3.1 In-car observation with a focus on the Vienna driving test

With respect to in-car observations, we want to cite Lajunen *et al.* (referred to by [31]), who postulated that the 'total set of variables is meant to be a reflection of the observed participants' driving behaviour, or driving style'. However, in-car behaviour observation does not cover the strategic part of behaviour, according to [32].

It does not tell us about planning processes, about behaviour at odd times (evening, night), about precautions taken before longer trips, etc. However, it gives insight into behaviour and communication shown during the observation period, and it allows us to assess and compare drivers with respect to this time frame.

The Vienna driving test (‘Wiener Fahrprobe’) is such an in-car-observation method. It was developed in order to determine the lacks of safety in traffic processes. The original idea behind the Wiener Fahrprobe was that it should be used as an instrument to analyse driving behaviour in order to make sure whether a person is apt for driving a car or not [5]. Work was based on earlier approaches to achieve this, e.g. the different versions of Quenault’s driving-test (Klebelsberg, 1982; in [5]), the Kölner Fahrverhaltenstest (Kroj *et al.*, 1973; in [5]) and other more or less well known, and more or less structured instruments. In the meantime, the concept of behaviour has expanded towards including interaction and communication aspects. The main assumption thereby is that looking at the individual driver’s behaviour as steered only/mainly by physical outside stimuli and by internal variables (thoughts, motives, opinions, attitudes) is a dead-end street. Communicative aspects (the social environment, communication clusters that develop in connection with the distribution of power, imitation, etc.) most probably influence car driver behaviour to a much higher degree. Moreover, due to the fact that the Wiener Fahrprobe so far was used mostly in urban areas, the interaction between car drivers and vulnerable road users, especially pedestrians, gained importance (see [33]).

By combining observation with special ways of conducting interviews it turned out that it is mostly the communication with car drivers that creates frustration in pedestrians and bicyclists. This communication is very much characterised by the fact that car drivers are those who have much more power [19]. Expansion took place also in another way, namely by adapting the observation instrument for the assessment of sections of the road network – intersections and links (see [17]) – by following drivers on rural roads without their knowledge; there, it could be assumed that many drivers would drive along the same route for a longer time so that drivers could be compared to each other.

10.3.1.1 Observers and observation contents

Behaviour is usually recorded by two observers, the so-called ‘coding’ and the so-called ‘free’ observer. The two observers register several behaviour variables in both a standardised and a non-standardised way. The ‘*coding*’ *observer* concentrates on standardised behaviour such as lane keeping, speed behaviour, distance keeping, as well as behaviour while turning and in bends, behaviour at crossings or usage of the indicator. He/she records the observation on a specific sheet that has to be filled in per section. The ‘*free*’ *observer* concentrates on three types of behaviour phenomena that are hardly predictable: violations, interaction/communication processes and traffic conflicts.

Violations are events such as neglecting a stop sign that represent ‘a severe offence of the law and/or cause danger’ [31]. *Interaction/communication processes* are assumed to appear since the car driver rarely is alone on the road. In the majority of cases, other car drivers, pedestrians, cyclists and others will be on the road as well and the driver has to interact with them. He/she is not independent of other road users. *Traffic conflicts* are discussed further, below.

10.3.1.2 Route

The method is applied in real time traffic. For the observation, a standardized route with a length of 25–50 km is used. It should include different road types typical for the place (city) where one carries out the studies, mainly urban roads but, as well, rural roads and motorway sections in the surroundings or passing the city. The route is divided into sections like motorway entrances or exits or sections between intersections. All characteristics like stop signs, public transport stops, zebra crossings, etc., should be noted.

10.3.1.3 Point in time

Especially within a scientific context, rides ought to be comparable; hence, observations should always be done on the same day time. Driving observations usually are made during weekdays. Therefore, it is utterly important to avoid rush hours in order to keep schedule because it is crucial to get indeed a comprehensive impression of the driving behaviour, not just the behaviour while the car is rolling or even standing still. In practice, an observation ride along a route with standardized lengths takes between 40 and 60 min, depending on the actual traffic situation and the driving style of the participant. The actual recording of the observation starts after about 15 minutes of the trip. The literature (Höfner, 1972; in [5]) and also the practice show that the presence of the observer is 'forgotten' after that while or does not influence the observed person's behaviour any more (see also Section 10.2.1).

10.3.1.4 Quality criteria

In social science, three criteria are very important: objectivity, validity and reliability. The value of an instrument always was and will be evaluated on those criteria. In the basic literature about the methodology of social science, many definitions can be found. Researchers working with the 'Wiener Fahrprobe' ([5,34]; in [31]) thought about those criteria as well, and the results they found will shortly be presented in the following.

If the outcome of an instrument is independent of the operating, coding and interpreting person, it can be characterised as *objective*. The objectivity concerning coding and interpreting the data can be enhanced through comprehensive observer training [5, 35]. Risser and Brandstätter (in [5]) calculated a congruence of 67% for six different observers. Hjälmdahl and Varhelyi [16] found a congruence of observers with a 'key'** observation of 90% after an intense training.

**Hjälmdahl and Varhelyi (2004) trained three students in a theoretical and practical way on the behaviour observation method. Afterwards "test" observations were carried out. The students observed the behaviour of a participant, followed by a video-assisted discussion about the ride. On the basis of those discussions between the observers, the trainers and the participants "key" observations have been developed. Subsequently, the observation sheets of the observers were compared to the "key" observation.

With a *reliable* instrument (next to), identical results should be derived in the course of repeated measurements. 'Reliability refers to the extent to which a test or other instrument is consistent in its measures' (http://www.alleydog.com/glossary). In order to test the reliability of the data collection method of the 'Wiener Fahrprobe' via counting violations, Risser and Brandstätter (in [5]) used Cronbach's alpha (α = 0.84) and the split-half method of Spearman-Brown (r = 0.81) and Guttman (r = 0.80).

An instrument that measures the concept it is supposed to measure can be seen as being *valid*. 'If the test does indeed measure what it is intended to measure, then we can say that the test is valid (or has validity)' (http://www.alleydog.com/glossary). Concerning the 'Wiener Fahrprobe', Risser and Brandstätter compared the amount of recorded violations of a driver with the amount of traffic conflicts he/she was involved in. They found a significant correlation of r = 0.65. This result means that the more violations a driver made, the more he/she was involved in critical incidents. According to this study, the same is true for the involvement in conflicts (r = 0.65). Moreover, the authors could find a significant correlation between the amount of violations and the crashes he or she was involved in in his/her driver's career so far.

Another validation study was carried out by Risser and Lehner [31,34] and Chaloupka and Risser [31,34]. Within this study, the data of the behaviour observation have been correlated with crash data of 51 road sections along a standardised route (police-registered crashes). The following significant validation coefficients were derived [36] (Table 10.1).

According to this study, all behavioural variables in the table above are to a significant degree correlated to crash figures, even if the relationship between keeping too short distance to the car ahead and crash occurrence is significant only

Table 10.1 Overview of the correlation of different observation variables with accident data

Observation variables	**r_{XY} with crashes**
Driving extremely on the left or the right side of the lane	0.42 (1%)
Inadequate overtaking	0.42 (1%)
Too small lateral distances	0.37 (1%)
Delayed lane change in case of obstacles	0.38 (1%)
Early change of lane before obstacles, reflecting anticipation	
Problems with lane choice (e.g. the wrong lane for proceeding after the intersection)	0.48 (1%)
Speed not exceeding the limit and well adapted to the situation	
Exceeding speed limits	0.46 (1%)
Early deceleration whenever deceleration becomes necessary	
Distance to the car ahead too short	0.29 (5%)

1% = highly significant, 5% = moderately significant.

at the 5% level. Concerning objectivity, reliability and validity, see also Turetschek [37].

10.3.2 Traffic conflict registrations on site

The Traffic Conflict Technique is a method to register and analyse conflicting interaction processes between road users. The first registration method for traffic conflicts was presented by General Motors in the USA in 1967 [38]. Since then, the Traffic Conflict Technique has been modified several times and is still used in many countries (e.g. the UK, Germany, France, Sweden, Austria, etc.) to enhance road safety. The Traffic Conflict Technique can be characterized as a method for observation, description and interpretation of behaviour. This technique leads to direct, systematic and objective information on behaviour and interaction processes between road users. Based on this information, assumptions concerning behaviour and interactions leading to critical situations can be made. Traffic conflicts occur more often than crashes; thus, more relevant data for traffic safety analyses are available. Assuming that there is an understandable relationship between traffic conflicts and crashes, conclusions with regard to existing traffic safety problems could be drawn easier and faster based on conflict data than with the analyses of crash data. Thus, in principle, the application of the traffic conflict technique enables one to evaluate the effect of safety measures directly after their implementation.

Traffic conflicts are described as events when there is a collision course of two or more road users, and at least one of the involved parties has to change the course by modifying the speed (accelerating or braking) or the direction (swerving) in order to avoid a crash. The later the evasive action is performed, the more the risk of a collision increases and the severity of the conflict will be higher.

Many studies (e.g. [24,39]) assume that there exists a continuum of dangerousness in traffic processes (see Figure 10.1). According to this way of looking at the processes, they can vary from being smooth (no risk, most probably safe), via representing potential and slight conflicts, up to resulting in serious conflicts or

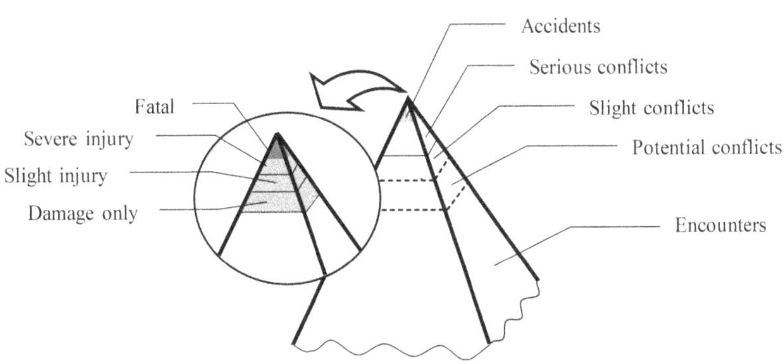

Figure 10.1 Degree of the seriousness of conflicts [24]

even accidents. From a more holistic – and psychologically relevant – point of view, it is interesting to see in what kinds of interactions traffic conflicts are embedded.

10.3.2.1 Types of traffic conflicts

Traffic conflicts can be subdivided, among others, into the following types, similar to crash types: rear-end conflict, left-turning to head-on conflict, right-angle conflict, head-on conflict, lane-change conflict, conflict with vulnerable road users (at intersections, mid-block, right angle or straight going pedestrian/cyclist with left or right turning vehicle).

10.3.2.2 Methodology

For evaluating the degree of seriousness of conflicts, different countries developed different techniques, which can be categorised as (cf. [40], pp. 53)

- semantic techniques
- time-to-collision techniques

Semantic techniques [40] are based on the human perception ability to evaluate the criticality of a situation, though of course with a subjective touch. No additional criteria for quantification are used, such as measuring the time till a potential collision. It is assumed that different experts/observers are able to train their perception sufficiently in order to estimate the seriousness of situations in a comparable way. The decision of whether a traffic conflict has happened or not, as well as the identification of the degree of severity, is based on semantic aspects; e.g. 'That was very close!' Closeness can, for instance, be pseudo-quantified on a Likert scale [41], most often as a five-point scale, or on two levels only, as frequently is the case: slight and severe. The semantic techniques are mainly used in Austria, Germany, France and the USA.

Other countries prefer time-to-collision (TTC) techniques. These techniques are characterised by the use of time measurements in order to get quantitative data of traffic conflicts. According to the Swedish method, which was developed at the technical University of Lund, the observer has to distinguish between three to five degrees of severity. The criterion is the time span from the identified beginning of the collision course to the potential collision that is prevented by at least one of the involved road users taking an evasive action [24].

Still another group of researchers connects post-encroachment time (PET) to the feature of a conflict. It is usually defined as the time between the first road user leaving the potential crash zone and the second road user arriving there. This method has been used in Canada and the USA (e.g. see [42]).

In general, traffic conflict studies are conducted within observation series in the field (respectively on the street) by specifically trained persons. Intensive training of the sensibility in order to identify risky traffic situations is indispensable, technical registration methods like video recording or electronic measure of time (see TTC) are only supplements to the human observer.

10.3.2.3 Quality criteria: objectivity, reliability, validity

Concerning quality criteria there is no clear picture pertaining to traffic conflict registration. With respect to the question of validity, Zheng *et al.* [43] refer to the process of how crashes occur and to the understanding of this process. According to them, there are different opinions as far as the position of the traffic conflict in this process is concerned. Some researchers look at traffic conflicts and collisions as two mutually exclusive outcomes, depending on whether an evasive action is taken or not and on the type of this action. According to this group, traffic conflicts and crashes are complementing events. Others argue that traffic conflicts precede evasive actions that can be successful or not. In the case of many crashes, there would be no evasive action at all due to a lack of time. Zheng *et al.* [43] conclude that 'although there is no agreement on the process model validity, the latter argument gained popularity'. Thus, the question of validity is subject to the point of view of researchers, where the latter version mentioned above seems to have more support, while no figures seem to be available.

The same is true concerning the objectivity and reliability of traffic-conflict registrations. No measurements and no figures are available. However, there is a strong belief among those who practise traffic-conflict registration that these two quality criteria can be achieved by thorough training of observers, supported by video and accompanied by extensive discussion. A high inter-observer correlation concerning the identification and labelling of traffic conflicts is considered an indicator that reliability and objectivity are satisfactory.

10.3.3 On-site observation of behaviour and interaction of road users

Behaviour observation in road traffic is carried out in order to gather data on road user behaviour and interaction. Road users do thereby not know that they are observed. Such observation is carried out in order to learn about behaviour and interaction processes in different situations that reflect frictions of different severity. Such frictions might indicate that traffic safety is at risk, but also that other problems – lack of comfort barrier effects, etc. – are present or imminent.

10.3.3.1 Procedure

On-site observation of road users' behaviour and their interaction is characterised by the aim of observation, place of observation, methods used to collect data and methods used to analyse and interpret data [10, 44].

(a) Aim – usually the aim is to explore, operationalise, describe and quantify defined behaviour. In some cases, it is not only visible behaviour that needs to be explored, but also motives that lie behind this behaviour need to be understood. This is important, because one motive can lead to many different behaviours, and one behaviour can have many different motives. The appropriate approach is to use qualitative methods of data collection (focus groups, interviews) to deal with research questions connected to motives. Such methods provide the best results in connection with behaviour observation.

(b) Place – we need to specify very precisely the observation or spot characteristics so that the observed behaviour at these spots have a potential to answer relevant questions in connection with the study aims. Usually, there are two main strategies: to select more spots with the same (or as similar as possible) characteristics in order to collect homogenous data; or to select different spots in order to collect data that are shaped by different infrastructure characteristics.
(c) Methods used to collect data – usually data are collected by human observers on the spot or recorded by cameras. When collecting data with human observers, we need to make sure that we define what the observers are exactly looking at selected area, selected behaviours and selected road users. It is also necessary to make sure that data are coded in a way that makes it possible to merge them afterwards. It is always handy to do a pilot study before the full data collection, in order to make sure that all instruments are working well.
(d) Methods used to analyse and interpret data – basically both qualitative (e.g. narrative analysis of both behaviour descriptions and interview results) and quantitative methods of data analysis (e.g. statistical methods) may be applied, depending on the case. A frequently used approach is to combine both in the frame of a so-called mixed-methods design. For instance, when referring to (c. Methods used to collect data, see above) one can draw lists of possible behaviour and interaction phenomena at a certain site, with the help of an unstructured series of observations (possibly completed with short on-site interviews) and then design a standardised observation sheet based on the qualitative analyses. Then, both qualitative and quantitative data analysis methods need to be applied. For the interpretation and better understanding of data expert interviews or expert workshops with relevant stakeholders are recommended.

10.3.3.2 A case study in Olomouc

In the further text, an example of a mixed methods design used in connection with on-site observations of pedestrians' and drivers' interactions at non-signalised crossings [25] will be described. The study was carried out according to the following steps:

1. **Exploration of potential pedestrian and driver interactions and identification of problems in the frame of these interactions**: Four focus groups were conducted, two groups of pedestrians and two groups of car drivers. The focus groups with pedestrians concentrated on the following aims: identify pedestrians' needs and problems according to crossings; identify their strategies in connection with crossing (wait/go/communication/signs given) and the factors that influence their decision to cross or not; identify what they consider as dangerous (road design, drivers' behaviour) and what they would like to change or to be changed. The focus groups with drivers concentrated on the following aims: identify their strategies when approaching a crossing (wait/go/communication/signs given) and the factors that influence their decision to yield or not; identify what they consider as dangerous in pedestrians' behaviour or what is misleading communication in their eyes; describe possible types of

conflicts that they consider a problem or a safety issue. For both groups: identify research sites.

2. **Pilot study:** During the pilot study (4 h of field observation), all the materials and processes were tested (the sites chosen for observation, the interview questionnaire, the observation sheet and the quality of the camera recordings).
3. **Data collection:** on-site observations (four sites – zebra crossing, in an urban area, three observers, 50 h at each site), camera recordings (24 h), speed measurement, measurement of car and pedestrian densities and interviews (brief on-site interviews with pedestrians). The on-site observation should focus on pedestrians' behaviour in connection with crossing the road (before crossing, while crossing), pedestrians' communication with drivers and vice versa and drivers' behaviour (mostly focused on the decision on whether to yield or to go).

   Four observation sites were selected, all with zebra crossings and in the urban area. The sites were chosen on the basis of information from pedestrians and drivers from focus groups, traffic data (accidents, densities) and feasibility, like the possibility of having cameras installed, availability of suitable spots for observation, etc. Requirements for the sites in this special case were that they should be non-signalised, with marked pedestrian crossings, on two-lane streets (one lane in each direction) and with high densities of pedestrian flow. For the direct observation, three methods were used: camera recordings, on-site observation (pedestrians and drivers) and measurement of speed and volume of both cars and pedestrians. The situations under observation were those when a car was approaching a crossing where a pedestrian was present (waiting), approaching or crossing the road. Camera recordings were used as a back-up for the identification of frictions and near-misses (traffic conflicts). They provided support for thorough and detailed descriptions of these events.

   The interviews were carried out on-site by a third observer after a pedestrian crossed the road, while the first and second observers were registering pedestrians' and drivers' behaviour and interactions. Pedestrians were asked to give a short interview (five questions, approximately 3–4 min). The response rate was approximately 30%. We used brief on-site interviews with pedestrians as a method for exploration. The aims were to identify pedestrians' decision factors about starting to cross the road, to understand how pedestrians understand drivers' signs or other communication and to learn what pedestrians do to let drivers know their intention to cross or force drivers to yield.

   The data from the interviews was linked to the data from the on-site observations and the camera registrations, which made it possible to acquire a better understanding of the underlying motivational factors involved in the pedestrians' behaviour. The objective of the combination of all these methods was to identify factors which lead to conflicts.
4. **Exploration and development of recommendations**: The last stage of the data exploration was an expert workshop, where all the information from the research was presented and discussed. The workshop proposed interpretations of the results and further steps in the research.

10.3.3.3 Quality criteria: objectivity, reliability, validity

The concept of objectivity in social research is referring to the principle that, as far as is possible, researchers should remain distanced from what they study. Findings need to depend on the nature of what was studied rather than on the personality, beliefs and values of the researcher [45]. In on-site observations, noting that the measurement is done by humans, full objectivity cannot be achieved. To ensure the maximum objectivity possible we should use more than one observer and make sure that their observations are free from bias as far as possible. To achieve this, we need to train them and deliver only as much information about the purpose of the study as needed. It is beneficial when observers are rather 'professionals' in observing rather than experts in the research topic. The reason is that interpretation should not interfere with the empirical task.

Validity expresses the degree to which any measurement measures what it purports to measure. Validity of an on-site observation of behaviour and of interaction of road users can be established by using a panel of experts who explore theoretical constructs, or by carrying out focus groups with the target populations, in order to explore the field of observation. This form of validity reflects how well the idea of a theoretical construct is represented in an operational measure – what is observed, what are the categories, what type of answers are used etc. Validity can also be established by carrying out pilot studies in the form of field tests, thus examining how well observations relate to one or more external criteria, based on empirical constructs.

Reliability refers to the degree to which the results obtained by any measurement or procedure can be replicated. In the case of on-site observation, the lack of reliability may arise from divergence between observers and their assessment of the situation. To ensure that reliability suffices we need to train observers and to calibrate their assessment. Reliability of the design of observation procedures can be tackled by carrying out pilot tests, as well.

10.3.4 FOT/natural driving studies (NDS)

During the last few years, several field studies have been carried out under the label FOT. Such studies are sometimes set equal to NDS. However, as Lietz *et al.* [46] formulated,

> 'there has to be made a clear distinction between an NDS and an FOT. An NDS is the observation of drivers in naturalistic settings (during their regular, everyday driving) in an unobtrusive way. The essential driver behaviour is what is of interest in these studies, usually in relation to crashes. The purpose of an FOT, however, is to evaluate functions of ICT Solutions under normal operating conditions, using quasi experimental methods. So, an FOT is always related to some form of technology that is to be tested in the field',

and the driving there is most often not naturalistic, but registered in the frame of an artificial setting.

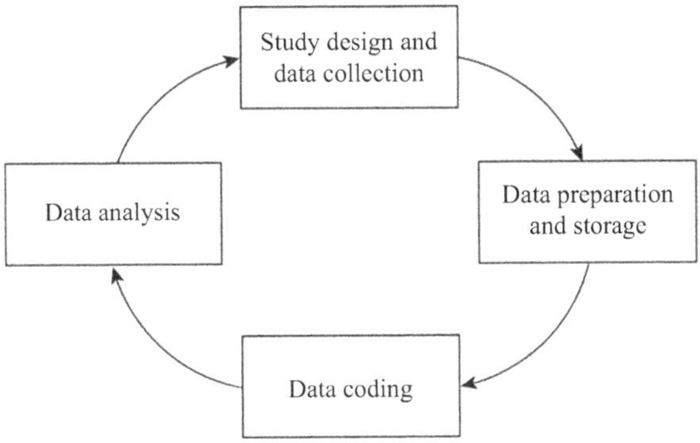

Figure 10.2 Life cycle of naturalistic driving studies [81]

In the frame of NDS 'Naturalistic' refers to the fact that the natural, every-day behaviour of drivers/road users is registered over a longer period of time. The quality of data registration and also of interpretation can theoretically be improved continuously (see Figure 10.2). According to the theory lying behind such studies, the data that is collected with different instruments reflects the 'real' behaviour of the road users, consisting of a large number of different situations the road user gets involved in every day. The test persons know that their car is instrumented. The assumption is, though, that they forget this fact after a short time and that they then show their real, or their usual, behaviour. FOT differ from designed experiments in that they are naturalistic and lack direct manipulation of most test conditions and independent variables. Thus, experimental control lies in the commonality of the test vehicles driven and the ability to sample driving data from the data set on a within-subjects basis. The within-subjects experimental design approach, in which drivers serve as their own control, is powerful in that it allows direct comparisons to be made concerning individual drivers on how the vehicles are used.

10.3.4.1 Technical equipment and registered data

The instruments referred to above are: video cameras to the front, to the sides and to the rear of the car; sensors that register distance to the car in front or position of the car/vehicle in relation to the driving lane; meters to register speeds and speed variations, mechanisms that recognise jerks, i.e. abrupt longitudinal or lateral decelerations or accelerations (that might reflect a traffic conflict). Among many others, Barnard *et al.* [47] and Dingus *et al.* [48] have provided overviews of the combinations of data registration methods applied in the frame of NDS viz. in FOT.

Sophisticated methods and instruments are necessary in order to cope with the extremely large amount of data collected in this way, while reconstructing the interaction of the road user with the environment and especially with other road users in a more holistic way constitutes an extra challenge. One attempt to meet this challenge was the EU project INTERACTION. It had the goal to combine what can be learned from naturalistic-driving-data collection on the one hand and from behaviour registration by human observers on the other hand. This should help to interpret automatically collected data, not least with respect to the detection of critical events (see [28]).

10.3.4.2 Objectivity, reliability, validity

The study by Kaufmann and Turetschek [28] could be seen as an indirect approach to tackle the issues of objectivity and validity: The data are objective, and they give insight into 'real life' road use, but only if one is able to interpret them correctly. Important impulses for this project came from the groups of researchers that constituted the HUMANIST Network of Excellence that later on became the HUMANIST-Virtual Centre of Excellence (VCE; see https://www.humanist-vce.eu/). In the frame of the collaboration of this group, among other things, two studies were carried out where data collected with the help of the Viennese Driving Test and by instruments, in parallel, were compared. The driving test results were used to interpret the instrument data in a holistic way. With the help of this approach, among other things, 'jumps' in the system, reflected by hard forward, lateral and vertical acceleration/deceleration could be placed in a behavioural frame. With the help of these results, descriptions of behaviour, including emergency actions, could be associated with data clusters produced by the instruments, thus improving understanding of data.

To our knowledge the issue of reliability has never been taken up explicitly. Maybe there is no reason to doubt that the same type of registered data always reflects the same type of behaviour or interaction. However, as the appropriate interpretation of the huge amount of information that comes in the frame of FOT/NDS seems to be tricky, the interpretation part could pose problems with respect to the reliability of the method. This would, of course, also affect the objectivity and validity of the method [49].

10.4 Some results of studies with the help of observation methods

In the following, some selected cases are presented including the results they delivered and the conclusions that could be drawn from these results – measures/countermeasures, restructurings of infrastructure, adaptations of HMI, etc.

10.4.1 In-car observations: some applications and results

The Wiener Fahrprobe[††] was used in the frame of many projects in order to study car driver behaviour with and without certain equipment viz. before and after certain measures, both at the individual and at the infrastructure side (e.g. [37]). So far, the Wiener Fahrprobe has been used several times in numerous national and also international studies. Short summaries of the various application areas are given in

[††]Lately, the label 'ride-along observation' has emerged for all types of studies where observers accompany subjects in the car that the subjects are driving.

Risser and Lehner [31] as well as in Kaufmann and Risser [50]. Several mutations of the Wiener Fahrprobe exist. The original version was used, e.g. in an evaluation study of a dual-mode route-guidance system (HOPES project; cited in [34]), for the evaluation of the AICC-system [51] as well as for the evaluation of the effects of an active accelerator pedal [52]. Moreover, the method was adapted several times to the research needs. In some studies, the participants did not know about the observation. Bruhning *et al.* [83], for instance, used the Wiener Fahrprobe to observe car drivers along a transit route from Germany through Austria to the border of Slovenia from a following car. In 1992, Risser *et al.* investigated the interactive behaviour between pedestrians and cyclists in Vienna with a modified version of the instrument. As the application of two observers is rather time- and cost-intensive, in some evaluation studies, only one observer was involved (Gstalter 1991, Almqvist *et al.* 1991, 1993, Almqvist and Nygard 1997, all cited in [4]). Risser *et al.* [53], for example, analysed the driving behaviour of electric mini-car users in Vienna with only one observer (see [4]). In a further study ('Sicher mobil mit Handicap'), the driving behaviour of disabled people was observed in order to assess their ability to take part in road traffic (http://www.clubmobil.at). In 2007, FACTUM OHG, together with CDV, ran a study in Vienna and Brno with the aim of combining the observation data of the 'Wiener Fahrprobe' with the data of a Black Box [28]. The most recent study, where behaviour of drivers of half-automated cars with a special focus on their vigilance was observed, was carried out in a driving simulator at the Technical University of Graz [54].

10.4.2 Traffic conflict registrations: some applications and results

The traffic conflict technique is applied worldwide. For instance, Uzondu *et al.* [55] investigated quite generally and in an exploratory way unsafe behaviours and interactions at intersections in Nigeria with the help of the traffic-conflict technique. Guo *et al.* [56] compared simulated and field-measured conflicts for the safety assessment of signalized intersections in Australia. Van der Horst *et al.* [57] evaluated speed management measures in Bangladesh, among other things, based upon the DOCTOR traffic conflict observation method, a special type of traffic-conflict measurement developed in the Netherlands (Dutch Objective Conflict Technique for Operation and Research).

Other studies dealt with bicyclists, motorcyclists and pedestrians, viz. with interactions among these and between these groups and car drivers. van der Horst *et al.* [58] analysed events on bicycle paths, very often registering quite severe conflicts between cyclists and pedestrians. Marizwan and Manan [59] observed motorcycles entering from access points and merging with traffic on primary roads in Malaysia. They aimed at understanding behavioural and road environment influence on the occurrence of traffic conflicts. Cafiso *et al.* [60] observed conflicts between busses and pedestrians with what they called an in-vehicle stereo-vision system, which in principle constitutes the combination of a ride-along observation and traffic-conflict registrations. Chebanyuk *et al.* [61] connected an analysis of how the presence and behaviour of pedestrians at intersections influence the traffic flow parameters (by 'traffic' meaning 'car traffic' and thus obviously not considering pedestrian movement as 'traffic') with road safety indicators, among which traffic conflicts play an important role.

For an overview concerning the studies mentioned here and also other projects connected to traffic conflict techniques, see Tarko [62].

10.4.3 Some results derived from on-site observation studies

A typical result of on-site observation studies is the description and quantification of specific aspects of behaviour. As mentioned above (section 3.3), when the aim of the research is road safety and road user behaviour, it is crucial to understand also the motivation for the behaviour. Only knowledge of motivation together with the real (observed) behaviour can provide sufficient information needed for the wishes for change of the behaviour (usually to support road traffic safety). Below, we will show as an example the results of an on-site observation study focused on vulnerable road users [25]. This research studied encounters between pedestrians and drivers at pedestrian crossings in urban settings. The aim was to describe drivers' and pedestrians' behaviour and their mutual communication, as well as providing an account of situations that occur, with a particular view to conflict situations. Subsequently, interviews with pedestrians were held in order to identify their needs and subjective feelings of safety in relation to crossing the road at the sites under consideration. The results were interpreted on the basis of the infrastructural characteristics of the crossings.

The most relevant predictors of pedestrians' and drivers' behaviour when they encounter one another at pedestrian crossings are the density of car traffic, the density of pedestrian flows and the speed of approaching vehicles. The majority of the pedestrians who were interviewed found it fairly safe to use the marked crossings under study (60%), while 40% of the respondents do not find it safe to traverse the road at these crossings. 46% of the pedestrians require drivers to stop before the crossing (not only slow down) for them to feel safe to cross. Conversely, only 17 % of the drivers actually did so (and 47% slowed down). The present study supported the following factors as influencing pedestrians' wait/go behaviour: the speed of the approaching car, the distance of the car from the crossing, traffic density, whether cars are approaching from both directions, various signals from the driver (eye contact, waving, flashing of lights), and the presence of other pedestrians.

Pedestrians indicate their intention to cross a road by standing at the roadside waiting, indicating forward movement or actually stepping into the roadway, by non-verbal signals, by making eye contact and by looking around.

The factors influencing drivers' yield/go behaviour were as follows: speed (higher speed leads to lower willingness to yield), traffic density (higher density leads to lower willingness to yield) and driving in a platoon (greater willingness to yield); the driver's willingness to yield increases where there is a group of pedestrians.

The probability of conflict situations increases with cars travelling at higher speeds, a higher traffic density and pedestrians being distracted by different activities while crossing, like talking on a mobile phone.

The congruence in the pedestrians' perceived safety of the crossings (subjective degree of threat) and objective safety (the number of conflicts at the given site) suggests that pedestrians are good at estimating the real risk. As for infrastructure-specific factors, the width of the crossing proved to be the most evident predictor of safety. Supporting the safety in numbers principle, the pedestrians rated the sites with a more balanced pedestrian/car ratio as safer.

The assumption that a great proportion of drivers (36%) fail to yield to pedestrians on marked crossings, although this is illegal, was confirmed. The drivers who do give way to pedestrians prefer lowering their speed instead of bringing their car to a complete standstill. This is the opposite of what pedestrians find safe: they prefer the car to stop rather than just to slow down for them to feel safe. A major difference between drivers and pedestrians in their respective needs for mutual communication was demonstrated. Unlike the drivers, a great majority of the pedestrians were proactive in seeking eye contact or tried to establish other forms of explicit communication with the drivers. Conversely, only a small proportion of the drivers (approximately one-third) communicated explicitly with the pedestrians.

10.4.4 FOT/NDS: some applications and results

10.4.4.1 FOT

Lyu *et al.* [63] produced a comprehensive overview of FOT, partly in combination with questionnaires. The T in FOT refers to the fact that the effect of certain measures or equipment on driver behaviour is tested. In Table 10.2, one can see from studies from all over the world that the equipment, or systems, tested were most often forward collision warning (FCW), lane deviation warning (LDW), autonomous cruise control (ACC), time headway warning (THW), but also lane-change assistance and lane-change warning, curve-speed warning, general hazard warning, braking advice and side-blind-zone alert.

10.4.4.2 NDS

In the last 4 or 5 years, many NDS have been carried out, or also studies making use of formerly gathered naturalistic driving data. Van Nes *et al.* [8] generally discussed the potential of naturalistic driving for an in-depth understanding of driver behaviour. Pantangi *et al.* [64] investigated whether high-visibility enforcement programs would affect aggressive driving behaviour. Kovaceva *et al.* [65], as well, were interested in aggressive driving and worked with an attempt to identify such behaviour from naturalistic data in car-following situations. Precht *et al.* [66] were after the identification of the main factors contributing to driving errors and traffic violations with the help of naturalistic driving data. Both Wijayaratna *et al.* [67] and Kreusslein *et al.* [68] used naturalistic driving data in order to assess the impact of mobile phone use in different situations. Morgenstern *et al.* [69] tried to answer

Table 10.2 Examples of FOT (source: [63], modified by the authors, 2020)

ADAS type	Major parameters	Country	Sample	Method
FCW	Speed, THW	Israel	15 M, 15 F	FOT
FCW	THW	Israel	29 M, 14 F	FOT
LDW	Mean max. lateral deviation	Germany	18 M, 12 F	FOT
FCW	Speed, THW, no. of alarms, alarm length, reaction time, workload	Italy	20	FOT questionnaire
FCW, LDW, Lane-change/ merge warning, curve-speed warning	Acceptance, lane departure rate, THW, turn signal use rate	USA	180	FOT questionnaire
FCW, ACC	THW, TTCi	China	26 M, 10 F	FOT
FCW, LDW, ACC and braking advice	Speed, THW, journey time, braking and acceleration forces, yaw rates	15 M, 15 F	UK	FOT
FCW, LDW, side blind zone alert	Acceptance, alert per 100 km	China	22 M, 11 F	FOT questionnaire
FCW, LDW	Acceptance, THW, FCW counts, PJ1.5, LDW counts, SDLP	Korea	26 M, 26 F	FOT questionnaire
Lane change, driver assistance system	Parameters of longitudinal adjustment and gap acceptance, lane change kinematics	U.S.A.	3 M	FOT

the question of whether drivers reduce their speed when texting on highways with the help of European naturalistic driving data. Itkonen *et al.* [70] attempted to characterise the motorway driving style of different subjects using naturalistic driving data. Wang and Xu [71] compared self-reported driving behaviours and data derived from naturalistic driving behaviour observation. Ismaeel *et al.* [72] used NDS data to deal with the question of whether drivers self-regulate their engagement in secondary tasks at intersections. Feng *et al.* [73], Kovaceva *et al.* [74] and Rasch *et al.* [75] analysed how drivers overtake bicyclists, viz. how they overtake pedestrians. Orlovska *et al.* [76] investigated the effects of the usage of automated driver assistance systems (ADAS) on drivers' behaviour with NDS data. One may assume that many other studies have been carried out that cannot yet be found in the peer-reviewed literature and that quite a few studies are going on at his moment. There is a strong belief among many researchers that the registration of road-user behaviour in the frame of natural-driving-observation settings delivers ultimate know-how about and understanding of road-user behaviour. This would allow to take efficient steps to modify non-wished-for behaviour.

10.5 How to understand traffic behaviour with help of the model of Michon

Which types of data, or so to say 'knowledge', do behaviour observations provide? From a psychological point of view, it appears appropriate to answer this question by using the model of [77]. This model refers to operational, tactical and strategic levels of behaviour, originally explaining behaviour of car drivers, but it can also be used in other fields of the traffic domain (Figure 10.3). In the case of car drivers, the strategic level consists of mode choice, route planning, reflections on one's own status, weather conditions and conditions of the car and the like. The tactical level involves manoeuvres related to approaches to places, interactions with other road users, anticipation concerning the physical and social environment, decisions concerning overtaking, yielding and the like. On the operational level, the control of the vehicle is taken care of like changing gears, braking, steering and alike.

This figure also indicates what types of methods are appropriate to assess different aspects of driving behaviour: psychophysical tests, behaviour observation, personality tests and exploration (interviews or focus groups). So what can the different behaviour observations tell us about observed persons and settings? In the following text, the focus will be on the different methods of behaviour observation and the ability to out of them derive information about individual drivers' behaviour on the respective level of Michon's model.

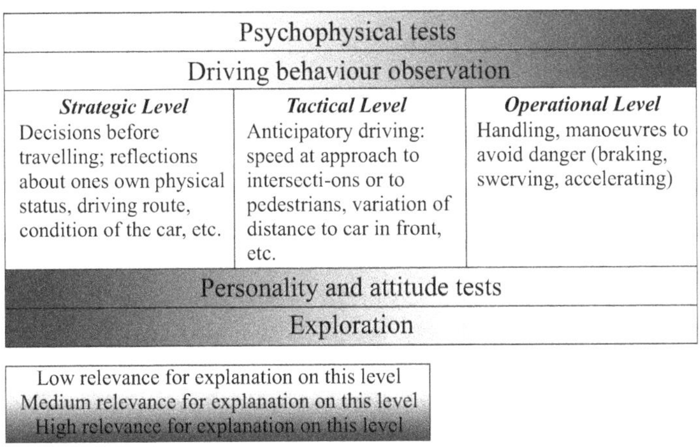

Figure 10.3 Michon's behaviour levels and the respective relevance of psychological methods to assess them [82]

10.5.1 Ride along and road-user following

Both ride-along and (unobtrusive) road-user-following observations provide information concerning the tactical level of behaviour. Typical elements of tactical behaviour are lane use, ways to change lanes in the presence of other road users, lane choice before one has to turn left or right at intersections. One most important aspect is the adjustment of behaviour (speed, attention, signalling) when approaching sites where there is a change in the physical setting of the road (a narrowing, a curve, an intersection, etc.) or when other road users come in sight or have to be dealt with. Such adjustment can range from perfectly anticipatory to plain reactive, ending up in an operational manoeuvre to avoid a crash. One element of such evasive actions is obviously that the observed person might have been caught by surprise. For the observer, it is difficult to decide whether an observed person relies on his/her good reaction abilities and thus takes risks consciously, waiting till the last moment to (re)act, or if he/she is (regularly) caught by surprise. But, in the ride-along situation, one can hear the person's comments (anger, insults) and observe the person's non-verbal and para-verbal expressions, like sounds that reflect surprise or a shock, or head and body movements indicating that developments of situations had not been expected. In any case, in the frame of ride-along observations, one has also access to the operational level, assessing handling and the behavioural context of manoeuvres to avoid danger, not least those that are connected to traffic conflicts, i.e. evasive actions. Thus, it is possible to decide whether a person has a non-anticipatory driving style under control or not.

When following road users without them knowing it, such information is not available. However, the ride-along observer cannot be sure of his/her interpretation of what he/she observed, either. To improve understanding in this respect, it is necessary to talk to the observed persons. Therefore, interviews of different types are usually carried out in connection with ride-along observations, which is obviously easier to arrange than in connection with road-user following studies. Interviews with the observed persons allow one to develop some legitimate assumptions concerning the strategic behaviour of the observed road users, which observation alone does not do.

In contrast to observations on site, ride-along and road user-following observation provide information about the observed individuals in the first place, especially in combination with interviews. The most extensive procedure to assess the individual driver is to combine behaviour observation, in-depth interviewing (exploration) and testing (psychophysical and personality tests). This is done in some countries (e.g. Germany, Austria or Czech Republic) in the frame of psychological assessment of driver fitness to drive and is usually prescribed by law there [78–80].

Apart from what has just been stated, it should be repeated that if larger numbers of observation rides are carried out along the same route, conclusions concerning possible impacts of route- and site characteristics can be drawn, as well. Also, the methods – ride along and road-user following – are appropriate to compare behaviour before and after the implementation of any measure in the infrastructure or in the vehicle. For example, to carry out comparisons of behaviour without and with certain equipment of the vehicle, like ADAS or in-vehicle-information-systems.

10.5.2 Behaviour observation on-site and traffic conflict registrations

Behaviour observation on-site, carried out by human observers and backed up by video, helps us watch and register interactions between road users and, in case there are no other road users present, how road users make use of the observed spot. This could include breaking the rules like driving against red, over stop signs, using the wrong lane for turning left or right and alike. Concerning interactions with others, one would register how the right of way is negotiated, how lane changes and turning manoeuvres in the observed area are handled in the presence of others, and whether there are frictions, there. The interaction of car drivers with pedestrians and cyclists is usually also an important part of what can be observed. In sum, behaviour observations on-site provide data that mostly describe the tactical part of behaviour of larger numbers of persons both in the frame of interaction processes and of dealing with the physical environment. For example, with the design details of the site, and to a lesser part the operational behaviour of single individuals. By observing from outside, we do not learn so much about the skill concerning steering, changing of gears or use of mirrors. But we learn about abiding by the rules and the types of interactions (communication between road users) that can be observed at the spots selected for studies. To a certain degree, one can also register to which degree the behaviour of the road users at the studied site is anticipatory. A lack of such behaviour could certainly also be due to design elements, and not only due to the anticipation habits of the observed road users.

Again, without interviewing the observed persons, we do not learn about the way they look at their own behaviour (e.g. driving, walking, cycling) and interacting (communicating with others), nor about the motivation that lies behind different types of activities: breaking or respecting rules, respecting other people's right or not taking care of them. We do not learn about the strategic behaviour of individuals. However, for all types of observations on-site it is more important to note what types of problematic behaviour and interactions can be registered frequently or systematically, there. This, then, allows well-funded assumptions concerning which features of the observed spot enhance such problems. Observations on-site are not appropriate to assess the behaviour of specific individuals.

10.5.3 FOT

FOTs yield information concerning how (especially) drivers solve certain tasks or react to certain measures, similar to ride-along and road-user following. Depending on the set-up, different sections of tactical and operational behaviour are registered, and strategic behaviour is hardly covered at all. The question is to which degree a holistic view of the interaction processes is produced by data derived from instruments. But certainly, behaviour variables that can be measured are registered most exactly with the help of technical instruments.

Also, similar to ride-along and road-user following observation by human observers, researchers have direct contact with the observed persons, and thus, it is easy to talk to them in one or the other scientifically funded way. Again, this would allow us to learn more about the observed person's view on different aspects of road use and to get a more complete picture of their driving behaviour in the sense of Michon's model. This also helps to interpret behaviour, its changes, and different changes of behaviour by different persons or groups of persons, in connection with certain changes in the vehicle (or other traffic mode) or in the environment. As Michon [32] puts it: 'traffic conditions can trigger changes in tactical or strategic choices'. That needs to be understood.

10.5.4 NDS

NDS at first sight seem to be candidate number one for providing data that give a comprehensive picture of all levels according to Michon's model (driving behaviour or behaviour of other road users). Data from the road users involved are registered during 24 h 7 days a week whenever they use their vehicle. Thus, the strategic part is covered to a certain degree, motivational and decision-making aspects to be covered with the help of interviews. However, the question of to which degree a holistic view of interactions with other road users is provided by NDS remains to be analysed more in detail and more frequently. More studies like the one by Kaufmann and Turetschek [28] are recommended, who connected black box data and driving behaviour observation by human observers in order to achieve a better understanding of what types of behaviour and interaction data provided by vehicle instrumentation reflect.

10.5.5 Conclusion – Understanding what road users do and why they do it

The goal is to not only see or register what road users do but to understand why they are doing it. A swerve could be owed to something a person observes in the infrastructure, it could be unintentional, but it could also be a signal to another road user. The context might tell the researcher what exactly was registered or what he or she really saw. However, this remains speculation until the observed person tells what it was. Interviewing person is difficult and getting reliable data that reduce insecurity concerning such conclusions is not trivial. We will not go into further detail, but we want to underline the importance of talking to road users (subjects) that have been observed in one or the other way (e.g. rapid on-site interviews, in-depth interviews, group discussions). Sound understanding will allow more accurate conclusions concerning what leads the road users to behave in a certain way. This is of course important when we analyse the impact of infrastructure or equipment. We always need to know what the motivation for the observed behaviour is.

But another goal that behaviour observation can help us to reach is to find ways to influence road user behaviour in a positive way. As cited in the introduction, traffic behaviour and especially driving a car is highly influenced by emotional aspects and extra-motives such as fun and thrill. The importance of measures for changing antisocial or risky behaviour – especially behaviour that puts others at risk – should be obvious when looking at the high numbers of injured and deaths in traffic crashes. When talking about changing the behaviour of traffic offenders, a widespread knowledge base is necessary. There exist important psychological and social theories about traffic as a system and the reasons for erroneous or even dangerous behaviour. Effective measures to achieve changes in attitudes and, ideally, sustainable changes in behaviour must build on this and on a sound analysis of what must be improved. This chapter contains psychological, social as well as concrete practical background information intending to make a contribution to a safer and more social traffic environment.

References

[1] Turetschek Ch. and Risser R. (2004). Behaviour observation in connection with the evaluation of ISA, STORM and ACC, Presentation in the frame of the NoE HUMANIST conference 'Drivers Needs in Relation to IST', Lisbon.

[2] Rothengatter T. (1997). Psychological aspects of road user behaviour, Applied psychology: An international review – Special issue. *Traffic Psychology*, 46(3), 223–234.

[3] Singh V. and Dickson J. (2002). Ethnographic approaches, in: Partington D. (eds.) *Essential Skills for Management Research*, London: Sage Publications.

[4] Chaloupka-Risser C., Risser R. and Zuzan W.D. (2012). *Verkehrspsychologie: Grundlagen und Anwendungen*, Wien: Literas Universitätsverlag.

[5] Risser R. (1985). Behaviour in traffic conflict situations. *Accident Analysis & Prevention*, 17(2): 179–197.

[6] Rumar K. (1988). Collective risk but individual safety. *Ergonomics*, 31(4): 507–518.

[7] Vollrath M. and Krems J. F. (2011). *Verkehrspsychologie: Ein Lehrbuch für Psychologen, Ingenieure und Informatiker*. Kohlhammer Verlag.

[8] Van Nes N., Bärgman J., Christoph M. and van Schagen I. (2019). The potential of naturalistic driving for in-depth understanding of driver behaviour. *Safety Science*, 119, 11–20.

[9] Elvik R. and Mysen A.B. (1999). Incomplete accident reporting: Meta-analysis of studies made in 13 countries. *Proceedings of Annual Meeting of the Transportation Research Board No78*. National Research Council, Washington, DC, USA, pp. 133–140.

[10] Van Haperen W., Riaz M. S., Daniels S., Saunier N., Brijs T. and Wets G. (2019). Observing the observation of (vulnerable) road user behaviour and traffic safety: A scoping review. *Accident Analysis & Prevention*, 123, 211–221.

[11] Riaz, M. S. (2016). Naturalistic Behaviour Observation of Road Users: A Scoping Review. Master of Transportation Sciences-Traffic Safety), Universiteit Hasselt, Belgium. Retrieved from http://hdl.handle. net/1942/22284.

[12] Höfner K.J. (1967). Fahrverhalten und Persönlichkeitsbeurteilung junger Mopedfahrer. Eine Gegenüberstellung von wissentlicher und unwissentlicher Beobachtung des Fahrverhaltens. Dissertation zur Erlangung des Doktorgrades an der Philosophischen Fakultät der Universität Wien.

[13] Yerkes R.M. and Dodson J.D. (1908). The relation of strength of stimulus to rapidity of habit-formation. *Journal of Comparative Neurology and Psychology*, 18(5), 459–482.

[14] Kiegeland, P. (1999). Pkw-Nutzung in der Stadt?-Natuerlich nicht!!. In 37. BDP-Kongress Fuer Verkehrspsychologie.

[15] Rathmayer, R., Beilinson, L., Kallio, M., and Raitio, J. (1999). Kokeenjohtajan ja näkyvien mittalaitteiden yhteys ajokäyttäytymiseen instrumentoidussa autossa. [The observers' and the visual instruments' effect on driving behaviour when driving in an instrumented vehicle]. VTT, Finland.

[16] Hjälmdahl M. and Várhelyi A. (2004). Validation of in-car observations, a method for driver assessment. *Transportation Research Part A: Policy and Practice*, 38 (2), 127–142.

[17] Brühning, E., Chaloupka, Ch., Höfner, K. J., *et al.* (1989). *Sicherheit im Ferienreiseverkehr, Ausländische Kraftfahrer –insbesondere Deutsche – in Österreich, Unfall- und Sicherheitsforschung Straßenverkehr, Heft 75*, BASt, Bergisch Gladbach.

[18] Regan M., Triggs T., Young K., *et al.* (2006). *On road evaluation of ISA, following distance warning and seat belt reminder systems: Final results of the TAC Safecar Project*. Monash University Accident Research Centre, Clayton.

[19] Risser R., Schmidt L., Snizek S., Hulmak M. and Scheidl M. (1992). Konflikte Fußgänger — Radfahrer in Wien. Untersuchung im Auftrag der Stadt Wien, MA18 Stadtstrukturplanung, Werkstattbericht Nr. 1.

[20] National Academies of Sciences, Engineering, and Medicine. (2007). Impact of Behavior-Based Safety Techniques on Commercial Motor Vehicle Drivers. Washington, DC: The National Academies Press. https://doi.org/10.17226/23193.

[21] Falkenstein M. and Karthaus M. (2017). *Fahreignung im höheren Alter. Sensibilisieren – erfassen – fördern*, Verlag Kohlhammer.

[22] Risser R., Steinbauer J., Amann A., *et al.* (1988). *Probleme älterer Menschen bei der Teilnahme am Straßenverkehr*. Wien: Literas Universitätsverlag.

[23] Várhelyi, A., Kaufmann, C., and Persson, A. (2015). User-related assessment of a Driver Assistance System for Continuous Support – A field trial. *Transportation Research Part F: Traffic Psychology and Behaviour*, 30, 128–144.

[24] Hydén Ch. (1987). The Development of a Method for Traffic-Safety Evaluation: The Swedish Traffic-Conflicts Technique, Dissertation at the Technical University of Lund, Department for Technology and Society (formerly: Traffic Planning and Engineering).

[25] Šucha M., Dostal D and Risser R. (2017). Pedestrian-driver communication and decision strategies at marked crossings. *Accident Analysis and Prevention*, 102, 41–50.

[26] Basil M. (2011). Use of photography and video in observational research. *Qualitative Market Research*, 14, 246–257.

[27] Guo F. (2019). Statistical methods for naturalistic driving studies. *Annual Review of Statistics and ITS Application*, 6(1): 309–328.

[28] Kaufmann C. and Turetschek Ch. (2008). Connecting black box data and driving behaviour observation for better understanding of driving behaviour, *22nd ICTCT Workshop*, Leeds, UK.

[29] Lawrence B.M. and Oxley J.A. (2019). You say one route, we observe four: Using naturalistic observation to understand route-choices in cyclists. *Safety Science*, 119, 207–213.

[30] Meuleners L. and Fraser M. (2015). A validation study of driving errors using a driving simulator. *Transportation Research Part F: Traffic Psychology and Behaviour*, 29, 14–21.

[31] Risser, R. and Lehner, U. (1998). *Acceptability of Speeds and Speed Limits to Drivers and Pedestrians/Cyclists, EU-project MASTER*, FACTUM OHG, Austria.

[32] Michon J.A. (1985). A critical view of driver behaviour models. What do we know, what should we do? In: L. Evans and R. C. Schwing (eds.) *Human Behaviour and Traffic Safety*. New York: Plenum Press, pp. 485–520.

[33] Chaloupka, Ch., Risser, R. and Roest, F. (1991). Methode zur Identifizierung gefährlichen Verhaltens im Straßenverkehr, Abschlußbericht über das FWF-Projekt 6656-Soz, Wien.

[34] Chaloupka, C. and Risser, R. (1995). Don't wait for accidents—Possibilities to assess risk in traffic by applying the 'Wiener Fahrprobe', *Safety Science*, 19(2–3), 137–147.

[35] Hjälmdahl M. and Varhelyi A. (2000). Validation of Wiener Fahrprobe – A method description. Paper presented at *13th ICTCT Workshop*, Corfu, Greece.

[36] Risser, R. (ed.) (1997). *Assessing the Driver, Contributions to a Workshop on Driver Diagnostic and Selection, March 17, 1997 in Basel*. Braunschweig: Rot-Gelb-Grün.

[37] Turetschek Ch. (2009). Effects of ISA-like IST – Applications on the communication behaviour of young car drivers in urban areas, Doctor's thesis at the University of Vienna, Department of Sociology.

[38] Perkins, S. R., and Harris, J. I. (1967). Traffic conflict characteristics: Freeway curve and exit area F1, December, 1966.

[39] Svensson Å. (1998). *A method for analysing the traffic process in a safety perspective. Department of Traffic Planning and Engineering, Lund Institute of Technology*. (Bulletin/University of Lund, Lund Institute of Technology, Department of Traffic Planning and Engineering).

[40] Risser, R., Zuzan, W. D., Tamme, W., Steinbauer, J. and Kaba, A. (1991). *Handbuch zur Erhebung von Verkehrskonflikten mit Anleitungen zur Beobachterschulung. Kleine Fachbuchreihe Band 28*. Wien: Kuratorium für Verkehrssicherheit.

[41] Likert R. (1932). A technique for the measurement of attitudes, *Archives of Psychology*, 140, 1–55.

[42] Kassim A., Ismail K. and Hassan Y. (2014). Automated measuring of cyclist–motor vehicle post encroachment time at signalized intersections. *Canadian Journal of Civil Engineering*, 41(7), 605–614.

[43] Zheng, L., Ismail K. and Meng X. (2014). Traffic conflict techniques for road safety analysis: Open questions and some insights. *Canadian Journal of Civil Engineering*, 41, 633–641.

[44] Šucha, M. (2014). *Road users' strategies and communication: Driver-pedestrian interaction*. Transport Research Arena (TRA).

[45] Payne, G. and Payne, J. (2004). *Key Concepts in Social Research*. London: Sage.

[46] Lietz H., Petzoldt T., Henning M., *et al.* (2008). Methodische und technische Aspekte einer Naturalistic Driving Study, Forschungsvereinigung Automobiltechnik FAT, *FAT-Schriftenreihe* 229.

[47] Barnard Y., Innamaa S., Koskinen S., Gellerman H., Svanberg E. and Chen H. (2016). Methodology for Field Operational Tests of Automated Vehicles. *Transportation Research Procedia*, 14, 2188–2196.

[48] Dingus T. A., Klauer S. G., Neale V. L., *et al.* (2006). *The 100-Car Naturalistic Driving Study, Phase II —Results of the 100-Car Field Experiment Virginia Tech Transportation Institute*, Blacksburg, VA, NHTSA DOT HS 810 593.

[49] Carmines E. G. and Zeller R. A. (1979). *Reliability and Validity Assessment*, vol. 17. London: Sage.

[50] Kaufmann, C. and Risser, R. (2009). Verkehrspsychologische Untersuchungen. Validierung mithilfe von Fahrverhaltensbeobachtungen/Psychological test-validation with help of a driving test. *Zeitschrift für Verkehrssicherheit*, 55(4), 11–18.

[51] Chaloupka, Ch., Risser, R., Antoniades, A., Lehner, U. and Praschl, M. (1998). Abschätzung von reaktiver Anpassung an fahrzeugtechnische Veränderungen, Berichte der BASt, Heft M 84, Bergisch Gladbach

[52] Várhelyi, A., Hjälmdahl, M., Hydén, C., and Draskóczy, M. (2004). Effects of an active accelerator pedal on driver behaviour and traffic safety after long–term use in urban areas. *Accident Analysis & Prevention*, 36(5), 729–737.

[53] Risser, R., Stockinger, M. and Liedl, K. (1993). Studie über das Fahrverhalten von LenkerInnen von Elektrokleinmobilen 1/60. Report Arbeitsgruppen für Unfallmechanik Wien/Zürich, Universität Zürich.

[54] Eichberger A. (2022). *Müdigkeit am Steuer präzise und prädiktiv erkennen: Das Forschungsprojekt WACHsens*. Technische Universität Graz, Institut für Fahrzeugtechnik.

[55] Uzondu Ch., Jamson S. and Lai F. (2019). Investigating unsafe behaviors in traffic conflict situations: An observational study in Nigeria. *Journal of Traffic and Transportation Engineering*, 6(5), 482–492.

[56] Guo Y., Essa M., Sayed T., Haque M.M. and Washington S. (2019). A comparison between simulated and field-measured conflicts for safety assessment of signalized intersections in Australia. *Transportation Research Part C: Emerging Technologies*, 101, 96–110.

[57] Van der Horst A.R.A., Martijn C.T., Vet J.M. and Rahman A.K.M.F. (2017). An evaluation of speed management measures in Bangladesh based upon

alternative accident recording, speed measurements, and DOCTOR traffic conflict observations. *Transportation Research Part F: Traffic Psychology and Behaviour*, 46, 390–403.

[58] Van der Horst A.R.A, de Goede M., de Hair-Buijssen S. and Methorst R. (2014). Traffic conflicts on bicycle paths: A systematic observation of behaviour from video. *Accident Analysis & Prevention*, 62, 358–368.

[59] Marizwan M. and Manan A. (2014). Motorcycles entering from access points and merging with traffic on primary roads in Malaysia: Behavioral and road environment influence on the occurrence of traffic conflicts. *Accident Analysis & Prevention*, 70, 301–313.

[60] Cafiso S., Di Graziano A., and Pappalardo G. (2017). In-vehicle stereo vision system for identification of traffic conflicts between bus and pedestrian. *Journal of Traffic and Transportation Engineering (English Edition)*, 4(1), 3–13.

[61] Chebanyuk, K., Prasolenko, O., Burko, D., *et al.* (2020). Pedestrians influence on the traffic flow parameters and road safety indicators at the pedestrian crossing. *Transportation Research Procedia*, 45, 858–865.

[62] Tarko A.P. (2019). *Measuring Road Safety Using Surrogate Events*. Amsterdam: Elsevier Publishers.

[63] Lyu N., Deng C., Xie L., Wu C. and Duan Z. (2019). A field operational test in China: Exploring the effect of an advanced driver assistance system on driving performance and braking behavior. *Transportation Research Part F: Psychology and Behaviour*, 65, 730–747.

[64] Pantangi S.S., Fountas G., Anastasopoulos P.C., Pierowicz J., Majka K. and Blatt A. (2020). Do High Visibility Enforcement programs affect aggressive driving behavior? An empirical analysis using Naturalistic Driving Study data. *Accident Analysis & Prevention*, 138, 105361.

[65] Kovaceva J., Isaksson-Hellman I. and Murgovski N. (2020). Identification of aggressive driving from naturalistic data in car-following situations. *Journal of Safety Research*, 73, 225–234.

[66] Precht L., Keinath A. and Krems J.F. (2017). Identifying the main factors contributing to driving errors and traffic violations – Results from naturalistic driving data. *Transportation Research Part F: Traffic Psychology and Behaviour*, 49, 49–92.

[67] Wijayaratna K.P., Cunningham M.L., Regan M.A., Jian S., Chand S. and Dixit V.V. (2019). Mobile phone conversation distraction: Understanding differences in impact between simulator and naturalistic driving studies. *Accident Analysis & Prevention*, 129, 108–118.

[68] Kreusslein M., Morgenstern T., Petzoldt T., Keinath A., Krems J.F. (2020). Characterising mobile phone calls while driving on limited-access roads based on SHRP 2 naturalistic driving data. *Transportation Research Part F: Traffic Psychology and Behaviour*, 70, 208–222.

[69] Morgenstern T., Schott L. and Krems J.F. (2020). Do drivers reduce their speed when texting on highways? A replication study using European naturalistic driving data. *Safety Science*, 128, 104740.

[70] Itkonen T.H., Lehtonen E. and Selpi B.R. (2020). Characterisation of motorway driving style using naturalistic driving data. *Transportation Research Part F. Traffic Psychology and Behaviour*, 69, 72–79.

[71] Wang X. and Xu X. (2019). Assessing the relationship between self-reported driving behaviors and driver risk using a naturalistic driving study. *Accident Analysis & Prevention*, 128, 8–16.

[72] Ismaeel R., Hibberd D., Carsten O. and Jamson S. (2020). Do drivers self-regulate their engagement in secondary tasks at intersections? An examination based on naturalistic driving data. *Accident Analysis & Prevention*, 137, 105464.

[73] Feng F., Bao S., Hampshire R.C. and Delp M. (2018). Drivers overtaking bicyclists—An examination using naturalistic driving data. *Accident Analysis & Prevention*, 115, 98–109.

[74] Kovaceva J., Nero G., Bärgman J. and Dozza M. (2019). Drivers overtaking cyclists in the real-world: Evidence from a naturalistic driving study. *Safety Science*, 119, 199–206.

[75] Rasch A., Panero G., Boda Ch-N. and Dozza M. (2020). How do drivers overtake pedestrians? Evidence from field test and naturalistic driving data. *Accident Analysis & Prevention*, 139, 105494.

[76] Orlovska J., Novakazi F., Bligård L-O., Karlsson M., Wickman C. and Söderberg R. (2020). Effects of the driving context on the usage of Automated Driver Assistance Systems (ADAS) – Naturalistic Driving Study for ADAS evaluation. *Transportation Research: Interdisciplinary Perspectives*, 4, 100093.

[77] Michon J.A. (1979). *Dealing with danger (No. VK 79-01)*. Haren: Verkeerskundig Studiecentrum, Rijksuniversiteit Groningen.

[78] Fastenmeier W. and Strohbeck-Kühner P. (2019). Driving behaviour observation as the 'gold standard' of the performance check? *Blutalkohol* 56/3, 185–190.

[79] Brenner-Hartmann J., Wagner T., Seidl J., Mußhoff F., Hoffmann-Born H. and Löhr-Schwaab S. (2014). Assessment of personal resources for safe driving – The principles of medical psychological assessment in Germany. *Schriftenreihe Fahreignung*.

[80] Wagner T., Uhlmann L. M., DeVol D., and Rethfeldt I. (2018). Factors influencing the outcome of the Medical-Psychological Assessment of speed-affine drivers in Germany. *Transactions on Transport Sciences*, 9(2), 57–66.

[81] Klauer, S. G., Perez, M., and McClafferty, J. (2011). Naturalistic driving studies and data coding and analysis techniques. In *Handbook of Traffic Psychology* (pp. 73–85). New York: Academic Press.

[82] Chaloupka-Risser C. (2005). What are we allowed to ask, what can we know – traffic-psychological analysis of driver behaviour. In *Proceedings of the ICTCT Extra Workshop*, Campo Grande, Brazil.

[83] Brühning E., Chaloupka C., Höfner K. J., *et al.* (1989). Sicherheit im Fernreiseverkehr. Ausländische Kraftfahrer – insbesondere Deutsche – in Österreich. Unfall- und Sicherheitsforschung Straßenverkehr Heft 75, BASt: Bergisch – Gladbach.

Chapter 11

Road Safety Inspection and Road Assessment Programme

András Várhelyi[1]

Road Safety Inspection (RSI) and Road Assessment Programme (RAP) are specific tools in the framework of Road Infrastructure Safety Management. Other tools are Black-spot Management, Network Safety Management, Road Safety Impact Assessment and Road Safety Audit. This chapter is devoted to RSI and RAP.

11.1 Road Safety Inspection

Road Safety Inspection (RSI) is a systematic on-site review of existing roads conducted by independent road safety expert(s) with the intention of identifying any potential hazards, faults or deficiencies that may lead to serious accidents. Systematic means that it is carried out in a methodical way following a formal procedure. Potential hazards are to be considered from the point of view of all types of road users. Independent experts are road safety professionals with experience in road infrastructure safety management, accident analysis, traffic engineering, road design and road user behaviour, especially trained to carry out RSI. RSI is pro-active, aiming at preventing severe accidents through the identification of safety deficiencies for remedial action rather than reacting after accidents occur. In some older references, RSI procedures were covered by the term Road Safety Audit. RSI of existing roads is a today's concept that has been adopted because this term appears to be more appropriate when associated with existing roads. Road Safety Audit (a formal safety performance examination of planned roads or intersections) and RSI have several similarities, but also differences, especially regarding the procedures.

The EU directive 2008/96/EC [1] requires the establishment and implementation of procedures relating to, among others, RSI by the Member States. Hence, RSI is mandatory as a minimum on Trans-European Road Network (TEN-T) roads, while every country can decide if RSI is also applicable to other roads. Many countries have established their own procedures and manuals for RSI as well as a database on certified inspectors. A follow-up directive of the European

[1]Department of Technology and Society, Faculty of Engineering, LTH, Lund University, Sweden

Parliament and of the Council, Directive 2019/1936 amending Directive 2008/96/EC on road infrastructure safety management [2] states that the 'Directive shall also apply to roads, which are not part of the trans-European network (but situated outside urban areas)'.

11.1.1 Why RSI?

The largest part of the existing roads was designed and constructed many years ago, initially for different amount of traffic, vehicle fleet or road user types. Very often, the human factor, i.e. road users, is blamed as the main cause of road accidents. However, in a system built in a way in which mistakes can lead to accidents, accidents will occur [3]. People have always made mistakes and will make mistakes, and information campaigns cannot make mistakes disappear. The most cost-efficient measure is not necessarily the one directed to the main factor, the human road user. Human behaviour can often be more easily influenced by engineering measures than by education, legislation or surveillance.

Road agencies (state, private or municipal) do not have sufficient knowledge of road safety to identify or analyse the risks built into their road network. The idea with RSI is to provide a systematic method that will help the road agencies improve their knowledge of the safety of their road network through inspection visits made by someone from the outside who has a fresh look. Such visits are made by qualified, specially trained professionals, called auditors, whose task is to identify safety problems and report them. Subsequent actions to eliminate the identified safety problems are the responsibility of the road agency that is familiar with the local context. RSIs provide the road agency with a tool to improve the safety of the road network by prevention and contribute to the development of a culture of 'attentiveness to safety issues' of the road network.

11.1.2 Benefits and costs of RSI

RSI minimises the risk and severity of road accidents that may result from deficiencies in the road infrastructure and hence minimises losses to human health and the economy. Expert knowledge and the systematic procedure of RSI provide the possibility to reduce the number and severity of traffic accidents by improving the safety performance of existing roads. Since it considers all types of road users as equally important, it elevates consciousness of the exposed situation of an often-neglected group in traffic planning and engineering: vulnerable road users. Elvik *et al.* [4] refer to two evaluation studies in the USA and conclude that RSI yielded a 12.5%–40% reduction in accidents. Elvik [5] compiled the expected effects of various treatments on injury accidents as a result of RSIs (Table 11.1).

For RSI to improve road safety, a number of criteria need to be fulfilled [5]:

- RSI should be targeted at hazards known to be risk factors for injury/fatal accidents.
- RSI should identify treatable traffic hazards and propose corrective measures with known efficiency.

Table 11.1 Expected effects on injury accidents because of RSIs [5]

Treatment	Accidents affected	Expected accident reduction (%)
Correcting erroneous signs	All accidents	5–10
Removing sight obstacles	All accidents	0–5
Guardrails along embankments	Running-off the road	40–50
Providing clear recovery zones	Running-off the road	10–40
Flattening side slopes	Running-off the road	5–25
Signing hazardous curves	Running-off the road in curves	0–35
Collapsible lighting poles	Vehicles striking lighting poles	25–75
Guardrail end treatments	Vehicles striking guardrail ends	0–10

- It should be made sure that the measures proposed by an RSI are actually implemented.
- The measures implemented should not elicit behavioural adaptations that may offset the intended effects of the measure.

The costs (auditor fees and costs for treatments) associated with RSIs are overwhelmingly compensated by the benefits from saved lives and prevented injuries.

11.1.3 When should RSI be carried out?

As the EU directives prescribe, RSI should become a routine process '... inspections shall be sufficiently frequent to safeguard adequate safety levels for the road infrastructure in question'. A rule of thumb is that an adequate frequency is around every third to fifth year. Although the basic features of the roads, such as alignment and cross sections, may not change for many years, their function, traffic volumes and traffic composition may change, which calls for their check-up. Also, since changes in some road elements (e.g. surface condition, road markings, signs and lighting) can occur more often than in others, a focused RSI limited to these elements can be carried out.

Other reasons to initiate an RSI:

- A road section or site (intersection, pedestrian or cycle crossing) is identified as a high-risk location based on accident records.
- Serious safety problems are reported by police or maintenance personnel, local administration like city government, etc.
- A reconstruction or rehabilitation project is planned by the road agency or municipality.
- Possible other causes relevant to road safety.

11.1.4 Partners involved in the RSI process

The partners involved in the RSI process are the Client and the Auditor. The Client (usually the road authority, road operating company or local municipality) is the institution responsible for road safety. The Auditor is the independent road safety expert, a team or organisation who will conduct the inspection. The task of Auditors is to identify road safety problems and present the findings, deficiencies and recommendations in a formal written report. Then, it is the Client who finally decides whether recommendations are to be adopted or not. Possible courses of action are in the competence and under the responsibility of the road operator who is familiar with the local context. All recommendations should be given due consideration, and the road operator should provide an implementation plan of the corresponding countermeasures. Safety problems identified by the Auditor but considered to be insignificant or the recommended solutions considered to be not suitable must be addressed by a formal response. This formal response must give reasons why the recommendations are not accepted. The response to the audit report will become part of the RSI project documentation which is the formal responsibility of the superior authority of the road operator.

11.1.5 Who can perform RSI?

The term Auditor is used for a specially trained and qualified road safety professional who performs RSI. If an RSI is performed by more than one Auditor, the team is called Inspection Team, where the members are Road Safety Auditors. The Inspection Team should comprise expertise and qualification from various road safety relevant fields, such as accident analysis, traffic engineering, road design and road user behaviour.

The identified safety problems and recommendations from RSIs are based on the auditors' knowledge and experience. Hence, parallel inspections performed by different auditors at the same site may result in somewhat different recommendations. Still, since these recommendations build on basic safety principles, they should be duly considered. Anyway, when inspecting large objects, it is advantageous if the inspection team has more than one pair of eyes and all necessary skills and backgrounds are covered by the team members.

An Auditor should, apart from a university degree or comparable education, have participated in a specialised RSI training course and then should take part in subsequent follow-up training, regular seminars, and workshops to keep his/her knowledge up to date. Although there are working aids in the form of checklists, they cannot replace the knowledge and experience of an Auditor.

11.1.6 Organisational issues

A central national organisation with overall responsibility for road safety should administrate the RSI process, provide guidelines for RSI procedures, regulate the training activities of Auditors, certify Auditors, maintain a central database of certified Auditors, maintain a database of RSI cases, initiate regular meetings where Auditors can exchange experience and discuss interesting cases. Regular follow-ups of RSI results help identify frequently occurring deficits, which can give valuable input into updating RSI procedures, guidelines, training curricula and RSI checklists.

11.1.7 Performing an RSI

When starting an RSI, the first step is to identify the object – a road section or a road element (e.g. intersection, pedestrian crossing, etc.). This task is performed by the Client (road agency or local municipality). The object of the RSI will be outlined in an agreement between the Client and an RSI Team (or Auditor) procured by the Client. The agreement should describe the object to be inspected, timelines, who pays for what and who should contribute with what.

Preparatory work

RSI starts with preparatory work in the office. Background information about the object (the function of the road, traffic volumes, traffic composition, accident data, possible land use changes, previous audit, inspection reports, etc.) should be obtained. Information from residents might prove useful and can be obtained through face-to-face discussions or a questionnaire. The Client usually can provide this information.

Site visit

When conducting the site visit, the auditor should place himself in the position of the various road users (motorists, cyclists and pedestrians) who might be using the inspected object. The Auditor should drive (with the eyes of a car driver/truck driver), cycle (with the eyes of a cyclist, a rider of a powered two-wheeler) and walk (with the eyes of a pedestrian, a child, an elderly person, a person in a wheelchair) through the object from all directions. The inspection should consider a range of human factors related to possible road user errors induced by the road environment. Identified safety deficiencies should be photographed for documentation, and the photos to be linked to the location of the deficiency. The Auditor should wear safety equipment/reflective vest.

During the site visits, RSI checklists are used to cover the most important issues. Checklists are assigned for road sections, intersections, traffic signals, roundabouts, pedestrian crossings, etc. The content of the checklists reflects the prevailing relevant types of hazards that may be encountered. Using a checklist is less likely to overlook problems. However, they cannot be a substitute for expertise.

New technology, in the form of software, allows to improve the quality and efficiency of the inspections, as well as reduces the risk of the Auditor being exposed to traffic on the site. These tools are run either on a notebook computer (when the inspection is performed by car) or on a mobile device when the inspection is performed on a bike or on foot (e.g. cycle paths, footpaths, etc.). They allow to register safety deficiencies during the inspection, automatically locate the identified deficiencies using GPS and facilitate post-processing of the registered issues.

RSI should be performed in various selected periods of time so that the most relevant traffic situations are covered: during winter and summer, day and night, dawn or dusk in East–West aligned roads. Also, site-specific matters should be taken into consideration, e.g. locations, where extraordinary events could cause site-specific circumstances or if the road passes a school; site visits also should take place when school children go to and from the school or locations.

The RSI report

The main content of an RSI report is the identified safety deficiencies and recommendations for their treatment. The report should only list safety-related deficiencies. It starts with general information about the RSI carried out, such as the general characteristics of the inspected object, RSI team members (Auditor) and their role during the RSI and sources of information/data used for the RSI. Then, the identified road safety deficiencies are described, and recommendations are made on how to improve the situation. The auditor formulates for alleviating the problem but does not necessarily design the countermeasure. According to international practice, the comments on identified safety deficiencies are structured on two levels:

- Problems – conditions that have been documented entail accident risk.
- Remarks – conditions that experience has shown should be given attention, but for which there is no documented evidence that they entail accident risk to road users.

Auditor comments do not need to refer to norms, standards or regulations. It is important to remember that an RSI is not an inspection in the literal sense of the word, i.e. an external control of whether the inspected object is in line with standards or norms. Norms and standards may be outdated, and Auditors should base their recommendations on the latest knowledge.

Appendices to the report may include accident data, maps and illustrations (photos or sketches of countermeasures).

To guarantee that every RSI is free of subjective elements, irrespective of the performing inspecting team, a standardised report form should be used. To achieve objectivity, standardised text for the most common hazards describing a number of typical frequent situations should be employed. A standardised report form has the advantage of being easy to read and allowing to compare different reports.

Completion meeting

After receiving the RSI report, the Client must consider the indicated safety deficiencies and proposed countermeasures and decide on whether to accept or not the proposed measures. Sometimes, some additional clarification and/or discussion is needed to ensure that both the Auditor and the Client understand the issues pointed out in the same way. This can be done at a clarifying meeting between the Client and the Auditor. In certain cases, it can be helpful to invite parties who are not part of the formal RSI procedure but have to do with road safety issues concerning the inspected object, e.g. the traffic police, public transport company, local communities, NGOs, etc.

11.1.8 Good RSI practice guidelines

Based on the analysis of RSI practice in 11 European countries, Elvik [5] listed the following guidelines for good RSI practice:

1. The elements included in RSIs should stand as risk factors for accidents or injuries.
2. Inspections should be standardised and designed to ensure that all elements included are covered and assessed in an objective manner.
3. Checklists for RSI should include the following core of recognised important elements:
   (a) The quality of traffic signs, with respect to their need and whether they are correctly placed or legible in the dark.
   (b) The quality of road markings, whether they are visible or are consistent with traffic signs.
   (c) The quality of the road surface characteristics, with respect to friction (macro and micro-texture) and evenness.
   (d) The adequacy of sight distances and the absence of permanent or temporary obstacles that prevent timely observation of the road or other road users.
   (e) The presence of roadside traffic hazards, near the carriageway, such as trees, exposed rocks, drainage pipes and culverts, steep high embankment slopes, etc.
   (f) Aspects of traffic operation, in particular if drivers' speeds are adequate to local conditions and to the function of the road. This also includes items such as the suitability of the road to its function, the adequacy of space for current traffic and separation between motorised and vulnerable road users.
4. For each element included in an inspection, a standardised assessment should be made by applying the following categories:
   (a) The item represents a traffic hazard that should be treated immediately. A specific treatment should then be proposed.
   (b) The item is not in perfectly good condition or deviates slightly from current standards, but no short-term action is needed to correct it. Further observations are recommended.
   (c) The item is in good condition and in accordance with current standards.
5. RSI should state their findings and propose safety measures by means of standardised reports.
6. Auditors should be formally qualified for their job. They should meet on a regular basis to exchange experiences and to ensure a uniform application of safety principles in the inspections.
7. There should be a follow-up of RSI after some time to check if the proposed measures have been implemented or not.

These guidelines are general, and each country should define its own national administrative, regulatory legal and policy specificities and procedures for RSI.

11.1.9 Typical safety deficiencies identified by RSIs

Experience from several RSIs identified the following typical safety deficiencies (photographic examples of these are shown below): road function, road alignment: predictability, consistency, clarity, credibility, visibility, roadside features, location of intersection and traffic signals.

Road function

A basic safety principle for the road network is mono-functionality of roads as either through roads, distributor roads, access roads or local roads in a hierarchically structured network. Mixing these functions creates hazards, most often for vulnerable road users (Figure 11.1).

Road alignment – predictability

The road alignment should guide the driver in a predictable manner and not create any surprise. For an adverse example, see Figure 11.2.

Consistency

Road users should encounter uniform design and arrangements at all sites of the same kind, so they are conditioned to act in the expected way. For an adverse example, see Figure 11.3.

Clarity

All signing and instructions must be clear for road users so they can be certain about what is required in terms of correct behaviour. For an adverse example, see Figure 11.4.

Credibility

Information and instructions to road users should be 'believable' so they can rely on that the received messages are a true representation of what will occur ahead.

Figure 11.1 Main arterial street (also functioning as distributor street and local street) violating the principle of mono-functionality (photo: A. Várhelyi)

Figure 11.2 *A straight secondary road crosses a main road. The stop-regulated junction is not salient enough, and it is difficult to detect the driver in good time (photo: A. Várhelyi).*

Figure 11.3 *Straight wide carriageway giving the impression of being the main road, hence own priority, but the right-hand rule is in force at the junction ahead, creating confusion (photo: A. Várhelyi)*

Figure 11.4 *Delineation should be clear to guide the traffic past the work zone so that the driver does not end up inside the construction site (photo: A. Várhelyi)*

One aspect of credibility is the agreement between the posted speed limit and road alignment. For an adverse example, see Figure 11.5.

Visibility

Road users, expected to interact with each other, have to be visible to each other during the whole interactive process. For an adverse example, see Figure 11.6.

Roadside features

Hard objects on the roadside should be eliminated or shielded with guard rails. For an adverse example, see Figure 11.7.

Figure 11.5 Inconsistency between the posted speed limit of 70 km/h and the sudden sharp curve – it is impossible to take this curve at a speed of 70 km/h (photo: A. Várhelyi)

Figure 11.6 A right-turn lane on the main road creates visibility problems since a right-turning vehicle from the main road may hide a straight passing vehicle approaching the intersection for the driver attending to enter the main road (photo: A. Várhelyi)

Figure 11.7 Unshielded trees on the roadside outside a curve (photo: A. Várhelyi)

Figure 11.8 An undue T-junction in a curve on a hilltop (photo: A. Várhelyi)

Location of intersection

Intersections should not be located at places where they may surprise the approaching driver. For an adverse example, see Figure 11.8.

Traffic signals

Incorrect programming of traffic signals can create a trap for vulnerable road users. For an adverse example, see Figure 11.9.

11.2 Road Assessment Programme

A RAP is a set of tools to assess the safety level of a road, i.e. how forgiving it is based on its features. It involves the collection of data on road characteristics with the aim of determining the level of protection the road environment provides for the road user when a crash occurs. The first RAP, the EuroRAP (European Road Assessment Programme), was founded by the British, Dutch and Swedish governments, working in partnership with leading European automobile clubs and safety

Figure 11.9 The traffic signal head on the near-site signal post indicates red for pedestrians, while the far-site signal head indicates green, which might lure pedestrians to begin crossing (photo: A. Várhelyi)

charities [6]. The focus of RAP, like the NCAP (New Car Assessment Programme), was to create a market for safety by raising consumer awareness. The programme spread globally, becoming iRAP (International Road Assessment Programme), which is an umbrella organisation for regional RAP programmes including Australian, Brazilian, Chinese, European, Indian, New Zealand, South African, Thai, the US and RAP programmes [7].

A RAP is a proactive approach to road safety management, and it is particularly useful if crash data are of poor quality or are not geo-referenced, making traditional crash-data-based approaches, such as Black-spot Management or Network Safety Management, difficult. The available tools in the framework of RAP are Star Rating, Risk Mapping, and Performance Tracking.

11.2.1 Star Rating

Star Rating is based on site inspections along the road, and it is used to show how well the road infrastructure would protect from death or serious injury when a crash occurs. It provides a simple measure of the 'built-in' level of safety for road users of the road section in question. Star Rating considers the safety of car occupants. It assumes that cars have a rating of at least four stars according to EuroNCAP, their drivers obey speed limits and that occupants wear seat belts [6]. Star Rating is especially useful when crash data is unavailable or unreliable since it can be carried out without reference to crash data. Star Rating is based on Road protection scoring, and its results are often presented cartographically in publications of motorist organisations to inform drivers/riders about the safety level of different road sections. Road protection scoring involves the collection of road characteristics data that are known to influence the severity of three main crash types: run-off road crashes, head-on crashes and crashes at intersections. The observed elements are aspects of the road environment known to influence either the likelihood of crash occurrence or the severity of those crashes that do occur. The safest roads can attain five stars, while the least safe roads get one star. As an example, a road with a speed limit over 70 km/h is scored as safe – with respect to running-off-the-road crashes – if it has a safety zone of at least 10 m [6] or an appropriate safety barrier. A safety zone is a level area along the road, which does not contain hard objects (rocks, trees, etc.) that may cause injury in case of a crash. For assessing the protection score of a road environment for head-on crashes and crashes at junctions, specific criteria of relevance are used.

Site inspections can be undertaken using specially equipped vehicles to collect digital images or videos of the road environment during driving. The video-based road-element data is geo-referenced and coded later in an office by a team using specialised software to record details of the road. Additional data on actual driven speeds and traffic volumes can also be entered into the analysis tool.

To analyse the collected road element data, special analysis tools can be applied (see, e.g. https://vida.irap.org/en-gb/home). These tools can

- Estimate the number of casualties that would be expected based on the condition of the road elements.
- Propose suitable countermeasures that would treat detected safety deficits.
- Estimate the number of casualties that would be expected on the basis of the proposed treatments.
- Appraise the economic benefits and costs of the proposed treatments.

Although Star Rating basically considers the safety of car occupants, National EuroRAP programmes increasingly include in their analysis the safety performance of roads for motorcyclists. Evaluating the road environment with respect to unprotected road users helps to elevate the need for measures designed to protect riders on higher-risk sections, e.g. motorcyclist-friendly crash barriers and cushioning around lampposts.

11.2.2 Risk mapping

If detailed crash data is available, colour-coded Risk Maps can be produced. A Risk Map represents the actual number of deaths and injuries on a road network, and it provides an at-a-glance view of where crash risk is greatest. Risk Maps are a useful tool for road safety policy-makers and professionals since they provide easily interpretable visual information for decisions on priorities for road safety action. The risk is calculated by comparing the frequency of road crashes resulting in death and serious injury on every stretch of road with how much traffic each road is carrying. For example, if there are 10 injury crashes per year on a road carrying 5,000 vehicles a day, the risk on this road is twice that of a road with the same number of crashes but with a daily flow of 10,000 vehicles.

Different Risk Maps can be produced depending on the target audience. The risk of the individual road sections can be calculated by relating the number of killed/injured to either [6]:

- crashes per kilometre
- crashes per kilometre travelled

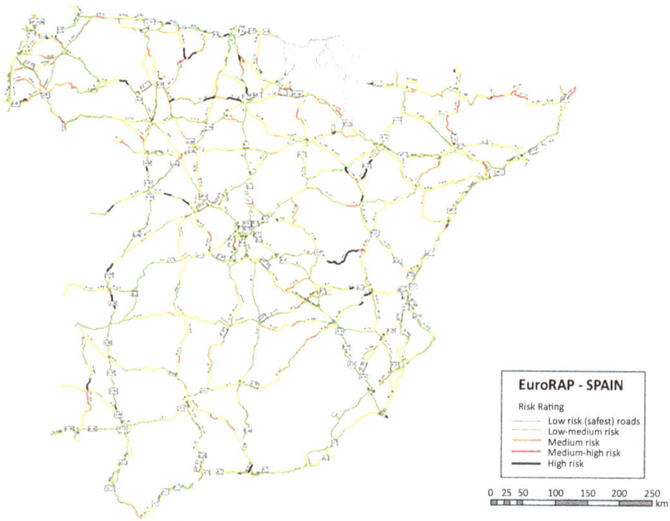

Figure 11.10 Risk Map of the road network of Spain (by courtesy of EuroRAP: https://www.eurorap.org/risk-mapping-the-spanish-national-road-network)

- crashes per road user group or crash type
- crash costs per kilometre or kilometre travelled
- potential crash savings per kilometre and per kilometre travelled

The estimated risk of the individual road sections can be shown in colour-coded bandings from high risk (black) to low risk (green). For an example of a Risk Map, see Figure 11.10.

11.2.3 Performance tracking

An important part of Road Safety Management is monitoring safety performance over time towards the achievement of road safety targets. RAP star rating offers safety performance indicators for road infrastructure. Performance tracking identifies whether the applied countermeasures have been effective and fewer people are being killed or seriously injured on a certain road section over time. Regular risk mapping or star rating of a road network provides insight into the success of road safety investments. Trends in performance can be monitored and early action can be taken to address any newly detected issues. An example of visualising the evolution in risk distribution of the road network of a country is shown in Figure 11.11.

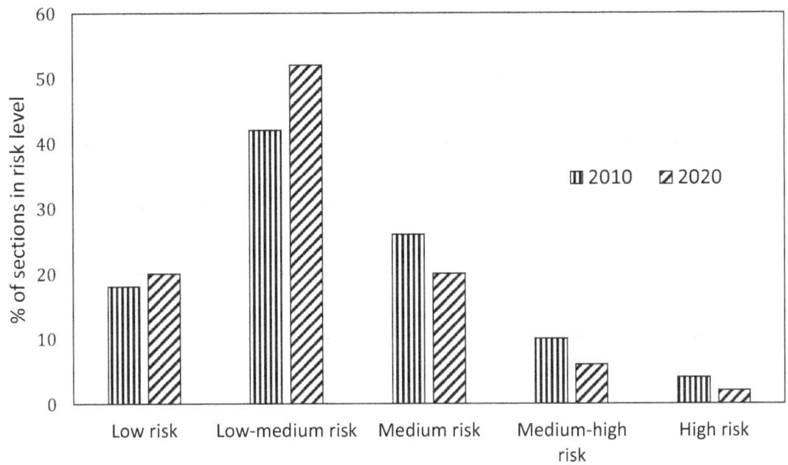

Figure 11.11 Example of a diagram presenting the evolution in risk distribution of the road network of a country from 2010 to 2020 (arbitrary data)

11.3 Concluding remarks

RSI and RAP are efficient tools in Road Infrastructure Safety Management. To reach their full potential, it is important that they are done systematically on a regular basis, well documented and carried out by trained professionals. Their mandatory utilisation should be stipulated by national legislation. Also, legislation should point out the authority with responsibility to supervise the process of carrying out these activities, issue authorisation for certified experts who can carry out these activities and keep an open database of certified experts. A national regulatory framework should establish procedures and responsibilities for initiating, carrying out, supervising, administrating and documenting these activities.

References

[1] EC (2008) Directive 2008/96/EC of the European Parliament and of the Council of 19 November 2008 on road infrastructure safety management. Official Journal L 319, 29/11/2008, pp. 59–67.

[2] EC (2019) Directive 2019/1936/EC of the European Parliament and of the Council of 23 October 2019 amending Directive 2008/96/on road infrastructure safety management.

[3] Perrow, C. (1999) *Normal Accidents – Living with High-Risk Technologies*. Princeton University Press, Princeton, NJ.

[4] Elvik, R. Höje, A., Vaa, T., and Sörensen, M. (2009) *The Handbook of Road Safety Measures*. 2nd edn. Elsevier, Amsterdam.

[5] Elvik, R. (2006) *Road Safety Inspections: Safety Effects and Best Practice Guidelines. TØI Report 850/2006*. Institute of Transport Economics, Oslo, Norway.

[6] RISMET (2011) Recommendations for the development and application of Evaluation Tools for road infrastructure safety management in the EU. Deliverable Nr 7 of the Road Infrastructure Safety Management Evaluation Tools (RISMET) project.

[7] iRAP (2020) iRAP home page. www.irap.org/about-us/?et_open_tab=et_pb_tab_3#mytabs [3 Accessed 20 Jan 2020].

Useful sources on RAP

AusRAP Australian Road Assessment Programme – https://web.archive.org/web/20080704075126/http:/www.ausrap.org/.

EuroRAP – European Road Assessment Programme https://www.eurorap.org/.

iRAP international Road Assessment Programme – https://www.irap.org/.

New Zealand Road Assessment Programme – http://www.kiwirap.org.nz/.

ViDA – The iRAP online software to help create a world free of high risk roads – https://vida.irap.org/en-gb/home.

Chapter 12

Safety performance indicators

Wouter Van den Berghe[1], Victoria Gitelman[2] and Shalom Hakkert[2]

12.1 Introduction

12.1.1 What are safety performance indicators?

Indicators are numeric values that give information about a relevant characteristic of an object, an activity or a phenomenon [1]. For instance, the average temperature and the annual rainfall in a country are indicators of climate; the average life expectancy in a country is an indicator of the performance of the welfare system, etc. Indicators help to capture the complexity of phenomena in relatively simple terms and facilitate monitoring these phenomena and communicating about them [2]. Presently, indicators are used in a wide range of contexts and policy areas [3], including public health and transport. Similarly, in the field of road safety, current policies and interventions strongly rely on a systematic monitoring of the road safety situation, with a variety of indicators in use [4,5].

When monitoring road safety progress, in the past, road safety was mainly assessed in terms of accidents, injuries or their social costs. However, more than two decades ago, it was recognised [6] that counting accidents or injuries only does not offer sufficient insight into the factors that affect safety and the current state of the system, and thus, safety performance indicators are required to monitor the road safety situation in more detail.

The place of safety performance indicators (SPIs) in a safety management system was originally described in a New Zealand Land Traffic Safety Authority paper [7] and then elaborated in a report prepared by a group of European road safety experts [6] and by the SUNFlower report [8] (Figure 12.1).

The model presents essential components of a safety management system in the country in the form of a road safety pyramid defining the layers of background characteristics (structure and culture); safety interventions (measures and programs); safety performance indicators as intermediate outcomes; accidents, fatalities and injuries as final outcomes and the social costs of accidents/injuries (at the top).

[1]Tilkon Research & Consulting, Belgium
[2]Technion, Transportation Research Institute, Israel

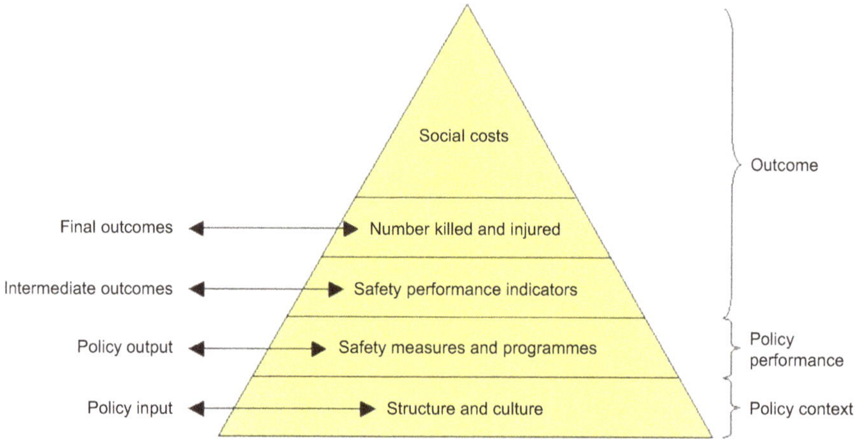

Figure 12.1 The road safety pyramid (source: Koornstra et al. [8])

Furthermore, the European project SafetyNet provided methodological fundamentals for SPI development and showed their application in various road safety areas [9]. The detailed approach strengthened the role of SPIs in reflecting operational conditions of the road traffic system, which are related to the system's safety performance, and outlined the common use of the term *safety performance indicators* as indicators at the intermediate level of the road safety pyramid (Figure 12.1). Yet, a distinction was made between SPIs and direct outputs of specific safety interventions (Section 12.2.1).

In line with the traditional taxonomy of factors affecting road safety and guiding road injury prevention, i.e. road infrastructure, vehicle conditions and road-user behaviours (e.g. [10]), SPIs are intended to reflect the operational conditions in those areas, but focusing on the characteristics which are known to be causally related to accident/injury occurrences. For example, regarding road users, SPIs may reflect risky behaviours such as speeding, distracted driving, driving under the influence (of alcohol or drugs) or non-use of protective equipment (e.g. seatbelts, helmets).

To note, a broader view of SPIs is possible which may also include, for example, characteristics related to traffic exposure [11]. As vehicle traffic volume was proven to be a contributory factor for accidents, indicators related to traffic volume can be considered as SPIs.

12.1.2 The role of SPIs in road safety management

As mentioned above, road safety monitoring and evaluation are the basics of road safety management [4, 5, 12]. Traditionally, road-safety authorities followed up the frequency and socio-economic costs of road accidents and injuries to assess the safety of transport systems. However, such values are incomplete as they do not offer sufficient insight into the underlying processes that lead to accidents. Typically, accidents or injuries are only the tip of the iceberg, because they occur as the 'worst case' of unsafe operational conditions of the road traffic system [9].

In addition, the number of accidents or injuries is subject to random fluctuations, whereas their reporting is frequently incomplete [6].

At the same time, managing road safety needs to take into account as many factors that affect safety as possible or, at least, those factors that can be adjusted or controlled and that have a tangible impact on road safety. Hence, SPIs are required for road safety management to provide a means for monitoring the effectiveness of safety actions applied [13].

Unlike traditional safety outcomes, SPIs can give a more holistic picture of the safety performance of a transport system and help in recognising the issues that are related to accident occurrences and/or consequences. Regular recording and monitoring of SPIs can help to reach a better understanding of the processes that lead to accidents. This understanding, in turn, can facilitate the identification of targeted interventions to bring positive safety change [14]. Furthermore, SPIs can play an essential role in implementing the Safe System approach to road safety [10]. They delve into the causes of accidents and help track progress, enabling process-oriented interventions that address various road-safety issues and reduce road users' risk of death or serious injury [14].

The primary objective of using SPIs is to enhance road safety, addressing specific safety concerns. A particular value of SPIs can be used for monitoring road safety plans and strategies. For example, they can be helpful in monitoring progress towards final safety targets, to set specific targets to be achieved in the safe system conditions (e.g. a reduction of the number of speeders by 50%), to assess whether the policy measures implemented have led to the desired results, etc. [4,15,16]. Without SPIs, policymakers and other stakeholders will experience difficulties in assessing the efficiency of interventions applied or the identification of measures that still need to be taken.

Other purposes of SPIs in the continued road safety management may include [9,11,13,17], for example:

- monitoring changes in accident/injury contributing factors (e.g. the use of mobile phones by drivers);
- detecting emerging trends in road-user behaviours (e.g. shifts in drink-driving);
- identifying policy measures/interventions that need to be taken (e.g. increase the number of roads with section control for speeding);
- comparing road traffic systems (e.g. differences between countries in the percentage of motorcyclists wearing a helmet);
- providing a focus for the efforts of the many actors involved, etc.

12.2 Methodological rules for using SPIs

12.2.1 The basic model

The model of road safety management (Figure 12.1) allocated SPIs to the level of intermediate outcomes. Further developments of the SPI theory [9] underlined that a core issue of SPIs is that they should be able to reflect unsafe operational conditions of the road traffic system and should therefore be of a more general nature

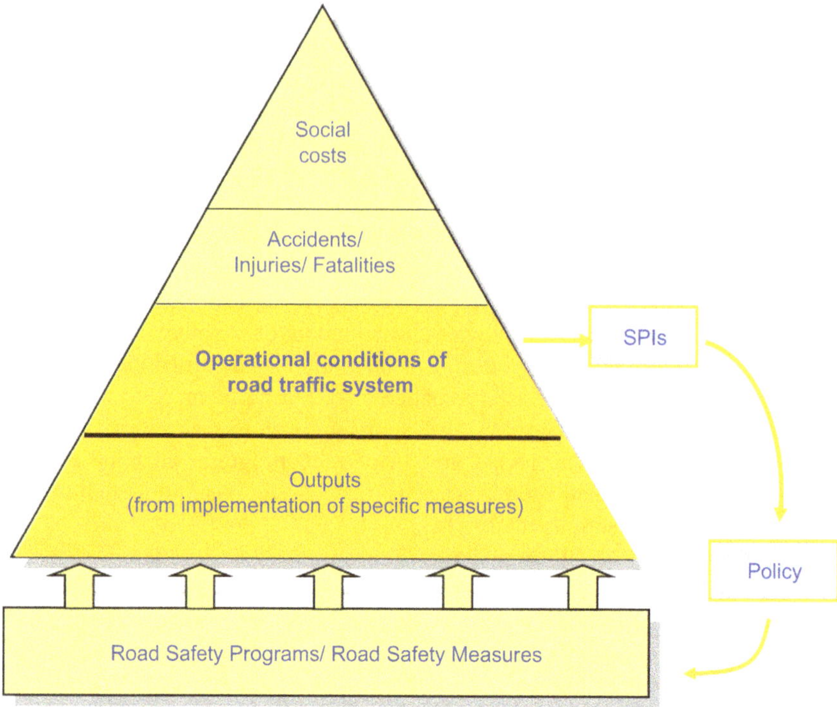

Figure 12.2 Place of SPIs in road safety management (source: Hakkert et al. [9])

than direct outputs of specific safety interventions. To demonstrate a more general character of SPIs and their independence from interventions, the layer of 'intermediate outcomes' was further divided into 'operational conditions of the road traffic system' and 'outputs' from specific measures (Figure 12.2).

Ideally, SPIs should reflect the unsafe operational conditions of the road traffic system and be sensitive to their changes, while the latter are caused by outputs of road safety programmes and interventions. For example, in the case of speeding, the unsafe operational conditions of the system (i.e. speeding) are affected by outputs from specific safety measures such as speed enforcement. The outputs of the intervention are the amount of speed cameras installed, the number of vehicles checked, speed tickets produced, etc., whereas the outcomes are in improving the operational conditions, e.g. lower level of speeding, which can be measured by SPIs. The improved operational conditions (lower speeding) will result in accident and injury reductions, whereas the whole process should reduce the social costs.

Consequently, SPIs were defined as follows [9]: 'safety performance indicators are the measures (indicators), reflecting those operational conditions of the road traffic system, which influence the system's safety performance'. The SPIs' purpose is

- to reflect the current safety conditions of a road traffic system (in the context of safety problems or safety gaps but not specific measures);

- to measure the influence of various safety interventions (as a combined effect reflected in the operational conditions);
- to compare between different road traffic systems or over-time developments of one system.

Furthermore, a common framework was suggested for SPI development across different road safety areas [9, 18]. In a certain safety area, SPIs should reflect the factors contributing to road accidents and/or injuries and characterise the scope of the problem identified. Thus, their development should begin with a definition of the problem, i.e. the operational conditions that are unsafe and lead to accidents/ injuries, and continue with a conversion of this information into measurable variables. A direct indicator is preferable but when it cannot be realised, for example, due to a lack of appropriate data, indirect indicators can be suggested or the problem can be divided into several sub-problems, with indicators established for each of those. When the measurement is possible only for certain conditions, the limitations of SPIs should be clearly stated. This way, the difference between the ideal and the realisable SPIs is recognised [9].

12.2.2 Criteria and conditions for using SPIs

Using a common framework, the SafetyNet project demonstrated a practical way for SPI development in the pre-defined safety areas [9]. In each area, first, based on a literature survey, a relationship between the area and road safety was shown, and the scope of accident/injury reduction potential associated with better system's operational conditions (e.g. lower speeding, better passive vehicle safety) was stated. Second, characteristics of the system's performance, user behaviour, etc., were analised in order to select those which can be measured and quantified. Using a literature survey, examples of SPIs in use by different bodies (countries, authorities) and/or research studies were considered. Third, the SPI concept in a specific problem area was developed, which stems from the structure of the area considered, the available experiences with measurements of similar characteristics, etc. Once the SPIs were defined, the data for estimating SPIs were collected via national experts, enabling country comparisons (e.g. [19]). The above framework actually reflected the main criteria, which should be satisfied in the SPI development, that is the proven relevance to road safety, measurability and applicability by authorities.

Additional literature [2,11,20,21] discussed further theoretical and practical criteria, which should be considered when developing or selecting SPIs. The report [1] suggested a comprehensive list of SPI criteria, which includes

- *Causal link* with road accidents and injuries, i.e. proven association between SPIs and road safety outcomes based on previous research, with no or limited impact of confounding factors.
- *Quantifiability* of the phenomenon, i.e. a possibility of capturing the phenomenon in a quantitative way that should be sufficiently sensitive (i.e. changes in the phenomenon lead to observable changes in the values of SPIs).

- *Relevance for policymaking*, which can be seen, for example, in the scale of accident/injury risks linked to the SPIs describing the phenomenon (e.g. not wearing a helmet by riders). In addition, the phenomenon should be integrable in policy measures and influenceable by interventions that can reduce road injury risks, while SPIs can be specified in the scope of target values and are easy to understand.

Furthermore, to be applicable, SPIs should meet several criteria, which refer to the methodology of data collection and processing, such as [1]

- *Representativeness* for the geographical area or period considered, which can be attained by applying random sampling for observations/measurements, with appropriate weighting factors for samples (e.g. traffic exposure).
- *Reliability* of the values, which depends on accurate initial registration of data, sufficient sample sizes, reproducible methodology of measurement and quality assurance of data collection and analysis.
- *Comparability* of the values, for example, over-time, with other geographical areas or other indicators.
- *Feasibility* of the methodology, which will preferably use measurements not restrained by legal obstacles (e.g. use of images, legislation on random breath testing); data providers and/or authorities willing to cooperate (e.g. police, local authorities, national agencies); access to all the data needed; availability of sufficient financial resources; a possibility for repetition of data collection on a regular basis in the future.

In order to meet the above criteria, appropriate sampling strategies need to be applied for SPI data collection and analysis (Section 2.4).

In current road safety policies, the ability to estimate and compare SPIs is essential both within a country (across regions) and across various countries since it allows for meaningful comparisons and benchmarking to identify areas for improvement. However, such comparison is frequently impossible due to the lack of data, insufficient quality of the data or incomparability of the data due to different measurement circumstances [14]. To facilitate international comparisons, recent European projects Baseline (www.baseline.vias.be) and Trendline (www.trendlineproject.eu) have been focused on developing common methodological guidelines for estimating SPIs in selected safety areas.

12.2.3 SPIs' subdivisions

In general, a large number of potential SPIs is possible [6]. However, not all of them are equally important. The importance of an SPI is higher when it has a stronger relationship with accident or injury occurrence and can be influenced by road safety measures. A natural starting point would be the main behavioural indicators, e.g. speeding, drink-driving and seat-belt use, which were proven to be related to road traffic injury and its severity [5,9,10]. Furthermore, quality indicators concerning other components of the road traffic system such as road networks and vehicle fleets should be added as well as the post-crash treatment of road accident victims. Thus, originally, within the SafetyNet project, SPIs were developed for seven problem areas, which are alcohol and drugs; speed; (use of) protective systems – seatbelts and child restraint systems (CRS) in cars, and safety helmets by riders; (use of) daytime running lights (DRL); vehicles (passive safety); roads and trauma management [9,18]. To note, the DRL area was included in the SPIs as it was a popular road safety measure in that period and also suggested by ETSC [6].

In all the seven areas, during the SafetyNet project, data were collected for the European countries and international comparisons of SPIs were performed [13,19]. However, the process was not easy in many cases. For example, while, for the use of protective systems (seatbelts, helmets), reasonably good data were available in many countries, in the case of alcohol and drugs' SPIs, legislative restrictions, lacking data and insufficient data quality flawed country comparisons [22]. For speed SPIs, the possibility of direct international comparisons was limited due to variability in the ways the countries conduct their surveys as well as differences in road classifications and speed limits [9]. For passive vehicle safety, trauma management and roads, new approaches were developed, and specific data collection undertaken for the SPIs' estimation. Nevertheless, it was demonstrated that the task of data collection for estimating SPIs is realisable, and meaningful country comparisons can be produced (e.g. [23,24]).

Nowadays, in Europe, eight SPI fields are defined for monitoring road safety progress [4], which include speed, safety belts' and CRS use in cars, protective equipment (helmet use) by riders, alcohol-impaired driving, drivers' distraction (due to mobile phones), vehicle safety, road infrastructure and post-crash care (response time). The newer list apparently reflects current safety concerns – termed as key performance indicators (KPIs). In general, it resembles the original list of SPI fields developed by the SafetyNet project, with *distraction* added but DRL removed from consideration.

A recent survey of current practices in (26) European countries [25] revealed that all the SPI fields are in use, with behavioural SPIs being the most common (Figure 12.3).

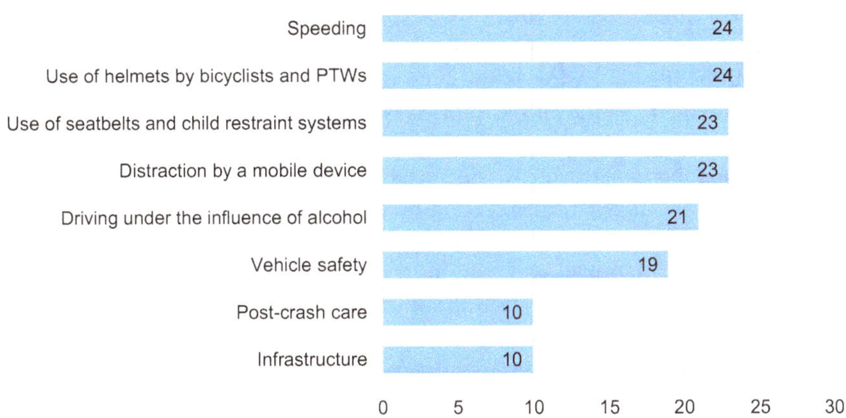

Figure 12.3 Number of European countries that measure SPIs (source: Mons et al. [25])

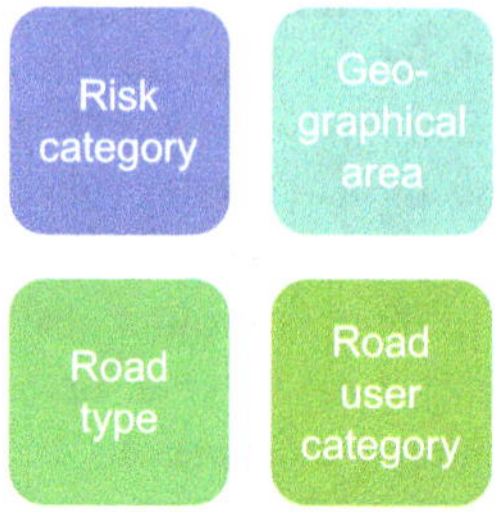

Figure 12.4 Four main dimensions of SPIs (source: EC [1])

In addition, the survey revealed that many countries use alternative formulations for SPIs than those defined as KPIs [4] and that some countries also apply SPIs in other areas, such as driving under the influence of drugs, red-light compliance, wearing of reflective materials by pedestrians or cyclists, fatigue, keeping safe distance between cars, etc.

For a general SPI classification, a four-dimensional typology was suggested [1], which considers the topic (risk category) addressed, the geographical area covered, the applicable road type and the road user category (Figure 12.4). For example, regarding speed SPIs ('risk category'), the percentage of car drivers ('road user') driving faster than the speed limit on rural roads ('road type') in France ('geographical area') can be estimated.

In general, the first dimension in Figure 12.4 refers to various SPI fields that were introduced above; the second dimension may range from local and regional to national areas (whole countries). Furthermore, SPIs that refer to risky behaviour of road users may differ between road types, while road categories and related speed limits can vary substantially between countries. Hence, when making international comparisons, broader road categories should be used [26]. Three broad road categories that are often used are urban roads (or roads in built-up areas), motorways and similar roads, and rural roads (or roads outside built-up areas).

The fourth SPI dimension is the type of road user, which enables focus on specific populations. While in the past SPIs mostly targeted car drivers, current SPIs are increasingly being deployed to refer to other types of road users, e.g. helmet-wearing by motorcyclists, distracted driving by truck drivers, seatbelt-wearing by car passengers, drink-driving by cyclists, etc.

For many SPIs, further breakdowns of the values are possible [16]. For instance, SPIs on safety-belt use may differentiate between vehicle types – cars, light trucks and heavy vehicles, and within cars between the driver, the front passenger and the rear passengers. When investigating drink-driving, it is useful to break down the data by age group and gender, as this information may help to target interventions. It may also be useful to calculate the values at sub-national (regional) levels since, in some countries, regions have considerable power to implement policy measures [1]. Conversely, SPI values, which are available at a sub-national level, may be aggregated to the national level by using appropriate weighting factors.

Other relevant SPI subdivisions may be by the time of day, the day of the week, month or season. This is justified when it is known that road safety risks vary considerably between different time periods. For example, driving under the influence of alcohol happens more frequently during weekend nights; hence, these time periods should be selected for SPI evaluation; there are fewer motorcyclists on the road during wintertime; thus, SPI-related data should preferably be collected in other seasons. Additional segmentations may help to identify the most efficient interventions. However, if it was decided to produce SPIs for a variety of time periods and road user groups, it may come at a price since a considerably larger and more complex data collection process can be required.

12.2.4 Principles for data collection and analysis for estimating SPIs

Recognising that a prerequisite for using SPIs lies in sufficient data quality while the data should also be collected in a harmonised manner to enable monitoring and comparisons, the processes of data collection and analysis for estimating SPIs should follow common methodological guidelines. The first SPI Manual was developed in the SafetyNet project [18]. This report summarised existing practices for the underlying data measurements, provided good practice examples for SPI evaluation and detailed the procedures, which are necessary to collect and process the required data for the SPIs' estimation on a national level. In particular, when setting up a survey for estimating behaviour SPIs, the issues detailed in the Manual concerned a definition of the survey population, a list of SPIs to be estimated, methods of data collection (survey instrument), sampling units and sampling design, for producing nationally representative SPIs.

In Europe, a recent road safety policy document [4] pointed to a number of methodological considerations to be applied when estimating country SPIs, such as

- *Geographical coverage* – the indicator should be representative of the whole territory.
- *Sampling* – a need for common methodology, with a random selection of units for measurement, pre-defined sample sizes and aggregation method aiming at weighting the results by traffic exposure.
- *Relationship with traffic rules* – to present SPIs in the appropriate local context, for example, regarding a blood alcohol content limit or the obligatory use of bicycle helmets.

The Baseline project conducted further development of a common international methodology for road safety SPIs [27], when, in each SPI area, it covered topics such as

- definition of the SPIs concerned;
- methods to be used for data collection;

- breakdowns requested of the SPI values – by road category, vehicle type, day of week, etc.;
- minimum sample of observations and observation locations required;
- methods for data processing and weighting;
- nature and format of data to be reported.

Methodological guidelines of all 8 SPIs (called KPIs in Baseline) are accessible from the Baseline website. Furthermore, the Trendline project brought together 25 European Union (EU) Member States and 4 other European countries as observers, for data collection, analysis and delivery of road safety SPIs and their use within road safety policies. The project is aimed at providing adaptations of the methodological guidelines, considering the experience gained. The updated methodological guidelines for all SPIs/KPIs can be downloaded from the Trendline website.

Below, we briefly present general principles on data collection and analysis, to be applied for estimating SPIs.

Data sources and data collection methods

Different data sources and data collection methods can be envisaged for estimating SPIs [1,14,16,28]. The common tools are

- *Data collection through roadside surveys, with observations made and coded by trained observers.* This is the most used method for data collection for behavioural SPIs, e.g. seatbelts' and CRS use, driver distraction and helmet use. In addition, it can be used for measuring speeds of selected vehicle types (vans, motorcycles, buses), particularly in densely urban traffic. The main advantages of roadside observations are objectivity and representativeness [29]. At the same time, they have limitations in the number and nature of variables that can be observed, provide little information about road users, require a sophisticated study design and protocol, and can be relatively expensive. Inadequate sampling design can be the cause of bias in the results. Regarding the use of child restraint systems, an in-car inspection is the 'golden standard', but it requires specific locations for the observations. The advantage of this method is that additional questions can be asked to the vehicle occupants, hence enriching the data.
- *Data collection through roadside surveys, with observations made by cameras or radars along the road.* In this case, large data samples can be collected, with semi- or fully-automated data processing, yet, when more complex behaviours are counted, the data coding is done afterwards by human operators. Speeding SPIs can be based on roadside observations using speed-detection equipment such as radars, lasers or cameras; they can also be made using roadside sensors such as Automatic Traffic Counters. Video detection followed by processing using AI technologies is becoming popular for local measurements of distraction by phone, seatbelt use (of front passengers) and helmet use, but not yet at national levels involving many observation locations.

- *Data collection through induction loops under the roads.* Such measurements are typically used for measuring traffic flow, but can also be used for speed SPIs. On cycle paths, induction loops can be used for measuring flow and speed of cyclists. A drawback of this method is that the locations of the induction loops, in general, are not compatible with the requirement of random sampling and may not fit free-flow conditions.
- *Data collection by the police when participating in the required survey design (with random sampling).* This method is currently used for SPIs on drink-driving in certain countries; it is often considered the 'golden standard'. Data for SPIs on drug driving can also be collected through a similar process. In general, the management and the logistics of this approach are considerable. The method is only applicable in countries where the police have the right to stop a vehicle without a specific reason and take a breath analyser test, even if there are no signs that the driver is drunk. An advantage of the procedure is that the police can ask additional questions (e.g. on the purpose of the route) and can gather personal data on the vehicle occupants (gender, age, etc.).
- *Data generated by in-vehicle sensors, tracing and GPS systems, or mobile phones of road users, including cyclists and pedestrians.* The analysis of such data makes it possible to identify unsafe road sections and risky behaviour on the road, most notably speeding and strong acceleration or deceleration. By using geospatial analysis techniques, such data can be used to systematically map road system risks, which eventually can become the foundation for the formulation of SPIs. The advantages of such an approach are the potentially high number of observations, access to historic data without the need to commission specific surveys, repeatable and reliable methodologies and access to data for a wide area of the road network that can be subject to random sampling. Challenges and drawbacks include the representativeness of the data within the vehicle fleet, the lack of a built-in free-flow speed measure, and availability in countries with small fleets of connected vehicles. Furthermore, data can be obtained through a large number of suppliers, and each will have their own strengths and weaknesses. Examples of studies with speed SPIs based on floating car data are [30–32].

All data sources and data collection methods have advantages and disadvantages. In all cases, care must be taken to identify a representative sample of locations, and to ensure consistency of data collection over time in order to be able to identify meaningful trends. Research is also needed to interpret data from these sources reliably, with attention paid to country-specific regulations on the use of personal data [33].

Sampling approach – basic requirements

SPIs for a particular behaviour, e.g. speeding, wearing a helmet, driving under the influence of alcohol, are usually expressed as the percentage of road users or of traffic volume that is behaving safely or respecting the regulation. For example, '97% of drivers of passenger cars are wearing a seatbelt'. Conversely, the percentage of road users *not* behaving safely can be indicated, e.g. '30% of vehicle drivers travel over the speed limit'.

In general, it is impossible to measure the performance of all vehicles and drivers at all times, nor would that be desirable or appropriate. In practice, SPI values are actually estimates based on a representative sample of vehicles and/or road users observed. In terms of sampling, this means that the statistical population to be considered is the total traffic volume of moving vehicles (typically expressed in kilometres driven) over a certain area (i.e. country or region) and a certain period of time (e.g. one year). Estimates are made by sampling individual vehicles (or road users) at particular locations and moments in time [34].

To estimate a national SPI, a representative sample of locations should be selected on the road network, where the behaviour of road users is observed. The sample design depends on the subdivisions of traffic exposure according to areas or road types and may also consider other factors that subdivide the population observed. For example, when seatbelt use' SPIs are estimated, the measurements are typically classified according to main road types and population groups. For survey design, a stratified sample can be applied in two stages: (1) an initial stratification according to urban/rural areas (or country regions), based on a subdivision of traffic exposure; (2) further stratification using additional characteristics, for example, road types, for rural area, and town types, for urban area. In the latter case, population size can serve as a substitute for traffic exposure subdivision.

At the end of the survey, the results of all observation sessions are weighted in order to calculate the overall SPI; the weighting factors strictly depend on the sample design. Methodologies for sampling and weighting the data to evaluate national SPIs were originally developed in the SPI Manual [18] and further elaborated in the Baseline and Trendline projects.

The amount of sites required for the survey and the number of observations per site depend on the statistical reasons, e.g. accuracy/sampling error and confidence level required [18] and cost/logistic considerations of the survey. As to the absolute minimum sample size for behavioural measurements, the Baseline and Trendline projects typically recommend a minimum of 2000 observations and at least 500 for each road category (i.e. urban, rural, motorway) [35]. Per stratum considered (a combination of a road category and a vehicle type), there should be at least ten observation locations selected via random sampling.

It should be mentioned that there is no statistical rationale for determining the required minimum sample size and the minimum number of locations to ensure the representativeness of the observations for an entire country. This mainly depends on the variance between locations within a country. The minimum requirements in the Baseline were set in order to make SPI data collection affordable, in particular for countries with little or no experience on the topic. Most countries participating in the Baseline and Trendline projects have exceeded the minimum requirements. In particular, sample sizes were much higher for speed measurements. Yet, there were also cases where it proved difficult to meet the minimum criteria, for example, for countries with a limited motorway network or observations related to cyclists and motorcyclists. More details on the actual sample sizes can be found in Van den Broek *et al.* [36] for speeding and in Yannis and Folla [37] for helmet use.

In general, the larger the sample size, the higher the number of observation locations, and the lower the variation of the observations, the smaller the confidence interval and the more precise the SPI estimation will be. Furthermore, in such cases, the differences between two SPI values are less likely due to chance or measurement error and more likely to represent genuine differences or trends in road safety performance. A smaller confidence interval provides more robust evidence that the observed differences in SPI values are meaningful and not simply the result of random fluctuations [14].

The selection of the month(s) when the observations are carried out is important as well. Periods with 'normal' or 'average' traffic are to be preferred. In many countries, therefore, data collection for behavioural measurements takes place in spring or autumn, outside school holidays. Some risky behaviours, such as driving under the influence of alcohol or drugs, tend to be more prevalent at night and, thus, observations should be conducted at night, despite lower traffic volumes [1].

Data analysis–need for weighting

For most behavioural SPIs, data are being collected during observations at different locations, which are selected in accordance with the sample design. It is common practice to define a number of strata, which could be based on one variable (e.g. road type) or a combination of variables (e.g. road type × vehicle type × time period). In general, as suggested in Baseline and Trendline [34], a stratum with less than 500 data points should not be considered for estimating separate SPIs, because the number of observations and/or observation locations is small, and thus, confidence intervals of SPI values will be too wide. Such strata can be combined with others.

As mentioned above, at the end of the survey, observations from various locations have to be weighed to evaluate the final SPI, e.g. the overall percentage of road users with a certain behaviour in the total population. The weighting procedure follows the sample design and applies weighting factors, which reflect the relative share of traffic exposure at the survey locations and the subdivisions of the survey population according to other characteristics, which were used to define the sample design. Typically, first, the data of each stratum are processed and then, aggregation of the SPI results of different strata is performed. Further details on conducting the calculations for estimating SPIs can be found in [34] and Hakkert and Gitelman [18].

12.3 Current practices on the use of SPIs

12.3.1 International recommendations

Today, SPIs appear in guiding documents on road safety promotion, which are published by leading international bodies. In particular, the Global Plan for the Decade of Action for Road Safety [5] includes the United Nations' (UN) voluntary global targets, which actually include SPIs in the fields of road infrastructure,

Table 12.1 The UN voluntary global targets on road safety as examples of SPIs

Area	Target formulation
Road infrastructure	• All new roads achieve technical standards for all road users that take into account road safety or meet a three-star rating or better. • More than 75% of travel on existing roads is on roads that meet technical standards for all road users that take into account road safety.
Vehicle safety	• 100% of new and used vehicles meet high-quality safety standards, such as the recommended priority UN Regulations, Global Technical Regulations or equivalent recognised national performance requirements.
Risky road user behaviours	• Speeding halves the proportion of vehicles travelling over the posted speed limit and achieve a reduction in speed-related injuries and fatalities. • Motorcycle helmets increase the proportion of motorcycle riders correctly using standard helmets to close to 100%. • Vehicle occupant protection increases the proportion of motor vehicle occupants using safety belts or standard child restraint systems to close to 100%. • Driving under the influence halves the number of road traffic injuries and fatalities related to drivers using alcohol and/or achieve a reduction in those related to other psychoactive substances.
Post-crash care	• All countries establish and achieve national targets in order to minimize the time interval between a road traffic crash and the provision of first professional emergency care.

vehicle safety, risky road user behaviours and post-crash care (Table 12.1). The target formulation outlines the way how the SPIs should be evaluated.

The global targets can be translated into a set of national indicators, as suggested in Van den Berghe *et al.* [16]. For example, regarding vehicle occupant protection, such SPIs may include % of motor vehicle drivers correctly wearing a safety belt, % of motor vehicle passengers correctly wearing a safety belt and % of children correctly fastened in CRS (or safety belts, from the appropriate age, weight and height).

Similarly, the EU road safety policy framework [4] encourages Member States to apply SPIs in eight safety fields, using SPI definitions as presented in Table 12.2. The work also briefly describes methodologies for the SPIs' estimation, in each field, some of them were further elaborated by the Baseline project. Further development of harmonised methodologies for SPI measurements has been conducted in the Trendline project. In addition, some complementary indicators are planned to be tested, such as

- Driving under the influence of drugs, helmet wearing by riders of personal mobility devices (e.g. e-scooters), alternative speeding indicators (e.g. speed variation) – as an extension of existing SPIs;

Table 12.2 SPIs recommended for use in the EU countries

Safety field	Definition of SPIs
Speed	Percentage of vehicles travelling within the speed limit
Safety belt	Percentage of vehicle occupants using the safety belt or child restraint system correctly
Protective equipment	Percentage of riders of powered two wheelers and bicycles wearing a protective helmet
Alcohol	Percentage of drivers driving within the legal limit for blood alcohol content (BAC)
Distraction	Percentage of drivers NOT using a handheld mobile device
Vehicle safety	Percentage of new passenger cars with a Euro NCAP safety rating equal or above a predefined threshold
Infrastructure	Percentage of distance driven over roads with a safety rating above an agreed threshold
Post-crash care	Time elapsed in minutes and seconds between the emergency call following a collision resulting in personal injury and the arrival at the scene of the collision of the emergency services

- Share of 30 km/h road lengths in urban areas, use of lights by cyclists in the dark, red-light negations by road users, compliance with traffic rules at intersections – as new SPIs.

Furthermore, indicators based on self-reported risky behaviour and attitudes towards risky behaviours will be examined as possible alternatives to traditional SPIs. Guidelines for these new indicators are available on the Trendline website.

One can expect that in the future, some of the new indicators will be incorporated in the recommended set of European SPIs.

12.3.2 Good practice in using SPIs

SPIs as part of national road safety policies and plans

Merely measuring SPIs is not sufficient for improving road safety. It is important to define objectives for the reduction in road trauma with regard to improvements in contributing factors, such as road user behaviours, vehicle quality, post-crash response, etc., as part of road safety plans [16]. Developing SPIs involves identifying and prioritising policy areas, selecting relevant indicators, continuously monitoring and evaluating progress, and aligning selected indicators with national road safety strategies [14].

The Baseline project revealed that all participating EU Member States include SPIs, in various ways, in their national road safety plans [27]. The consequent Trendline project strengthens the focus on the integration of SPIs within national policies. Detailed insights on incorporating SPIs in the policy process in different countries can be found in Mons *et al.* [25]. The survey showed that the vast majority of countries (24 of 26) use SPIs to monitor road safety performance and/or progress. Other aims that were mentioned are identifying and prioritising problems and seeking solutions, making information available for road authorities, target setting, providing background information for road safety activities, being able to make evidence-based decisions, evaluating the effectiveness of measures/policy and benchmarking/comparing with other countries.

As emphasised by a recent summary report [14], SPIs' use requires targets – quantifiable and specific goals outlining the desired level of safety performance within a certain time-frame (e.g. 1, 5 or 10 years). Setting a time-frame helps establish a sense of urgency and accountability in achieving the goal. Targets also provide a reference point for assessing progress in improving road-safety conditions. At the same time, SPI targets should be feasible. Unrealistic targets undermine the credibility of the entire process, while realistic targets motivate individuals and organisations to work towards achieving them [14]. The Trendline survey revealed [25] that only seven (of 26) countries had set targets for the EU-recommended SPIs on a national level. The remaining 19 countries have not (yet) set such targets but ten of them intend to do so. How to set appropriate targets is obviously challenging and many European countries would welcome support in doing so. Examining the experiences and practices of other countries is valuable when developing SPIs [14]. Shared focus areas across countries can provide insights and lessons that local authorities can adapt to their specific contexts.

Once targets are formulated for different time horizons, regular monitoring of the progress is important. Adopting periodic monitoring and evaluation processes for SPIs that help identify changes over time allows for tracking progress towards the targets, assessing the effectiveness of implemented measures, and making necessary adjustments to improve road safety performance. Analysing SPI trends will support the decisions on which actions are appropriate [14].

SPIs at the sub-national level

Although the idea of SPIs is generally applicable to sub-national – regional or local level, some national indicators may not be the most appropriate ones to use at the regional or local level [28,38]. One of the reasons is that local authorities need information on specific road safety issues on which they have the power to act. This often requires choosing SPIs and specific methodologies that are unique to the local conditions.

Furthermore, SPIs used at the sub-national level rely on the characteristics of the local road network and the data sources that are locally available and thus become country-specific. Examples of that can be seen in infrastructure-related SPIs – e.g. Deepdarshan and Kushnappa [39] and Viera Gomes *et al.* [40], which are useful within a country but cannot be extrapolated for international benchmarking.

The survey from Mons *et al.* [25] showed that only eight (of 26) European countries had examples of SPIs' use on the regional and/or local level. Among those, 'speeding' was the most common topic (reported in four countries), followed by infrastructure SPIs (in three countries), helmet use (in two), use of seatbelts and CRS (in two) and distraction (in one country). Additional SPIs in use (in two countries) referred to behaviours at pedestrian crossings (especially near schools) and cyclists' behaviours and equipment. One of the countries that measures speed

SPIs reported a difference in estimating SPIs at the regional vs. national level, which refers to the units of consideration: on the national level, those are the distance driven on roads or incoming traffic volumes at intersections, while on the regional/local level, it tends to be road length or the number of intersections of a specific type.

An important consideration when using SPIs at a sub-national level (e.g. a province) is that the sample used for the estimation is still large enough and representative of that area. It can be misleading to communicate and base countermeasures on SPI values that rely on a small data sample or a limited number of observation locations.

When SPIs are measured in the local context, it is also important that road safety targets in terms of SPIs are embedded within wider policy objectives [41].

12.3.3 Examples of SPI evaluations

When sampling strategies and data collection and analysis methods for SPIs are similar, it is possible to compare results from different countries. Figure 12.5 provides an example of country comparison by SPIs on the use of seatbelts in cars, which fit such conditions. The data were collected using a common methodology in the Baseline project for the EU Member States [42] and in the Westbelt project for Western Balkans countries [43]. The values in Figure 12.5 are the percentages of drivers and rear occupants wearing a seatbelt (at weekdays, during daytime). One can observe that in 10 out of the 14 EU countries, the seatbelt-wearing rate of drivers is above 90%, and some are above 95%. In contrast, in the Western Balkans region, only Serbia reaches the value of 90%. In all countries, the percentage of rear occupants using a seatbelt is lower (and often much lower) than that of drivers.

When collecting data for SPIs at regular intervals, it becomes possible to examine trends and assess whether policy measures have been effective. In Belgium, for more than two decades, data were collected for a range of SPIs, by Vias Institute (previously called the Belgian Road Safety Institute). Figure 12.6 shows an example of the over-time trend for one SPI – the percentage of car drivers that were driving with a blood alcohol concentration over the legal limit (0.5 g/l) between years 2003 and 2018 (based on [44]).

Despite the considerable number of drivers controlled – typically, eight to ten thousand in each observation year, the confidence intervals of the SPI values are relatively wide, leading to non-significant differences between the years. Hence, on the basis of the SPI values in Figure 12.6, one cannot state that a significant downward trend was observed over the years. Such results underline the need to have large samples in the SPI data collection; yet, this may not be feasible due to logistic and financial reasons.

12.3.4 SPIs as predictors of road safety outcomes

The causal link between risky behaviour on the road such as speeding, driving under the influence of alcohol, red-light negation, not wearing a seatbelt, distracted driving, etc. – and road injury risks has been amply documented [5,6,9,10]. In

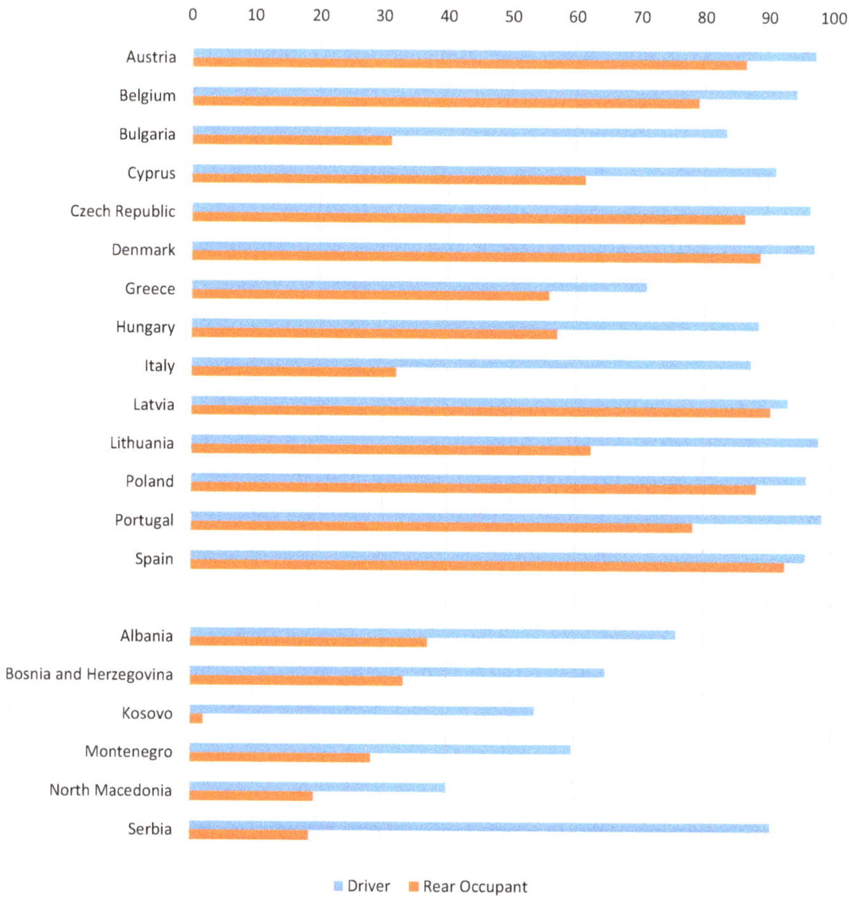

Figure 12.5 Country comparison by SPIs on the use of seatbelts in passenger cars (%): in EU Member States (in 2022) and the Western Balkans region (in 2023).

general, the more risky the behaviour of a road user, the higher the risk of road injury or fatality. Similarly, unsafe road conditions and worse vehicle quality would result in higher road injury. Those were the underlying assumptions in the development of SPIs (Section 1-2). Hence, one would expect that better SPI values in a country, when compared with another country, would be associated with better safety outcomes. Some research studies tried to explore this relationship, but the results were not always consistent.

For example, Holló *et al.* [45] examined the explanatory value of SPIs when considering over-time changes in road safety outcomes (fatalities and injuries) in several Central European countries, while the SPIs referred to the fields of seatbelt use, speeds and alcohol-impaired driving. Based on the numeric analyses shown, Holló *et al.* [45] concluded that SPIs only cannot provide a full understanding of

Figure 12.6 Over-time trend in the prevalence of driving under the influence of alcohol in Belgium.

road safety trends and can even be misleading if applied without the required background information such as changes in legislation, traffic exposure and socio-economic climate in the country.

Tingvall *et al.* [46] analysed linearity between SPIs' values and final safety outcomes, while considering over-time changes in seatbelt use and road fatalities in Sweden. The assumption of linearity was partly rejected, and the study concluded that a single SPI may not be a good predictor of safety outcomes. However, an additional analysis showed that 100% fulfillment of a set of SPIs, related to sober driving, non-excessive speed, seatbelt use and presence of divided roads on the road network, could lead to very few fatalities, demonstrating the importance of handling SPIs simultaneously.

Following similar lines, Van den Berghe and Christie [47] fitted a linear model to reflect a relation between a speed indicator – percentage of drivers that exceed speed limits in rural areas, and a safety outcome – fatality rate (per population) in 60 countries. Yet, unlike traditional SPIs, which are based on observations, self-reported speed behaviours were applied. The result showed (Figure 12.7) that the more drivers tend to speed in a country, the lower the fatality rate, which may appear to contradict the expectation and the current knowledge on the role of speeding in accident occurrences and injury severity (e.g. [10]).

The explanations suggested by this finding pointed to differences among the countries (which comprised high-, medium- and low-income countries together) in the quality of road infrastructure, the amount of traffic on the roads, the technical state of the vehicle fleet and the speed limits, whereas uncontrolled mixing of various conditions provided a misleading result. To note, this phenomenon was reported by previous research, e.g. Elvik [48], which demonstrated that mixing data from several road types (i.e. various conditions) would provide a negative relation between the increase in speed and accident rates (i.e. higher speed is associated

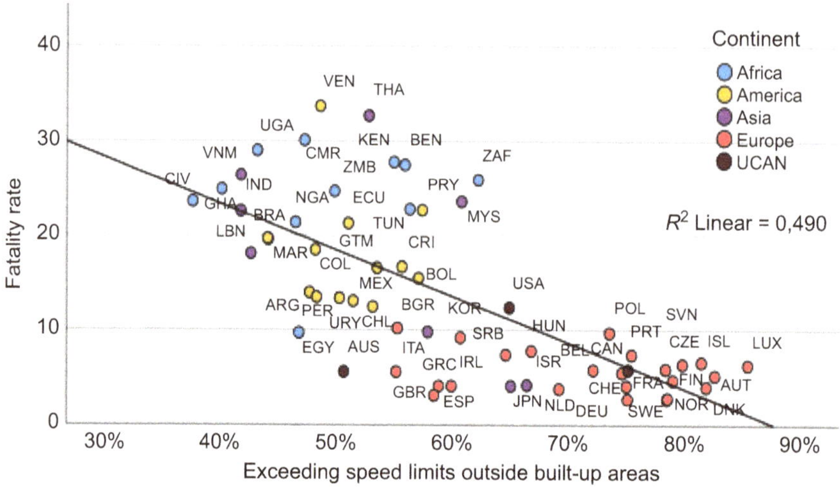

Figure 12.7 Accident fatality rate vs. the percentage of car drivers exceeding the speed limit in rural areas in 60 countries (source: Van den Berghe and Christie [47])

with fewer accidents), whereas a true relationship can be revealed only when the functions are fitted to homogeneous conditions.

Papadimitriou and Yannis [49] examined statistical relationships between various layers of the road safety pyramid (Figure 12.1) and road safety management based on data collected from 30 European countries. The SPI layer was presented by a composite index based on a weighted score of eight indicators from the fields of alcohol-impaired driving, seatbelt wearing and the quality of vehicle fleet. To explore the relationships multivariate regressions were applied. The results showed that increased SPI scores were related to lower fatality rates in the countries and that the (composite) indicator of road safety outcomes was mainly predicted by the layers of background country characteristics and SPIs. This suggests that a more thorough analysis conducted under homogeneous conditions may demonstrate a reasonable association between better SPIs and better safety outcomes in countries.

In general, it should be expected that the relationship between SPI values and road safety outcomes is complex since road accidents and injury are caused by a range of factors, while the context and the relative contribution of various factors may differ both within and between countries.

12.4 Discussion: challenges and future needs

Safety performance indicators are an emerging research area aiming at better measuring and monitoring the operational conditions of the road traffic system with a particular focus on factors contributing to accident occurrences and road injury [1,9,14]. SPIs' use provides wide opportunities for better planning and implementation of road safety policies and interventions, supporting a continued improvement of road safety conditions on the national (or sub-national) level. However, for the SPIs to be useful and effective, systematic data collection and processing should be conducted, following common methodological rules [9,34].

The use of SPIs in road safety management is not uniform yet among European countries (e.g. [25]) and ever lower outside Europe. However, evidence exists that many countries with better safety performance have a long tradition of using SPIs, at least in selected safety areas. Not surprisingly, leading international bodies encourage using SPIs in continuous road safety work and, particularly, in the framework of road safety programmes [4,5].

As with any evolving field, SPIs' development has a number of challenges both for research and for application. In relation to the *SPI data collection* and deployment of SPI methodologies, among issues needing attention can be mentioned [1,14]:

- The underlying dataset for the SPI may not be reliable or representative, which can lead to a bias in the SPIs.
- The data collection and analysis can be time-consuming and costly, leading to a lower sample size than what is desirable for sufficient accuracy and low confidence intervals.
- It can be challenging to define an appropriate sampling strategy and weighting approach (e.g. because of lack of traffic data).
- The sensitivity and accuracy of the SPIs may be low, making it difficult to detect real trends in a timely manner.
- It may not be possible to create desired segmentations for SPI estimation (e.g. by age group).

Concerning *the use of SPIs*, it is possible that the main purpose of SPIs – monitoring road safety performance with a view to undertake adequate measures – cannot be achieved. Issues needing attention in this context are [1,14,28]:

- There may be disagreements between stakeholders on the selection of SPIs and/or on the value and nature of the targets set.
- There is insufficient integration of the SPIs in road safety policy strategies and plans.
- The frequency of SPI measurements is not aligned with the policy monitoring cycle.
- SPIs cannot be used to identify possible countermeasures to be implemented.
- Topics for which no SPIs are available may be overlooked when deciding on policy measures and safety interventions.
- Authorities might be reluctant to adopt SPIs because they are held accountable for performance and for actions to improve it.
- The set of SPIs may insufficiently address evolving needs and new developments.
- Responsibilities for data collection, data analysis, policy integration and/or dissemination may be unclear.

Given the growing use of SPIs within road safety policies and current trends in mobility and technological developments, *future research* would be useful with regard to the following topics:

- The theoretical framework of SPIs, in particular, the relationships between risky behaviours, SPIs measuring and road safety outcomes.
- How SPIs can be used for implementing the Safe System approach.
- The potential and limitations of new technologies (e.g. floating car data, automated video analysis) and alternative data collection methods (e.g. self-reporting) for producing applicable SPIs, possibly at lower costs.
- The identification of preferable SPIs in relation to vulnerable road users, e.g. pedestrians, cyclists, mopeds, older people.
- The development of SPIs for new phenomena, e.g. micro-mobility, vehicle automation.
- The advantages and disadvantages of possible alternatives for estimating particular SPIs in terms of accuracy, cost and efficiency.
- The requirements for international comparability of SPIs and for international databases allow reliable country comparisons.

References

[1] European Commission. (2022). *Road Safety. Thematic Report – Road Safety Performance Indicators (RSPIs). European Road Safety Observatory*. European Commission, Directorate General for Transport. https://road-safety.transport.ec.europa.eu/system/files/2022-10/Road_Safety_Thematic_Report-Road_Safety_Performance_Indicators.pdf.

[2] Wegman, F., and Oppe, S. (2010). Benchmarking road safety performances of countries. *Safety Science*, 48(9), 1203–1211. https://doi.org/10.1016/j.ssci.2010.02.003.

[3] Eurostat. (2020). Sustainable Development in the European Union. Monitoring report on progress towards the SDGs in an EU context. 2020 edition. In *EUROSTAT. Statistical Books*. https://doi.org/10.2785/76551.

[4] European Commission. (2020). *Next Steps Towards 'Vision Zero' – EU Road Safety Policy Framework 2021–2030*. Publications Office. https://doi.org/doi/10.2832/391271.

[5] WHO. (2021). *Global Plan Decade of Action for Road Safety 2021–2030*. World Health Organization. www.who.int/publications/m/item/global-plan-for-the-decade-of-action-for-road-safety-2021-2030.

[6] ETSC. (2001). *Transport Safety Performance Indicators*. European Transport Safety Council. https://etsc.eu/wp-content/uploads/Transport-safety-performance-indicators.pdf.

[7] LTSA. (2000). *Road Safety Strategy 2010. A Consultation Document*. National Road Safety Committee. Land Transport Safety Authority.

[8] Koornstra, M. J., Lynam, D. A., Nilsson, G., *et al.* (2002). *SUNflower: A Comparative Study of the Development of Road Safety in Sweden, the United Kingdom, and the Netherlands*. SWOV.

[9] Hakkert, A. S., Gitelman, V., and Vis, M. A. (2007). Road Safety Performance Indicators: Theory. *Deliverable D3.6 of the EU FP6 Project SafetyNet*, 164. http://erso.swov.nl/safetynet/fixed/WP3/sn_wp3_d3p6_spi_theory.pdf.

[10] ITF. (2016). Zero Road Deaths and Serious Injuries. Leading a Paradigm Shift to a Safe System. In *Zero Road Deaths and Serious Injuries*. OECD Publishing. https://doi.org/10.1787/9789282108055-en.

[11] Funk, W., Orlowski, B., Braun, K., and Rücker, J. (2021). *Safety Performance Indicators im Straßenverkehr. Überblick und Erfahrungen aus der internationalen Praxis*. Berichte Der Bundesanstalt Für Strassenwesen.

[12] Muhlrad, N., Vallet, G., Butler, I., *et al.* (2016). Analysis of road safety management systems in Europe. In *Traffic Safety*. https://doi.org/10.1002/9781119307853.ch1.

[13] Vis, M., Weijermars, W., and Hakkert, S. (2014). Development of road safety performance indicators for the European countries. *Advances in Social Sciences Research Journal*, 1(4), 138–158. https://doi.org/10.14738/assrj.14.302.

[14] Sohu, V., and Zhang, X. (2023). *Using Safety Performance Indicators to Improve Road Safety. The case of Korea Case. ITF Case-Specific Policy Analysis Report No 126*. OECD Publishing. www.itf-oecd.org.

[15] Larsson, P. (2021). Management by objectives. *Baseline General Assembly 25th of November 2021*. https://www.baseline.vias.be/storage/minisites/experience-of-sweden-in-the-use-of-kpis-for-monitoring.pdf.

[16] Van den Berghe, W., Fleiter, J. J., and Cliff, D. (2020). *Towards the 12 Voluntary Global Targets for Road Safety. Guidance for Countries on Activities and Measures to Achieve the Voluntary Global Road Safety Performance Targets*. Brussels: Vias Institute and Genève: Global Road Safety Partnership. https://www.grsproadsafety.org/wp-content/uploads/Towards-the-12-Voluntary-Global-Targets-for-Road-Safety.pdf.

[17] Thomas, P., and Breen, J. (2022). Road safety strategies and safety performance indicators. *Developing and Measuring Safety Performance Indicators at Sub-National Level, Round Table 7 February 2022, February, 12*.

[18] Hakkert, A. S., and Gitelman, V. (Eds.). (2007). *Road Safety Performance Indicators: Manual. Deliverable D3.8 of the EU FP6 project SafetyNet*. http://www.dacota-project.eu/Links/erso/safetynet/fixed/WP3/sn_wp3_d3p8_spi_manual.pdf.

[19] Vis, M. A., and Eksler, V. (Eds.). (2008). *Road Safety Performance Indicators: Updated Country Comparisons. Deliverable D3.11a of the EU FP6 project SafetyNet*.

[20] Fosdick, T., Owen, R., and Staton, M. (2022). Developing and measuring safety performance indicators at sub-national level – Roundtable one.

Developing and Measuring Safety Performance Indicators at Sub-National Level – Roundtable One, 7th February 2022, 1–23.

[21] Van den Berghe, W. (2020). Using safety performance indicators for monitoring and improving road safety. In Juris Kreicbergs (Ed.), *Vision Zero for Sustainable Road Safety in the Baltic Sea Region, 2–3 December 2020 (Riga – online)* (p. 23). Riga Technical University.

[22] Assum, T., and Sørensen, M. (2010). Safety performance indicator for alcohol in road accidents-International comparison, validity and data quality. *Accident Analysis and Prevention*, 42(2), 595–603. https://doi.org/10.1016/j.aap.2009.10.005.

[23] Gitelman, V., Auerbach, K., and Doveh, E. (2013). Development of road safety performance indicators for trauma management in Europe. *Accident Analysis and Prevention*, 60, 412–423. https://doi.org/10.1016/j.aap.2012.08.006.

[24] Yannis, G., Weijermars, W., Gitelman, V., *et al.* (2013). Road safety performance indicators for the interurban road network. *Accident Analysis and Prevention*, 60, 384–395. https://doi.org/10.1016/j.aap.2012.11.012.

[25] Mons, C., Aarts, L., Van den Berghe, W., *et al.* (2023). The use of KPIs in the policy process. *Results of the first Trendline Policy Integration Questionnaire. Draft 2.2. Report produced as part of the Trendline project.*.

[26] Boets, S., Schumacher, M., Stelling, A., Jankowska-Karpa, D., and Pavlou, D. (2021). *Methodological Guidelines – KPI Distraction*. Brussels: Vias Institute.

[27] Baseline Project Team. (2023). *Baseline Policy Report. Conclusions and Recommendations*. https://www.esranet.eu/.

[28] Owen, R., Fosdick, T., Scott, S., and Staton, M. (2022). *Developing and Measuring Safety Performance Indicators at Sub-National Level Roundtables Summary Report*. May, 1–24.

[29] Vollrath, M., Schumacher, M., Boets, S., and Meesmann, U. (2019). *Guidelines for Assessing the Prevalence of Mobile Phone Use in Traffic* (Issue November). FERSI.

[30] Ambros, J., Usami, D. S., and Valentová, V. (2023). Developing speed-related safety performance indicators from floating car data. *IET Intelligent Transport Systems*, 17(3). https://doi.org/10.1049/itr2.12281.

[31] Bekhor, G., Lotan, T., Gitelman, V., and Morik, S. (2013). Free-flow travel speed analysis and monitoring at the national level using global positioning system measurements. *Journal of Transportation Engineering*, 139(12).

[32] Diependaele, K., Riguelle, F., and Temmerman, P. (2016). Speed behavior indicators based on floating car data: results of a pilot study in Belgium. *Transportation Research Procedia*, 14. https://doi.org/10.1016/j.trpro.2016.05.111.

[33] ITF. (2022). *The Safe System Approach in Action. ITF Research Report*. OECD. https://doi.org/10.1787/ad5d82f0-en.

[34] Bijleveld, F., Boets, S., Van den Berghe, W., and Elgner, J. (2024). Suggested approach for weighting sample data and calculation of statistics within Trendline. Document produced in the framework of the Trendline project.

[35] Moreau, N., Boudry, E., Zielinska, A., Ferrer López, S., and João Da Silva Barros, M. (2021). Methodological guidelines – KPI helmet use of cyclists and powered two-wheelers (PTWs). *Baseline Project*. https://www.baseline.vias.be/en/publications/methodological-guidelines-kpi/.

[36] Van den Broek, B., Aarts, L., and Silverans, P. (2023). Baseline report on the KPI speeding. *Baseline Project*. https://www.baseline.vias.be/en/publications/kpi-reports/.

[37] Yannis, G., and Folla, K. (2022). Baseline report on the KPI helmet use among cyclists and powered two-wheelers (PTWs). *Baseline Project*. https://www.baseline.vias.be/en/publications/kpi-reports/.

[38] Pešić, D., and Pešić, A. (2020). Monitoring of road safety performance indicators-current situation and trends in the Republic of Serbia. *Transportation Research Procedia*, 45(2019), 70–77. https://doi.org/10.1016/j.trpro.2020.02.064.

[39] Deepdarshan, K. P., and Kushnappa, B. K. (2020). Methodology to develop road safety index and evaluation for third world countries. *7*. www.jetir.org781.

[40] Viera Gomes, S., Cardoso, J. L., and Azevedo, C. L. (2018). Portuguese mainland road network safety performance indicator. *Case Studies on Transport Policy*, 6(3), 416–422. https://doi.org/10.1016/j.cstp.2017.10.006.

[41] Santacreu, A., and Samsonova, T. (2019). *Road Safety in European Cities – Performance Indicators and Governance Solutions* (International Transport Forum Policy Papers, No. 67). International Transport Forum.

[42] Van den Broek, B., Aarts, L., and Silverans, P. (2022). Baseline report on the KPI safety belt and child restraint systems. *Baseline Project*. https://www.baseline.vias.be/storage/minisites/baseline-kpi-safety-belt-and-crs.pdf.

[43] Transport Community. (2024). *Regional Thematic Report – Technical Assistance for Collecting Road Safety KPIs on Seat Belt and Child Restraint Systems in the Western Balkans Region*.

[44] Boets, S., Wardenier, N., De Vos, N., and Bouwen, L. (2023). *Nationale gedragsmeting "Rijden onder invloed van alcohol" 2021 – Drinken en rijden in België*.

[45] Holló, P., Eksler, V., and Zukowska, J. (2010). Road safety performance indicators and their explanatory value: A critical view based on the experience of Central European countries. *Safety Science*, 48(9), 1142–1150. https://doi.org/10.1016/J.SSCI.2010.03.002.

[46] Tingvall, C., Stigson, H., Eriksson, L., Johansson, R., Krafft, M., and Lie, A. (2010). The properties of safety performance indicators in target setting, projections and safety design of the road transport system. *Accident Analysis and Prevention*, 42(2), 372–376. https://doi.org/10.1016/j.aap.2009.08.015.

[47] Van den Berghe, W., and Christie, N. (2022). Differences in road safety "logics" between high income countries (HICs) and low income countries (LMICs). *Proceedings of the Transport Research Arena (TRA)*, Lisbon, 11–14 November 2022, *Transportation Research Procedia*. www.sciencedirect.comwww.elsevier.com/locate/procedia2352-1465.

[48] Elvik, R. (2011). Assessing causality in multivariate accident models. *Accident Analysis and Prevention*, 43(1). https://doi.org/10.1016/j.aap.2010.08.018.
[49] Papadimitriou, E., and Yannis, G. (2013). Is road safety management linked to road safety performance? *Accident Analysis and Prevention*, 59, 593–603. https://doi.org/10.1016/j.aap.2013.07.015.

Chapter 13

Insurance data

Magdalena Lindman[1] *and Anders Kullgren*[2]

13.1 Introduction

Data collected during the intricate management of insurance policies and claims can be used to develop informed strategies to improve traffic safety. Key application areas of insurance data include, each of which is assigned its own section/subchapter:

1. trends among injured traffic participants
2. retrospective studies on the effectiveness of safety features
3. consumer information on losses associated with used cars
4. refinements of insurance data through the creation of comprehensive crash databases

13.2 Analyzing trends of crashes and injured traffic participants to supplement official statistics

As seen in previous chapters of this book, there are various sources providing data on the number of fatalities and injuries in traffic accidents, each with its respective limitations that will influence the results when analyzing trends among injured traffic participants.

We will use Sweden and Europe as examples here, but similar organizations and structures can be found in many other countries.

13.2.1 Differences between insurance statistics and official road traffic statistics

In Sweden alone, there are at least five different references you can consult for the number of fatalities and injuries in traffic accidents, each of them using different data sampling and analysis methods.

[1]P&C Insurance, Stockholm, Sweden
[2]Folksam Insurance Group, Stockholm, Sweden

The Swedish Transport Agency supplies data from the police and emergency hospitals across Sweden, and the Swedish Transport Administration conducts in-depth studies on traffic fatalities. Transport Analysis, the agency responsible for transport policy analysis, publishes data based on part of the Swedish Transport Agency's statistics that originate from police records. Also, the National Board of Health and Welfare provides statistics on traffic accident injuries.

Insurance Sweden, the industry organization for insurance companies, provides information based on crashes covered by motor insurance. The traffic fatality data from insurance- and road traffic statistics have shown similar trends over the years, and the numbers remained at *roughly* the same levels. However, the data do not match exactly. Discrepancies can be attributed to the fact that the statistics cover different populations. Insurance statistics are based on vehicles that are required to have motor insurance, so crashes involving vehicles that do not require such insurance are not included in the insurance data. These accidents are captured in the national statistics collected by the Swedish Transport Agency if they occur within the Swedish road network. Conversely, the Transport Agency's statistics do not include accidents that occur, e.g., off-road or in private areas, but these crashes may be reflected in the insurance statistics. Furthermore, the national statistics do not include accidents that were neither reported by the police nor emergency hospitals.

When it comes to the number of injuries in traffic, the trends differ significantly. The discrepancies indicate that the official statistics underestimate the number of injuries, particularly in certain injury categories. Commonly, injuries in traffic accidents are overlooking a large group of serious injuries, which initially are classified as mild injuries, but can lead to long-term consequences for the person affected. For several reasons, these injuries may be significantly underreported in official statistics, which rely on reports from the police and emergency hospitals. Low-severity crashes rarely require emergency care and are often examined later, for example, at a health clinic. Additionally, they are not frequently reported to the police, as they often occur in city traffic and at low speeds, where the issue of liability is usually clear. As a result, it is likely that society's efforts to reduce the number of serious traffic injuries are missing a group of injuries.

13.2.2 Example of insurance data reports of trends for crashes and injured traffic participants

Insurance Europe is the European insurance and reinsurance federation. Through its 36 member associations, it represents all types and sizes of insurance and accounts for around 95% of total European premium income. In their webpage, a database with figures on the European insurance industry is available, including trends for motor third-party liability claims involving bodily injury [1]. Note that road and traffic-related factors stemming from demographics, terrain, and driving habits can affect the frequency of claims, as can the national regulatory environment and perceptions and attitudes to rules and obligations.

The National Insurance Associations [2] may provide more detailed information on the number of crashes. In Sweden, Insurance Sweden provides annual statistics for traffic insurance [3]. Based on data reported within the compulsory traffic insurance, it was shown that in 2023, about 25,600 people were injured in traffic. Over the last 10-year period, the number of people killed has decreased by percent, and the number of injured persons was reduced by 23%. Overall, the number of people killed in traffic has decreased since 1990, but persons injured in traffic have not decreased to the same extent. Here, it must be considered that the vehicle fleet has increased.

13.2.3 The advantage of a common long-term injury classification, a Swedish method

Differences in the classification of injury severity and diagnoses between data sources could have large consequences for the possibility of conducting studies, especially when studying long-term consequences of injuries.

In Sweden, all insurance companies are using the same system for the classification of permanent medical impairment. The procedure is that if an injured road user has not recovered within 3 years, the medical records from before and after the crash will be assessed by medical specialists to see if the injury and the symptoms can be traced to the crash. The assessment will be done according to the guidelines for Grading of Medical Impairment Degree [4]. The principles of grading medical impairment were developed at the beginning of the 20th century and established in consensus with physicians. The injury is given a degree of permanent medical impairment between 1% and 99%. Before a permanent medical impairment (PMI) is established, a preliminary forecast can be given because the long-term consequences of the injury can be difficult to estimate. Most PMI degrees are decided up to 5 years after the injury but in some cases even later. To consolidate the reliability of the judgment, all impairment degrees that receive a PMI of 10% or more are handled by a national medical committee. The PMI for each injured person is defined by this manual as physical and/or mental functional reduction, independent of cause and without regard to the occupation, hobbies, or other special circumstances of the injured person.

Since 2008, the definition in Sweden of a serious injury is an injury leading to PMI. The definition is based on insurance data to classify if an injured road user still had residual symptoms 3 years after a crash [5].

This common classification system and definition of a serious injury makes it possible to merge data from different data sources and to compare results from different studies based on the same injury classification.

13.3 Conducting retrospective studies on the effectiveness of safety interventions

Insurance data can be used to assess the effectiveness of safety initiatives by comparing crash rates before and after implementing measures such as traffic calming measures or enforcement of laws and by comparing crash rates in cars with and without a specific safety feature. This helps policymakers to determine which strategies are most successful in reducing accidents and injuries.

A current example includes advanced driver-assistance systems that are designed to avoid, or at least mitigate, crashes by alerting the driver in critical situations and perform countermeasures such as automatically braking or steering the vehicle. Evaluations based on data from real-world traffic are needed to improve the effectiveness of these systems.

Research until today in this area is mainly based on data from police-reported crashes and/or data collected by insurance companies. Various methodologies can be used for analyzing safety effectiveness, and examples can be found for regression analysis on crashes per insured vehicle-years [6–8], and for induced exposure methods based on data from police-reported crashes and/or data collected by insurance companies [6,9–11].

Note that the crash data sample can be associated with several confounding variables influencing bias, for example, those who choose to purchase vehicles with a certain safety technology are probably more concerned about their safety in the first place (i.e., selective recruitment).

Future applications in this area include safety impact evaluations of vehicles equipped with automated driving systems. Implementing and advocating for sound practices will improve the quality of this research, and experts across academia, insurance, and industry will collaborate to set up recommendations for performing and retrospective assessments of automated vehicles [12].

13.4 Providing consumer information on losses incurred by used cars

By considering insurance loss data alongside other factors such as safety ratings, individual vehicle drivers can make more informed decisions when purchasing a used car – that is, insurance data forms the basis for a sort of traffic safety consumer information.

13.4.1 Example IIHS, HLDI: insurance losses by make and model and driver death rates by make and model

In the US, the Highway Loss Data Institute [13] publishes losses in terms of percentages above or below average for passenger vehicles grouped by class and size. To minimize potential distortions from non-vehicle factors, the results are adjusted for variables such as operator age, calendar year, population density, model year, and state. Also, collision and comprehensive losses are adjusted for deductibles.

The insurance loss results generally serve as predictors for the performance of current versions of the same vehicle models. When automakers significantly redesign their vehicles, though, the performance of an earlier model may not predict the performance of the newer design.

Furthermore, IIHS [14] provides death rates based on insurance claims from US insurance companies. Beginning with 2020 models, they also calculate the rate at which drivers of other vehicles are killed in crashes with a given model. Both these rates provide perspectives on the relative safety of passenger vehicles across size and class. They take into account only the deaths of drivers, not passengers, since every vehicle that crashes has a driver, but not every vehicle has passengers. Rates are given as the number of driver deaths per million registered vehicle years.

13.4.2 Example: Folksam's How safe is your car?

Since 1983, the Swedish Insurance company Folksam has published car model safety ratings based on real-world crash and injury data. Over the years, it has been based on both national accident statistics and injury data from insurance claims. The rating methodology is based on pair comparisons using national data with two-car crashes where the injury outcome in a car model is compared with the injury outcome in the other vehicles it is colliding with. The output shows a relative injury risk in a car model (the injury risk in the model compared to the average injury risk in the car fleet) [Table 13.1 and (13.1)].

$$\text{The relative injury risk } R = (x_1 + x_2)/(x_1 + x_3) \tag{13.1}$$

When using this method, some factors will influence the outcome, and their influence needs to be adjusted for. The adjustments are described by Kullgren *et al.* [15].

The average injury severity in terms of risk for PMI is calculated for each of the included car models. The occupant injuries can be obtained from both national accident data and insurance data. The rating procedure and the methodology have been published in several studies [15,16].

The risk of PMI (RPMI) is used to measure the risk of long-term consequences. The risk of sustaining an injury leading to PMI of at least 1% and 10%, according to the procedures used by Swedish insurance companies, has been used. Impairment risks have been presented for each AIS level and body region and also for various degrees of PMI (more than 1%, more than 5% and more than 10%) [5], see examples of more than 1% and 10%, respectively, in Table 13.2.

In the Folksam car model safety ratings, the risk of PMI of at least 10% has been used since 1989.

Table 13.1 Number of impacts with different combinations of injured drivers in Car 1 and Car 2

		Driver of Car 2		Total
		Driver injured	**Driver not injured**	
Driver of Car 1	driver injured	x_1	x_2	x_1+x_2
	driver not injured	x_3	x_4	
Total		x_1+x_3		

Table 13.2 Risk of permanent medical impairment of at least 10%, for AIS levels 1–5, figures in percent (from [5])

	AIS level 1–5				
Body region PMI>10%	**1**	**2**	**3**	**4**	**5**
Head	2.5	8	35	75	100
Cervical spine	2.5	10	30	100	100
Face	0.4	6	60	60	n.a.
Upper extremity	0.3	3	15	100	n.a.
Lower extremity and pelvis	0.0	3	10	40	100
Thorax	0.0	0	0	15	15
Thoracic spine	0.0	7	20	100	100
Abdomen	0.0	0.0	5	5	5
Lumbar spine	0.1	6	6	100	100
External (skin) and thermal injuries	0.03	0.03	50	50	100

Table 13.3 Injury distribution for a specific car model

Body region	**1**	**2**	**3**	**4**	**5**	**6**
Skull/brain	44	13	5	2		2
Cervical spine	560	6				1
Face	16	8	1			
Arm	51	48	1			
Leg	6	38	9			
Thorax	26	38	17	6	3	
Thoracical spine	9	30	1			
Abdomen		13	2	1		
Lumbar spine	38	27	5			
External	760	9				

The RPMI for an occupant is calculated by multiplying the individual risks for each injury diagnosed with the highest AIS level in each body region according to (13.2), where p_i is the risk of sustaining a PMI from an injury of a certain AIS level to body region i. This is slightly conservative as it does not account for multiple injuries to the same body region

$$\text{RPMI} = 1 - \prod [1 - p_i] \quad (13.2)$$

Table 13.3 shows an example of injury distribution for a specific car model. The table shows the summary of the number of injuries sustained in all crashes the car model has been involved in.

To calculate the rating score for a car model, the relative risk of receiving an injury leading to fatality or PMI is obtained by combining the relative injury risk (with all adjustments, $R_{adjusted}$) and the RPMI (13.3). In the car model, safety ratings 95% confidence intervals for the relative risk of PMI are used

$$\text{Relative risk for PMI} = R_{adjusted} * \text{RPMI} \tag{13.3}$$

For the car model used in Table 13.3, $R_{adjusted} = 0.81$ and RPMI = 0.0241, which gives the relative risk of 0.0195 (0.0066, 95% CI).

13.5 Enrichment of insurance data: creation of crash databases and crash datasets

By collecting comprehensive data on traffic crashes and building a detailed crash database, we can identify unsafe details in vehicle design, unsafe traffic behavior, unsafe roadways and needs for traffic regulations. Crash databases also provide information that can be used for educating the public on safety issues (i.e., seat belt use, aggressive driving, and speeding) and estimating the efficiency of traffic safety measures. In this section, we explore some crash databases that are based on insurance claims information.

Again, we will use Swedish and European databases as examples, though similar actors and setups can be found in other countries.

13.5.1 UDB – the German Insurance Association crash database

German Insurers Accident Research (UDV), a division of the German Insurance Association, has access to all third-party vehicle insurance claims reported to the GDV. To support accident research, the UDV established a database called UDB, created from a representative cross-section of this extensive data pool [17]. The information collected from insurance claim files is processed for various use in the areas of vehicle safety, transport infrastructure, and traffic behavior.

Approximately 1,000 new cases are added to the UDB annually. The crash database only includes third-party vehicle claims involving personal injury and a total claim value of at least €15,000. Cases involving only property damage or less severe personal injury accidents (with a total claim value below €15,000) are excluded from the UDB. Each year, a stratified random sampling method is used, considering variables such as the type of traffic involvement, damage class, and time of year. Case-dependent extrapolation factors enable the UDB sample to be extended to the broader population of all claims in Germany, ensuring the results are representative of all claims handled by German insurers. However, single-car accidents are underrepresented, as cases without third-party injury or damage are not reported to the GDV.

13.5.2 VCTAD – Volvo Cars traffic Accident Database

Insurance claim inspectors from Volvia (part of If P&C Insurance), which insures most new Volvo cars, identify crashes with repair costs exceeding a certain threshold. Photos and technical details of the vehicles involved are then shared with Volvo Car's Accident Research Team [18].

Volvo Cars administers a detailed questionnaire to the car owner to collect information about the crash, the vehicle, and the occupants. This information includes crash circumstances such as weather, road conditions, speed limits, crash sequence, and collision objects. Occupant data, including seating position, age, gender, and restraint use, is also gathered – all of which is self-reported. With the occupants' consent, medical records are requested when applicable. Occupant injuries are then classified according to the Abbreviated Injury Scale and coded by a physician on Volvo's Traffic Accident Research Team. Based on photos from insurance claims settlement of the damaged vehicle and details from the questionnaire, Volvo experts code the car's deformation according to SAE recommended practice [19]. The repair cost threshold is adjusted over time to maintain consistent impact severity across different crash modes. For frontal impacts, the severity is estimated using the equivalent barrier speed.

Over the years, the database has played a key role in developing innovative systems like whiplash injury protection systems, side impact protection systems, and various child safety products. In the E.V.A. Initiative [20], Volvo Cars collects and shares some of the publications based on this data.

13.5.3 PAV – the People Around the Vehicle crash database by If

Information in the People Around the Vehicle crash database (PAV) is collected from insurance information at If P&C Insurance. Vehicle to active traveler crashes are identified using claims reported by the third-party liability motor insurance, which covers damage to property and personal injuries. The data cover crashes all over Sweden and include a wide range of car makes and models. Insurance claims data include crashes of all injury levels, from no injury at all to fatal outcomes, and so the dataset represents a broad spectrum of crash situations.

For each crash, information was annotated using crash descriptions from the persons involved. The vehicle claim report, the cyclist claim report, the police report and interviews with the drivers, cyclists and other eyewitnesses are reviewed in the case analysis process. A vehicle crash claim report was available for every case, and in most cases, a crash report was filed by the cyclist as well. The vehicle report contains information about the course of events from the driver's perspective. Also, estimated speed at impact, traffic environment, cyclist impact points, car damage, driver distraction elements and a description of injuries sustained are examples of information provided. For cases where the police were present at the accident scene, a police report including a sketch of the crash scene and witness statements from the driver, the cyclist, and any additional observers may be available as well. Additionally, further information such as notes from conversations with the cyclist and the driver are collected during the insurance claim handling process, and when relevant, this information is included in the database as well. Vehicle specification data adds car information to the database and map data provides detailed information of the crash scene, such as road geometry and roadside objects. Conflict Situation class is coded that describes the pre-crash maneuvers of involved participants in a crash [21]. In cases where photos of the car damage were available, car deformation following SAE recommended practice for Collision Damage Classification (SAE). If cyclists or pedestrians were injured, hospital reports and medical information are available for coding according to the Abbreviated Injury Scale (AAAM).

The PAV database was established in 2010 and about 350 new cases are added annually, providing a basis for interdisciplinary use in the areas of traffic safety.

13.5.4 Insurance datasets with information from crash recorders

Some insurance companies have been installing crash recorders in cars and created crash databases including information from crash recorders [22,23]. An important advantage of insurance claims data is that all crashes will be reported and could potentially be included in the dataset, irrespective of crash severity or injury outcome. When it comes to crash recorder data, this is a great advantage when creating injury risk curves.

The Folksam crash recorder database

The crash recorder fitments have been part of an insurance co-operation between several car manufacturers and Folksam Insurance Company in Sweden that started in 1992 and ended in 2018. In total, more than 300,000 vars were fitted with crash recorders. The main focus has been to study injury risk as a function of various crash severity parameters, such as change of velocity, and mean and peak acceleration, see, for example, Kullgren and Krafft [24], but also to study how crash severity varies in different crash types [25,26].

One important contribution to occupant injury protection is studies on injury and their link to impact severity. Studies have been published covering injury risks in both frontal and rear-end impacts. One main area has been whiplash injury risks in various impact directions.

Figures 13.1 and 13.2 show examples of risk for whiplash symptoms in rear-end impacts for change of velocity and mean acceleration [27].

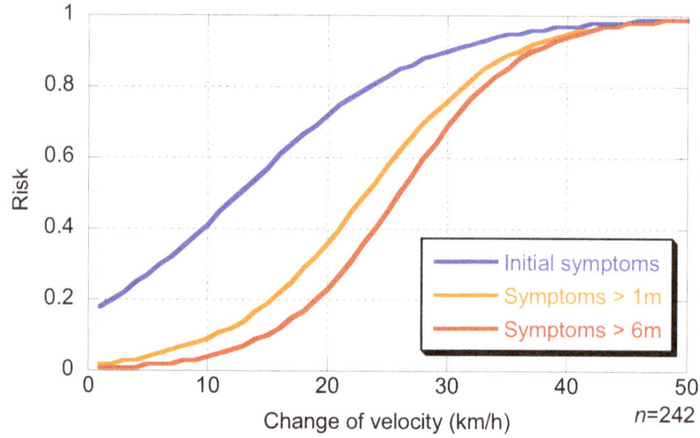

Figure 13.1 Risk of initial and long-lasting whiplash symptoms for change of velocity

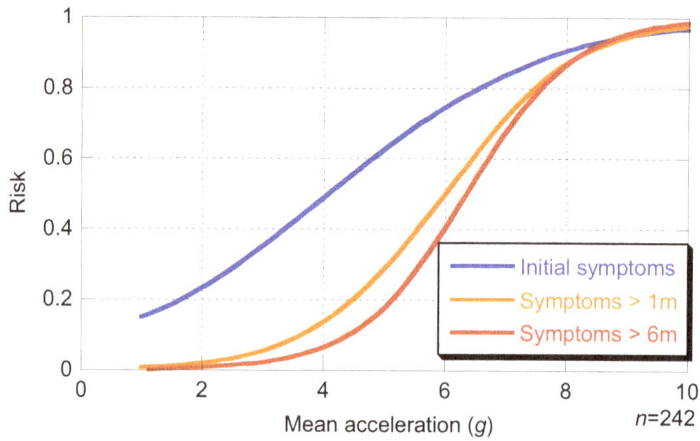

Figure 13.2 Risk of initial and long-lasting whiplash symptoms for mean acceleration

The AXA Winterthur crash recorder database

The AXA Winterthur Insurance Company in Switzerland launched a crash recorder in 2008 as part of an insurance solution for their policyholders. A large variety of vehicle models were equipped with the Crash Recorder. In 2016, more than 40,000 vehicles were equipped with crash recorders. Studies have been published based on this data. Figure 13.3 shows an example of whiplash injury risk versus change of velocity (Delta-V) based on 168 recorded crashes.

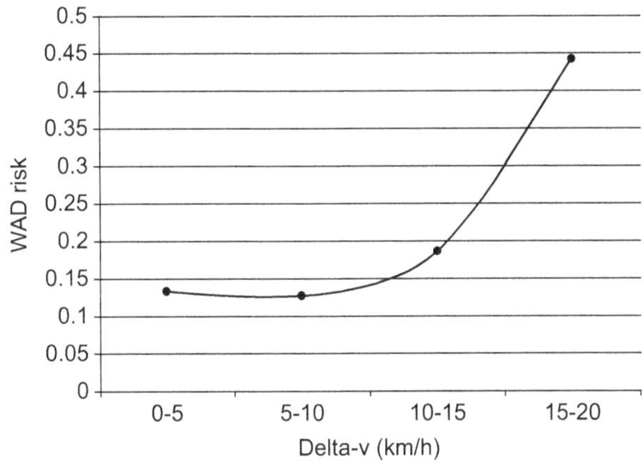

Figure 13.3 *Risk for initial whiplash symptoms in intervals of change of velocity (from [28])*

Outlook – insurance crash data enhancements with sensors and cameras

The future of traffic accident research based on insurance data may advance significantly with the widespread use of data from dashcams, telematics, and the increasingly sophisticated sensors and cameras integrated into modern vehicles. These technologies offer real-time and detailed data, including signals, videos, and vehicle performance metrics.

Dashcams capture high-resolution video of road conditions, driver behavior and the events leading up to accidents. This visual evidence allows for accurate accident reconstruction and helps identify the exact causes of crashes. Telematics systems add to collecting data on a vehicle's speed, braking patterns, GPS location, and long-term driver behavior. Modern vehicles, equipped with advanced sensors and cameras, continuously monitor and record various aspects of driving, the driver, and people inside and outside the vehicle, and environmental conditions. The integration of these data sources—dashcams, telematics, and vehicle-embedded systems—promises a more holistic understanding of traffic incidents; this, in turn, can inform the design of safer vehicles, the development of more effective safety regulations, and safer roads.

However, the collection and use of such data also present challenges. One of the most pressing concerns is the issue of personal privacy. This raises questions about how data should be used, stored, and shared. Data security is another challenge as vast amounts of data generated by telematics and vehicle-embedded systems are often transmitted wirelessly, making them vulnerable.

In summary, the adoption of dashcams, telematics, and vehicle-embedded sensors and cameras brings significant challenges but has the potential to revolutionize traffic accident research. By providing detailed data, these technologies will enable accurate accident reconstructions, deeper understanding of risk factors, and the development of more effective safety measures. The extent to which this data can be leveraged for research will depend significantly on the implementation rates into the vehicle fleet.

References

[1] Insurance Europe (2016), Statistics https://www.insuranceeurope.eu/statistics.

[2] Insurance Europe (2016), All members https://www.insuranceeurope.eu/about-us/2/who-we-are/members/all-members.

[3] Svensk försäkring (2016), Traffic and motor insurance. https://www.svenskforsakring.se/statistik/skadeforsakring/trafik-och-motorforsakring/.

[4] SverigesFörsäkringsförbund (2004), Medicinsk invaliditet. Gradering av medicinsk invaliditet 2004. Sveriges Försäkringsförbund, Stockholm, Sweden.

[5] Malm, S., M. Krafft, A. Kullgren, A. Ydenius, and C. Tingvall (2008), Risk of permanent medical impairment (RPMI) in road traffic accidents, *Annual proceedings of the Association for the Advancement of Automotive Medicine*, 52, 93–100.

[6] Cicchino, J. B. (2022), Effects of automatic emergency braking systems on pedestrian crash risk, *Accident Analysis & Prevention*, 172, 106686, https://doi.org/10.1016/j.aap.2022.106686.

[7] Isaksson-Hellman, I., and M. Lindman (2023), Estimating the crash reducing effect of Advanced Driver Assistance Systems (ADAS) for vulnerable road users, *Traffic Safety Research*, 4, 000036. DOI: https://doi.org/10.55329/blzz2682.

[8] Wakeman, K., M. Moore, D. Zuby, and L. Hellinga (2019), Effect of Subaru eyesight on pedestrian-related bodily injury liability claim frequencies, *26th International Technical Conference on the Enhanced Safety of Vehicles (ESV)*, Eindhoven, the Netherlands, 10–13 June 2019, https://www-esv.nhtsa.dot.gov/Proceedings/26/26ESV-000265.pdf.

[9] Leslie, A. J., R. J. Kiefer, M. R. Meitzner, and C. A. Flannagan (2021), Field effectiveness of general motors advanced driver assistance and headlighting systems, *Accident Analysis & Prevention*, 159, 106275, https://doi.org/10.1016/j.aap.2021.106275.

[10] PARTS (2022), Real-world effectiveness of model year 2015–2020 advanced driver assistance systems, Partnership for Analytics Research in Traffic Safety, https://www.mitre.org/sites/default/files/2022-11/pr%2022-3734-PARTS-real-world-effectiveness-model-year-2015-2020-advance-driver-assistance-systems_0.pdf. 22-3734.

[11] Kullgren, A., K. Amin, and C. Tingvall (2023), Effects on crash risk of automatic emergency braking systems for pedestrians and bicyclists, *Traffic Injury Prevention*, 24(S1), S111–S115, https://doi.org/10.1080/15389588.2022.2131403.

[12] Scanlon, J. M., E. R. Teoh, D. G. Kidd, *et al.* (2024). RAVE checklist: Recommendations for overcoming challenges in retrospective safety studies of automated driving systems, *Journal of Traffic Injury Prevention*, doi: 10.1080/15389588.2024.2435620.

[13] IIHS HLDI, Insurance losses by make and model (2016), https://www.iihs.org/topics/auto-insurance/insurance-losses-by-make-and-model.

[14] IIHS HLDI (2016), Driver death rates by make and model. https://www.iihs.org/ratings/driver-death-rates-by-make-and-model.

[15] Kullgren, A., H. Stigson, and A. Axelsson (2020), Developments in car crash safety since the 1980s. Paper presented at the Ircobi.

[16] Gustavsson, H., A. Hägg, M. Krafft, A. Kullgren, *et al.* (1990), *Folksam Car Model Safety Ratings 1989–90*. Folksam, Stockholm, Sweden.

[17] Kuehn, M., T. Hummel, and J. Bende (2009), Benefit estimation of advanced driver assistance systems for cars derived from real-live accidents. In Proceedings of the 21st International Technical Conference on the Enhanced Safety of Vehicles (ESV) (pp. 1–10). Stuttgart, Germany.

[18] Isaksson-Hellman I., and H. Norin (2005), How thirty years of focused safety development has influenced injury outcome in volvo cars, *Annu Proc Assoc Adv Automot Med*, 49, 63–77. PMID: 16179140; PMCID: PMC3217440.

[19] Collision Deformation Classification (2022), SAE J224-202205.

[20] Volvo Cars (2016), The E.V.A. Initiative. https://developer.volvocars.com/resources/eva/.

[21] Lindman, M., S. Jonsson, L. Jakobsson, T. Karlsson, D. Gustafson, and A. Fredriksson (2015, 9–11 September 2015), Cyclists interacting with passenger cars: A study of real world crashes *IRCOBI Conference*, Lyon, France. https://www.ircobi.org/wordpress/downloads/irc15/pdf_files/10.pdfAAAM (2015).

[22] Kamrén, B., M. v Koch, A. Kullgren, A. Lie, Å. Nygren, and C. Tingvall (1991), Advanced accident data collection – description and potential of a comprehensive data collection system. Paper presented at the 13th International Conference on the Enhanced Safety Vehicles (ESV), Paris.

[23] Sinzig, B. (2010), Crash Recorder: acceleration data of real world accidents. Paper presented at the 3rd International Conference Whiplash 2010 – Neck Pain in Car Crashes Munich, Germany.

[24] Kullgren, A., and M. Krafft (2008), Influence of change of velocity and mean acceleration on whiplash injury risk: results from real-world crashes. Paper presented at the 2nd International Conference on Whiplash: Neck pain in car crashes, Erding, Munich.

[25] Ydenius, A., and A. Kullgren (1999), Pulse shapes and injury risks in collision with roadside objects: results from real-life impacts with recorded crash pulses. Paper presented at the Int. IRCOBI Conf. on the Biomechanics of Injury, Sitges, Barcelona, Spain.

[26] Stigson, H., A. Ydenius, and A. Kullgren (2006), Variation of crash severity and injury risk depending on collisions with different vehicle types and

objects. Paper presented at the *Int. IRCOBI Conf. on the Biomechanics of Injury*, Madrid, Spain.

[27] Kullgren, A., and H. Stigson (2011), *Report on Whiplash Injuries in Frontal and Rear-End Crashes*. Folksam, Stockholm, Sweden.

[28] Jordan, B., K.-U. Schmitt, D. Bützer, and B. Zahnd (2016), Analysis of whiplash associated disorder claims using real-world data retrieved from event data recorders: a case-control study. Paper presented at the *Proceedings of IRCOBI Conference*, paper no. IRC-16-21.

Chapter 14

Traffic safety footprint and safe organisations

Matilda Magnusson[1] *and Carlos Viktorsson*[2]

A core principle of the Safe System approach is shared responsibility among stakeholders and road users. Traditionally, public actors have shouldered most of the stakeholder responsibility, and traffic safety has mainly been viewed as a duty of governments [1,2]. However, the United Nations and the World Health Organisation call on all stakeholders, both public and private, to take on the responsibility and move towards a world free from road traffic fatalities and serious injuries [3,4]. During the third Global Ministerial Conference on Road Safety, academic experts formulated nine key recommendations for the second decade of action for road safety. Two of these recommendations call on public and private organisations to incorporate road safety into sustainability reporting and to include traffic safety demands in procurement processes [1]. Among these calls for more stakeholders to take on the responsibility for a safe road environment, the concept of the traffic safety footprint has formed. The traffic safety footprint builds on the idea that all organisational activities within the road transport system have an impact on traffic safety and that this can and should be measured [5].

14.1 Traffic safety footprint

The concept of the traffic safety footprint is defined in FIA's (Federation Internationale Automobile) Road Safety Index as

> *The number of fatalities and seriously injured persons as a result of road crashes occurs within an organisation's entire value chain. All casualties resulting from relevant and significant activities, services and products should be included in the calculation (p. 31)* [6].

The value chain within the context of road traffic safety should contain all activities of an organisation related to road use, from any upstream sourcing to potential downstream use of products and services [1]. Employees, subcontractors, customers, clients and third parties that are victims of traffic accidents due to the

[1]Department of Technology and Society, Faculty of Engineering, LTH, Lund University, Sweden
[2]Sweco, Stockholm, Sweden

organisation's operations should be included [2,7]. However, the entire value chain can be vast for large organisations, and it can therefore be necessary to define which activities the organisation can realistically influence; this is the organisation's sphere of influence [6]. Measuring and improving the traffic safety footprint can be gradually implemented on the different parts of the sphere of influence, starting where the opportunity for impact is bigger and branching out [8].

14.2 Organisational traffic safety data

Nearly every organisation partakes in activities that utilise the road transport system in some way with activities like the delivery of goods, employee travel, commuting trips and customer travel [2,7]. It is at these points that organisations have the opportunity to influence traffic safety [2,3].

Depending on the organisation's specific context, there can be different entry points into working with traffic safety. For organisations that have an established process working with sustainability reporting, incorporating traffic safety can be an opportunity to benefit from existing sustainability tools and methods [8]. Procurement is an especially relevant approach for public organisations to influence traffic safety since public institutions spend large sums on procuring goods and services [1,9]. For organisations where employees travel during work Occupational Health and Safety (OHS), managing employee travel and systematic traffic safety work are other approaches for organisations to manage their traffic safety impact [5].

Regardless of the organisation's approach, the impact on traffic safety can be described through indicators. The relevant indicators will depend on the activities in the organisation's value chain [8]. The indicators can generally describe an organisation's influence on traffic safety in terms of outcome, performance or process.

Outcome indicators describe how many people are killed or severely injured due to the organisation's activities. Outcome indicators can be described using the same definition as the traffic safety footprint. For employees who use the transport system during their work, traffic safety is the employers' responsibility, as it is their work environment. Collecting data for the outcome indicators or the footprint, tends to be easier for internal activities. In most jurisdictions, employers are previously required to collect data on injuries and fatalities to employees during work as part of occupational safety and health. Sometimes this might also include commuting trips [10]. Collecting data for third parties, subcontractors or injuries and fatalities elsewhere in the value chain is not as straightforward, and strategies are still being developed [5,7]. For most organisations, these numbers will be low; it is therefore useful for organisations to work with additional indicators [6,8].

Performance indicators can be used to track and measure factors that are known to affect traffic safety outcomes and, through them, improve the organisations' traffic safety impact. For example, if an organisation has employees who travel with motorised vehicles during work, speed compliance can be a vital indicator to follow. Speed compliance directly impacts the risk and consequence of accidents and protects both employees and other road users from harm. Speed compliance can be evaluated at an aggregate level by using electronic driving logs [11]. Another example of a performance indicator is seat belt use, as road users not using seatbelts are overrepresented in fatal accidents. The indicator can be measured in employee surveys [8]. For bicyclists, the use of helmets and additional safety gear like high visibility vests are important. Initially, the availability of safety gear can be a relevant indicator, but expanding on it to verify if and how the employees use the gear is important. Another factor that influences traffic safety outcomes is safe vehicles, which can make a big difference in safety for both the vehicle occupant and other road users. Therefore, a relevant indicator to track can be how many vehicles in the fleet meet specified safety requirements, like five stars in EuroNCAP [6,8].

Lastly, process indicators can show how an organisation works with and handles traffic safety-related issues in their operations. These types of indicators can describe factors that contribute to shaping the conditions that promote or counteract traffic safety within the organisation [8]. Traffic safety should be reflected in policies, routines, requirements for procurement, scheduling and so on to create framework conditions that promote safety. For instance, what are the routines for maintenance or incident reporting [8]. It is also important to consider that there can be factors that counteract traffic safety. One example might be pay systems that encourage speeding or driving for too long periods by rewarding productivity [12,13]. The indicators can initially be evaluated on whether they take traffic safety into account or not. If they do, it might be useful to follow if the policies, routines, etc., are well-known in the organisation and if they are adhered to in the day-to-day operation [8].

14.3 Challenges and opportunities related to organisations and traffic safety

Understanding the impact on traffic safety might come more naturally for organisations where transport is a main aspect of the goods or services they provide. Many companies in the transport sector are already aware of traffic safety and might already work to improve their safety as a part of their work environment efforts or their sustainability reporting [5]. However, many other organisations might rely on the transport system but do not recognise the traffic safety impact they have both on employees and society in general [8]. At-home caregivers and craftsmen are examples of occupational categories that spend a significant amount of time in traffic [14]. It is essential for organisations to understand the extent of how the operations interact with and impact the transport system for them to improve traffic safety [2]. Increasing the awareness of how organisations impact traffic safety can initially be a challenge, as many individuals might not realise the impact or prioritise it [8]. Communicating that traffic safety is a sustainability issue and that all organisations have a traffic safety footprint that can be improved will be crucial for raising awareness [5].

Collecting relevant and accurate data on traffic safety from various parts of an organisation can be challenging. This can require coordination and collaboration between different teams or departments to ensure that the data is accessible and accurate [8]. Identifying and locating the data can be hindered by issues like data silos, where organisations often have multiple departments that operate independently and may store data in separate systems or databases, making it difficult to access and integrate data from different parts of the organisation. Inconsistent data formats and ensuring the quality and accuracy of data can also hinder data integration and create difficulties in obtaining a comprehensive and reliable view of the organisation's operations [15]. Collecting data from different parts of the organisation and third parties often requires cooperation and collaboration between various stakeholders. It can be challenging to align different teams, departments and external partners [16].

To overcome these challenges, organisations can implement strategies such as establishing a centralised data management system, implementing data governance frameworks, ensuring data quality and accuracy through data validation processes and fostering a culture of data collaboration and sharing [17,18].

Another challenge is complying with legal requirements regarding data collection for traffic safety. Organisations must comply with relevant data protection regulations and ensure that sensitive information is handled securely. An example of traffic safety is that the collection of speed compliance among drivers might not be compliant with GDPR [13]. However, there are examples of instances where the management, together with the trade union and drivers, have collaborated and come to an agreement for handling the data [5,11].

Having support from employees and trade unions is crucial for successful data collection within an organisation. Employees and trade unions can provide valuable insights and firsthand knowledge about the organisation's operations. Their support can ensure that relevant and accurate data is captured, enhancing the overall quality and completeness of data. Involving trade unions and employees in the data collection process also promotes transparency and builds trust within the organisation [19]. Employees are more likely to trust and participate in data collection efforts when they feel their input is valued and will be used for meaningful purposes.

Another challenge for organisations might be the issue of balancing productivity and safety concerns. Tight schedules, short delivery windows and narrow timetables are all examples of common conditions that can be counteractive to traffic safety [5]. There is a risk that organisations that take responsibility for traffic safety are not given any competitive advantages in today's market [13]. Demands set in the procurement processes are a powerful tool to create a driving force and an advantage for companies to work actively with traffic safety [5].

The commitment and interest of leadership play a crucial role in promoting and prioritising traffic safety within an organisation. Management leadership is crucial for building a safety culture, aka 'how things are actually done' [20]. Challenges may arise if there is a lack of leadership, as it can impact the allocation of resources, implementation of safety measures and overall organisational culture related to traffic safety [21]. There are, however, positive synergies between traffic safety and other areas that can be a selling argument for implementing traffic safety strategies. One example is the decreased risk of accidents, lower emissions and reduced costs that fuel-efficient driving can contribute to [5,22].

It should also be of interest to employees and trade unions to push for traffic safety, as another aspect of the current limited view of responsibility is that if a professional driver were to be involved in an accident, most likely only the driver would be held accountable with no regard for how the employer or potential transport contractor shaped the conditions under which the driver was working [13]. In addition, road traffic deaths that occur during work are not investigated as deaths at the workplace or as crimes related to the work environment by the police in Sweden [7]. Traffic safety is not only relevant for employees and trade unions when an accident occurs, traffic safety is often linked with other aspects of the work environment. A solution for one aspect can also improve other issues in the workplace. For example, introducing more flexible need-based schedules for at-home caretaking staff. This would improve both the general work environment by reducing stress levels and increasing the feeling of autonomy, control and influence, but also by increasing the traffic safety situation [14].

14.4 Future and potential

Engaging stakeholders who have not previously been involved opens new possibilities to address and mitigate accidents that have been hard to target earlier. In Sweden, nearly half of all traffic fatalities are work-related. 47 per cent of the deaths that occurred in 2019, either the fatally injured person or the collision partner was at work or commuting to their place of work [7]. In addition, for many organisations road transport is one of the leading causes of occupational fatalities and the biggest work environment risk [10,21]. This data suggests that there is substantial potential for organisations to improve their performance in the field of traffic safety.

There already exist some systematic frameworks for how organisations can work with traffic safety, such as ISO 39001 Road Traffic Safety Management Systems [2] and FIA Road Safety Index [6]. There have also been a couple of projects in Sweden exploring how indicators and traffic safety footprint can be integrated into a sustainability context for different organisations [8,23]. However, the established sustainability reporting standards like Global Reporting Initiative and UN Global Sustainability Goals Integrated Reporting, do not currently include explicit traffic safety indicators. To enable the integration of traffic safety within the sustainability context, there is a need for explicit inclusion of traffic safety in the reporting standards.

Exactly how traffic safety in the context of a broader involvement of stakeholders will take shape is still being established, but the collection and reporting of traffic safety data will be necessary for organisations to work systematically for continuous improvement [1]. With the information the data provides organisations can recognise the current traffic safety situation, identify the potential for improvement, make informed decisions and set goals and targets. By working systematically organisations can continuously improve their safety impact on society and protect their employees [13].

References

[1] Swedish Transport Administration, *Saving Lives Beyond 2020: The Next Steps — Recommendations of the Academic Expert Group for the 3rd Global Ministerial Conference on Road Safety 2020*, 2020.

[2] A. Lie and C. Tingvall, 'ISO 39001 road traffic safety management system, performance recording, and reporting', in *The Vision Zero Handbook: Theory, Technology and Management for a Zero Casualty Policy*, K. Edvardsson Björnberg, S. O. Hansson, M.-Å. Belin, and C. Tingvall (eds), Springer, Berlin, 2022, pp. 675–686. doi:10.1007/978-3-030-76505-7_26.

[3] United Nations, 'UN Resolution 74/299, Improving Global Road Safety. A/RES/74/299', 2020, *General Assembly*.

[4] World Health Organization, *Global Plan: Decade of Action for Road Safety 2021–2030*, WHO, Geneva, 2021.

[5] H. Wennberg and P. H. Mattisson, 'Traffic safety footprint in sustainability practises and reporting — Exploring the views of companies', *Sustainability*, vol. 16, no. 24, p. 10975, 2024.

[6] Federation Internationale de l'Automobile (FIA), 'FIA Road Safety Index: Framework', 2023. [Online]. Available: https://www.fia.com/fia-road-safety-index (accessed Sep. 13, 2024).

[7] A. Kullgren, H. Stigson, M. Rizzi, and C. Tingvall, 'Fatalities in value chains—An attempt to classify road traffic crashes in accordance with the United Nations General Assembly Resolution 74/299', *Traffic Safety Research*, vol. 5, no. Special issue, pp. 1–11, 2023, doi:10.55329/mcmr2018.

[8] C. Viktorsson, M. Hjälmdahl, T. Willstrand, M. Magnusson, A. Laureshyn, and M. Levin, 'Framtagning av verktyg för trafiksäkerhetsförbättrande arbete: Tre fallstudier om hur ett verktyg för ett trafiksäkerhetsförbättrande arbete kan integreras i en organisations hållbarhetsarbete (in Swedish) [Development of tools for improving traffic safety: Three case studies on how a tool for improving traffic safety can be integrated into an organisation's Sustainability practices]', Gothenburg, Sweden, 2024.

[9] L. Bidasca and E. Townsend, 'Reducing road risk at work through procurement: Integrating safety requirements in the transport sector supply chain', Brussels, Belgium, 2015.

[10] D. Adminaite, G. Jost, H. Stipdonk, and H. Ward, 'Tapping the potential for reducing work-related road deaths and injuries: PIN Flash Report 33', Brussels, Belgium, 2017.

[11] H. Lidestam, T. Dukic Willstrand, R. Bryman, and M. Levin, 'Uppföljning av trafiksäkerhetskrav vid upphandling av godstransporter (in Swedish) [Monitoring traffic safety requirements in the procurement of freight transport]', Gothenburg, Sweden, 2023.

[12] T. O. Nævestad, R. O. Phillips, and B. Elvebakk, 'Traffic accidents triggered by drivers at work – A survey and analysis of contributing factors', *Transportation Research Part F: Traffic Psychology and Behaviour*, vol. 34, pp. 94–107, 2015, doi:10.1016/j.trf.2015.07.024.

[13] A. Lie, C. Tingvall, and S. Eveby, 'Validering av FIA RS Index manual för bussbolag, speditionsföretag, hyrbilsföretag samt försäkringsbolag (in Swedish) [Validation of the FIA RS index manual for bus companies, freight forwarding companies, car rental companies, and insurance companies]', Stockholm, Sweden, 2023.
[14] A. Anund, M. Weidel, P. Henriksson, K. Brolin, and J. Ihlström, 'Hantverkares och omsorgspersonals resor i tjänsten Fokus på arbetsmiljö och säkerhet t (in Swedish) [Craftsmen and caregivers' work-related travel: Working conditions and traffic safety]', Linköping, Sweden, 2024.
[15] Z. Adams and A. Nelson, 'Breaking down silos: Integrating big data across organizational functions introduction', *International Journal of Network and Communication Research*, vol. 7, no. 1, pp. 81–89, 2022.
[16] H. Richter and P. R. Slowinski, 'The data sharing economy: On the emergence of new intermediaries', *IIC International Review of Intellectual Property and Competition Law*, vol. 50, no. 1, pp. 4–29, 2019, doi:10.1007/s40319-018-00777-7.
[17] R. Abraham, J. Schneider, and J. vom Brocke, 'Data governance: A conceptual framework, structured review, and research agenda', *International Journal of Information Management: The Journal for Information Professionals*, vol. 49, pp. 424–438, 2019, https://doi.org/10.1016/j.ijinfomgt.2019.07.008.
[18] J. E. Olson, *Data Quality: The Accuracy Dimension*, 1st edn. San Francisco, CA: Morgan Kaufmann Publishers Inc., 2003.
[19] K. M. Branch, 'Participative management and employee and stakeholder involvement', in *Management Benchmarking Study* (pp. 1–27). Washington DC, USA: Washington Research Evaluation Network, 2002.
[20] T. O. Nævestad, R. Phillips, I. B. Hovi, G. N. Jordbakke, and R. Elvik, 'Estimating safety outcomes of increased organisational safety management in trucking companies', *Safety*, vol. 8, no. 2, p. 36, 2022, doi:10.3390/safety8020036.
[21] B. Rowland, J. Davey, J. Freeman, and D. Wishart, 'Development of a proactive brief road safety intervention for industry', *Journal of Road Safety*, vol. 19, no. 4, pp. 27–35, 2008.
[22] T. O. Nævestad, 'Eco driving as a road safety measure: Before and after study of three companies', *Transportation Research Part F: Traffic Psychology and Behaviour*, vol. 91, pp. 95–115, 2022, doi:10.1016/j.trf.2022.09.012.
[23] M. Hjälmdahl, T. Willstrand, S. Andersson, and A. Nilsson, 'Trafiksäkerhetsavtryck: Om hur en kommun kan arbeta för att förbättra sin trafiksäkerhet (in Swedish) [Traffic safety footprint: How a municipality can work to improve its traffic safety]', Gothenburg, Sweden, 2022.

Chapter 15
Background data

Jenny Eriksson[1] and Maria Varedian[2]

By being in a traffic environment, you are exposed to it. Therefore, exposure is an important factor when explaining road safety and illustrating road safety problems [1]. Together with accidents and injuries from traffic, exposure can, for example, explain the risk in relation to traffic (volumes). By calculating this risk, it is possible to identify which groups are at the greatest risk of being killed or seriously injured in the traffic environment and should therefore be prioritised in the road safety programme. The risk of being killed or seriously injured in a road traffic accident is calculated by dividing the total number of injuries (x) by some measure of exposure (y), simply put:

$$\text{Risk} = \frac{x}{y}$$

A further dimension is the consequences of an accident, where traffic safety problems can be expressed as [1]

$$\textit{Traffic safety problem} = \textit{Exposure} * \textit{Risk} * \textit{Consequence}$$

In Figure 15.1, the expression is expanded to a three-dimensional picture for both fatalities and all injuries, where the products of risk bars are expanded to volumes. The volume describes the safety problem from an exposure, risk and consequence dimension.

An example of the traffic safety problem is illustrated in Figure 15.2. From the illustration, you can see that the consequence is highest for pedestrians and lowest for bus occupants. Also, car drivers have the highest number of injuries (the front area) and fatalities (the volume).

The most fundamental definition of exposure is to see it in terms of time and space. The concept of exposure is included in the definition to describe the road safety problem, using the exposure to calculate the risk of an accident occurring from different perspectives. The different perspectives can be the type of road, traffic environment (e.g. intersections), different types of road user groups, etc. [2].

[1]Research Department of Infrastructure, The Swedish National Road and Transport Research Institute, Linköping, Sweden
[2]The Swedish Transport Administration, Borlänge, Sweden

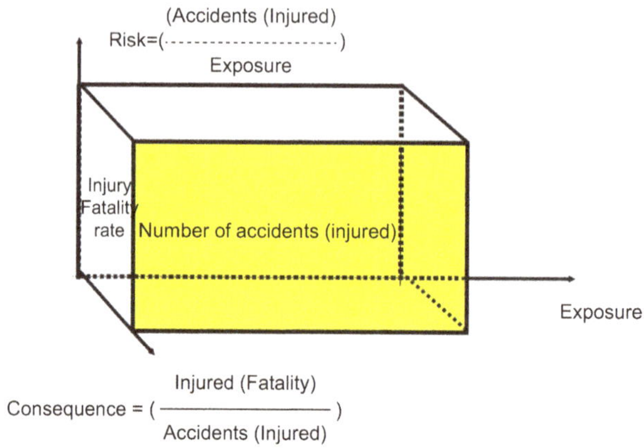

Figure 15.1 *The traffic safety problem = Exposure * Risk * Consequence. The volume is proportional to the number of injured (fatalities). Source: Nilsson [1].*

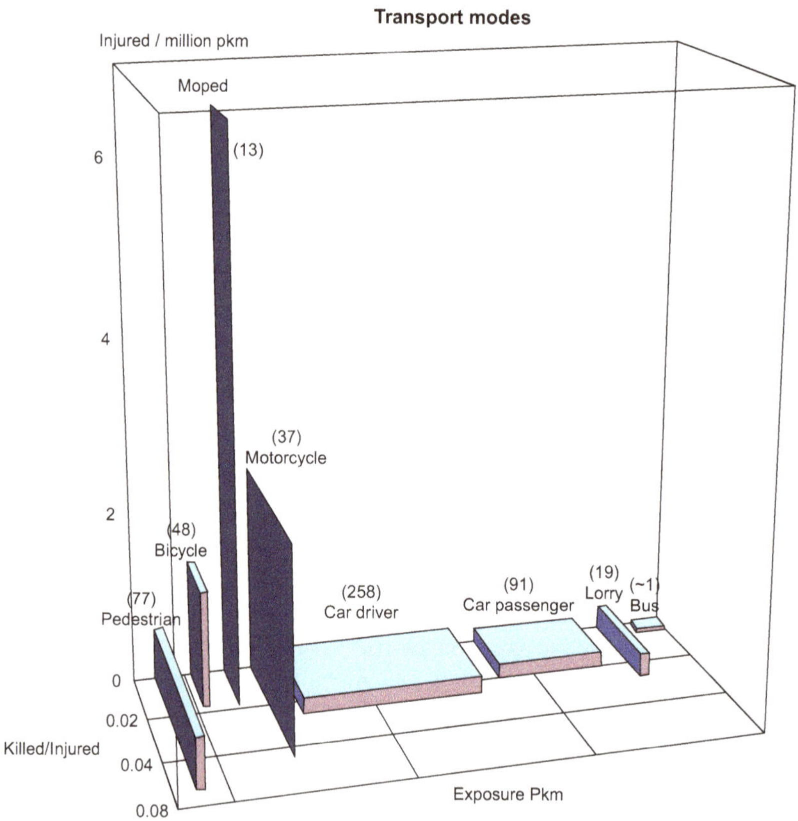

Figure 15.2 *An example of the traffic safety situation – average number of fatalities annually for different transportation modes in Sweden 1997–1999. Source: Figure 13 in Nilsson [1].*

For the definition of fatalities and injuries and how this is derived, see Chapters 2 and 4.

As regards the exposure measure, which is the subject of this chapter, it could be based on different data. The number of inhabitants is a common way of comparing different countries, as this is a reliable source, as is the number of registered vehicles. Even closer to the actual exposure is the calculation of vehicle mileage. There are three different sources for obtaining information from vehicle mileage: vehicle inspections, travel surveys and traffic counts [1].

- **Vehicle inspections**, data on kilometres travelled are usually obtained from the odometer during the inspection at regular intervals. However, we do not know where the journeys were made, neither in time nor space, and foreign road users are not captured (e.g. tourism).
- **Travel surveys** often consist of a random sample of a defined target population where travelling habits are surveyed. One advantage is that background data is collected such as gender, age, purpose of the trip, etc. The number of trips, length and travelled hours are also captured. Again, the problem remains that we do not know where the trips were made in space, or about the commercial traffic.
- **Traffic counts**, all traffic can be measured at a location or section regardless of who is travelling, but we do not get any background information, as we do from the travel surveys, for example.

It should be noted that the quality of data varies depending on the source. However, this is only dealt with in general terms in this chapter.

When the exposure rate is high (number of trials) and accidents are few (failures), the estimated number of accidents is assumed to follow the Poisson distribution. The estimated number of accidents is expected to be independent, but in practice, the estimators do not form independent and homogeneous trials for which risk remains constant independently of the number of trials. There are two problems in using accident rates in order to control for the effects of differences in exposure on the number of accidents: (1) Accident rate is not independent of exposure but tends to decline as exposure increases. (2) Computing and using accident rates arises in the case of composite exposure (i.e. exposure consisting of two or more traffic movements that both contribute to the risk of accident). There is a non-linearity, and there is more than just the exposure that can explain the accident, for example, a driver's lane change, behaviour, braking and overtaking [3]. Density is also something that can affect, and there are several encounter models that can be used. Johnsson *et al.* [4] were able to show that with a smarter use of the exposure, the relationship with accidents becomes almost linear and that the choice of a specific encounter model affects the outcome.

Traffic counts carried out today and used as input to risk models are still seen as a good method to calculate exposure from a holistic perspective. However, in the future when newer technologies can be used or existing technologies refined, the above details can certainly be captured to the extent required to cover the whole road network.

15.1 Total vehicle mileage – definition and theory

Total vehicle mileage is a measure of the total length of all journeys in time and space, expressed in millions of vehicle kilometres travelled. While the total vehicle mileage can be calculated in different ways, using the various sources mentioned above, this chapter focuses solely on traffic counts. The calculation is made for a specific road network and for a specific time. To calculate the total vehicle mileage, information on the vehicle passages on the entire selected road network is needed.

When the total vehicle mileage is calculated based on traffic counts, the annual average daily traffic (AADT) is usually used. AADT is a statistical parameter that describes the characteristics of the traffic on a road section or a location on the road. Thus, AADT is a measure that indicates the average flow per day for a year. For the conversion to total vehicle mileage on a stretch of road, the calculation is made by AADT × 365 × the length of the road [5]. The total vehicle mileage in an area is obtained by totalling all relevant stretches of the road. Thus, the basis for estimating total vehicle mileage is traffic counts that are planned to form a basis for the calculation.

15.2 Design of a study for the estimation of total vehicle mileage

Before estimating the total vehicle mileage, the following needs to be defined and decided:

1. the purpose of the study
2. define the time period
3. define the road network
4. strategy for the selection of road sections
5. strategy for measuring AADT
6. what types of vehicles to capture
7. measurements, by which technique
8. how to analyse the data

1. **Purpose of the study:** In this case, exposure data is collected to calculate injury and accident risks to measure the road safety problem. Therefore, traffic counts should be made in such a way that they can form the basis for the calculation.
2. **Define the time period:** Most commonly, a year is used, and often a calendar year, but it depends on the purpose of the study. Note that an average for the whole year should be used despite possibly large variations during the year. It is also important that the comparison is made within the same period for the accidents. Sometimes, it is a shorter period that may be of interest to investigate. If the AADT is not measured annually, it also needs to be adjusted over time, a so-called enumeration factor. There are different ways of calculating

AADT: e.g. index curves can be used for adjustment, factor method, regression model, and using information from similar roads, but all have the same purpose, to adjust for seasonal variations. If it is impossible to measure the whole year, it can be useful to learn how traffic varies over the year and to use this when estimating an annual average instead of just taking the average traffic counts on the days measured. Information on the length of the road and sections is obtained from a suitable database. It is important to get the length of the sections.

3. **Define the road network:** Determine a geographically defined area or the type of roads to be studied based on the purpose. First, before the investigation, the entire road network, motorways, certain speed signs on roads (e.g. 50, 80) and roads with different road safety classifications need to be defined. The selected road network is often taken from a national register. Road length is an important data item. Sometimes, the register contains incorrect data, which may be due to the fact that the specified road network is not updated in the database. This needs to be controlled and handled in some way.
4. **Strategy for selecting road sections:** The first step is to divide the selected road network into homogeneous sections where the traffic flow can be assumed to be constant. Then, the traffic flow is measured on all sections. A measurement site is chosen for each section where the AADT is measured. The location of this site should be randomly selected. Often, there is some variation in the flow of the section. If the location site with the highest traffic flow is always chosen, the total vehicle mileage may ultimately be overestimated. If the AADT is missing for some homogeneous sections, the traffic flow for these needs to be assessed. This could be done in several ways, such as modelling from traffic flow models.

   If impossible to measure all homogeneous road sections, a survey can be designed instead, measuring a probability sample to estimate the total traffic mileage. Such an approach does not have the same flexibility to estimate traffic volumes for different subdivisions of the road network, such as roads with certain speed limits. By randomly selecting homogeneous sections or points on the chosen network, statistical theory can be used to estimate the total vehicle mileage. To reduce the variance (uncertainty in the estimate), the network can be divided into different strata according to the expected flow. The number of sections or points to be selected depends on the purpose, budget and accuracy requirements of the estimation.
5. **Strategy for measuring AADT:** AADT can be measured throughout the year using long-term measurement techniques or for shorter periods using short-term measurements. If short-term measurements are used, a strategy is needed to estimate the average flow throughout the year. By using the information and assigning index curves or other assessments to the different road sections, temporary measurements can be carried out. The short-term measurements should be made on several occasions during the period.
6. **What types of vehicles to capture:** The most general approach is to collect the number of passages for all vehicles, but often, a breakdown by light or heavy

<table>
<tr><th colspan="4">Motor vehicle groups</th></tr>
<tr><th>1</th><th>2</th><th>3</th><th>4</th></tr>
<tr><td rowspan="10">MV Motor vehicle</td><td rowspan="6">LMV Light motor vehicle*</td><td rowspan="2">LMV 1</td><td>MC Motorcycle</td></tr>
<tr><td>MP Moped</td></tr>
<tr><td rowspan="4">LMV 2</td><td>PC Passenger car</td></tr>
<tr><td>LGV Light goods road motor vehicle</td></tr>
<tr><td>LB Light bus, Light motor coach</td></tr>
<tr><td>OLMV Other light road motor vehicle***</td></tr>
<tr><td rowspan="4">HMV Heavy motor vehicle**</td><td rowspan="4">HMV</td><td>HMV Heavy goods road motor vehicle</td></tr>
<tr><td>HB Heavy bus, heavy motor coach</td></tr>
<tr><td>RT Road tractor</td></tr>
<tr><td>OHMV Other heavy road motor vehicle****</td></tr>
</table>

<table>
<tr><th colspan="2">Coupled vehicle</th></tr>
<tr><th>1</th><th>2</th></tr>
<tr><td rowspan="2">WC With coupled vehicle</td><td>Vehicle train ≤ 24.0 m</td></tr>
<tr><td>Vehicle train ≥ 24.0 m</td></tr>
<tr><td>WOC Without coupled vehicle</td><td></td></tr>
</table>

* Gross vehicle weight rating (GVWR) <= 3,500 kg,
** Gross vehicle weight rating (GVWR) >= 3,500 kg
*** Light motor vehicles which do not belong to MC, MP, PC, LGV or LB
**** Heavy motor vehicles which do not belong to HMV, HB or RT

Figure 15.3 Vehicle classification in different levels. Source: From the Nordic Collaboration Project NorSIKT, Greijer and Danielsson [6].

vehicles is desired. Figure 15.3 provides a more detailed list of different types of vehicles that may be of interest to classify and at what level. How these are classified varies between countries, but there is a Swedish definition of all motor vehicles, and in some cases, they are also defined according to Eurostat, the statistical office of the European Union (see Appendix F in Greijer and Danielsson [6]).

Depending on the classification level chosen, it is also necessary to know whether the measuring equipment can detect vehicle classes with sufficient quality, see below.

7. **Measurements, by which technique:** It is recommended to measure AADT in both directions for bi-directional traffic, as well as to estimate the total vehicle mileage. As it is common to use short-term measurements to collect data for this purpose, pneumatic tubes or sometimes radar systems are common, but inductive loops could also be used [7]. When traffic flow is high, permanent measuring stations may be more appropriate as installing temporary equipment may hinder traffic or make the working environment for the installer dangerous.

   For permanent installations, induction is a common technique, but also magnetometers and other technologies. There is a constant development in the field. It is important to set requirements for detection capability. In an evaluation from 2014 [6], 27 different equipment from several manufacturers were tested in Amsberg, Sweden. The selected technologies were tested, sometimes combined, including pneumatic tubes, inductive loops, radar, video detection and magnetometers. The test included vehicle passage time, classification, direction and speed.

It should be noted that new technologies or other ways of analysing data should be considered. The technologies described above are in a current state. We know little or nothing about future technologies, but these are important to recognise. New techniques are always important to evaluate carefully, in terms of representativeness and other advantages and disadvantages before use. This may include measurement accuracy, integrity policies or issues, what types of traffic environments these are adapted to and their conditions. Other aspects to consider are the possible impact of weather on the detection capability (fog, snow, rain), darkness, and access to electricity, to name a few.

8. **How to analyse data:** First, before calculating the total vehicle mileage, data should be checked. There could be several sources of error, but the most common ones are measurement errors and non-response errors. Measurement errors occur when there is an error in the measured flow or if the measurement is taken at the wrong location. Often the degree of detection can vary, and it is common for the measurement equipment to under-detect to some extent, but over-detection can also occur. To some extent, measurement errors affect the uncertainty expressed in the confidence interval, as there are sometimes extreme values that may be more attributable to the capability of the measurement equipment than to the actual number of vehicles that passed. Non-response error occurs when observations are partially or completely missed. This needs to be assessed if it has a negligible or large impact on the estimate. Sometimes, whole or parts of measurement days may be missing. In such cases, these omissions need to be completed, and there are several ways to do this. One recommendation is to enlist the help of a person with statistical expertise for this. There are also road sections with missing data and, as mentioned earlier, these can be supplemented by assessments based on other measurements or traffic models.

   The estimation methodology differs depending on whether it is based on AADT or separate TA measurements. In the case of random sampling, statistical theory can be used for estimation and error margins.

15.3 Speed – approach and analysis

Correlations between traffic speeds and accidents are often requested. Speed is something that can both affect the severity of the accident and increase the risk of the accident occurring [8]. It may also be of interest to follow the speed tendency over time to be able to study if there are any differences. It is usually practical to measure speed when performing the flow measurements, as many techniques have that possibility. If speed is to be measured in addition to the flow, it requires equipment that can capture this and has sufficiently good quality, especially if it is to be divided into, for example, light and heavy traffic. Greijer and Danielsson [6] evaluated the speed detection for various technologies, but developments could have occurred after this study. Therefore, as a purchaser, it is important to make demands on the equipment supplier regarding the quality of speed data.

There are two different ways to present speed, at free flow or actual traffic. Both measure speed changes. The free flow speed estimates the speed the drivers use when they are not slowed down by traffic, winter conditions or other obstacles. Speed at actual traffic describes the prevailing speed regardless of what may have influenced it. There is no right or wrong choice between these two options. Both parameters are interesting but answer two different questions. The estimation of the free flow speed gives some answers to driving behaviour, which is not possible with the actual speed estimation. Actual flow describes the level of speed that occurs on the road regardless of queues or weather conditions. If free flow is chosen, you need to have some rules to decide when there is a free flow [7].

When measuring speed, a distinction should be made between space mean speed and time mean speed. The time mean speed is the arithmetic average speed of all vehicles for a specified period of time. The space mean speed is the average speed of vehicles travelling a given section during a specified period of time and is calculated using the average travel time and length of the section [9]. In this case, it is a question of obtaining speeds on a road network and not at a point, i.e. the space mean speed is preferable.

The relationship between traffic speed and accidents can be calculated using the power model or the exponential model. These two risk models estimate changes in accident risk according to aggregated speed measures (e.g. mean speed) and the number of accidents. The power model calculates the relative change in speed. Note that this has a relatively high effect on reducing the number of accidents at lower speeds compared to changes at higher speeds. The exponential model calculates the difference in speed effect before and after a change. A meta-analysis showed that the power model and the exponential model fit the data very well [8], but it is important to know the differences between them.

References

[1] Nilsson, G. *Traffic Safety Dimensions and the Power Model to Describe the Effect of Speed on Safety* [Doctoral Thesis (monograph)], Lund Institute of Technology and Society, Traffic Engineering, 2004.

[2] Chapman, R. The concept of exposure. *Accid. Anal. Prev.*, 1973; 5: 95–110.

[3] Elvik, R., Erke, A. and Christensen, P. Elementary units of exposure. *Transp. Res. Rec.: J. Transp. Res. Board*, No. 2103, 2009, pp. 25–31. https://doi.org/10.3141/2103-04

[4] Johnsson, C., Laureshyn, A., Dágostino, C. and De Ceunynck, T. 'The 'safety in density' effect for cyclists and motor vehicles in Scandinavia: An observational study.' *IATSS Res.* 2021; 45: 169–175. https://doi.org/10.1016/j.iatssr.2020.08.003

[5] Forman, G., and Brink, J. *NorSIKT 2. Rapport deluppdrag 1. Gemensam Nordisk definition av ÅDT.* Statisticon, 2013.

[6] Greijer, Å. and Danielsson, C. Analysis of measured data for the calculation of accuracy in vehicle classification and various measurement data.

Measurements in Amsberg, Sweden, August 2013. Trafikverket/NordFoU. Report ver. 1.0. 2014-05-07, 2014.

[7] Forsman, G. *NorSIKT 2. Kartläggning ÅDT trafikarbetsindex*. Statisticon, Report v. 3. 2014-03-25, 2014. https://docplayer.se/12636899-Norsikt-2-kartlaggning-adt-trafikarbetsindex-rapport-version-3-datum-2014-03-25-klass.html#google_vignette [accessed 7 May 2024].

[8] Elvik, R., Vadeby, A., Hels, T., and van Schagen, I. Updated estimates of the relationship between speed and road safety at the aggregate and individual levels. *Accid. Anal. Prev.*, 2019; 123: 114–122. https://doi.org/10.1016/j.aap.2018.11.014.

[9] Forsman, G., Haugen, B. and Nyfjäll, M. *NorSIKT IP Some notes on Time Mean Speed and Space Mean Speed*. Report nr 2019-13. NordFoU. 2021. https://api.vejdirektoratet.dk/sites/default/files/2022-08/NorSIKT_IP%20Some%20notes%20on%20Time%20Mean%20Speed%20and%20Space.pdf [accessed 23 August 2024]

Chapter 16

A brief introduction to research ethics

Ulf Görman[1]

16.1 Good research conduct

Science is a gigantic network of search for knowledge, where researchers set up and test hypotheses, and the results are made available to other researchers, mainly through publications. The outcome of the subsequent discussion defines the current standing of scientific knowledge.

For this endeavour to work, scientists must be able to trust the honesty of everyone's contribution. Several ethical conventions and international and national laws have been set up to describe how this aim should be achieved and protected. The result is often labelled *Good Research Conduct* or *Good Scientific Practice*, here briefly summarized (Good Research Practice 2017):

- You shall tell the truth about your research.
- You shall consciously review and report the basic premises of your studies.
- You shall openly account for your methods and results.
- You shall openly account for your commercial interests and other associations.
- You shall not make unauthorised use of the research results of others.
- You shall keep your research organised, for example, through documentation and filing.
- You shall strive to conduct your research without doing harm to people, animals or the environment.
- You shall be fair in your judgement of others' research.

When a researcher violates these rules, whether intentionally or by negligence, this may be considered a case of *scientific misconduct*, which may result in criticism or sanctions.

16.2 Ethics-related control

When you plan to do research on traffic data, you need to be acquainted with the general rules of research ethics and how these are expressed in laws, codes of ethics

[1]Center for Theology and Religious Studies (CTR), Lund University, Sweden

and quality requirements. These include many types of documents, such as international codes of research ethics, specific regulations for certain types of research and national legislation.

Before starting a research activity, you will have to be acquainted with such rules and their application in your country. Consult ethical expertise to learn the general principles and their application in your country and for your activities.

Most countries have also established control systems for ethics in research, in particular for research that involves human beings, animals, and other activities with specific risks. These are, to some extent, different for each country. You will have to investigate how the legislation and the control systems are set up in your country and plan your research accordingly.

The control of research that involves human beings is commonly organised through research ethics committees, in fact, a kind of peer review system. Such committees can be arranged in many ways. They are often set up by specific universities but sometimes as public authorities.

Their field of responsibility can be defined in different ways, and they can have different levels of authority. In many cases, they have the authority to approve specific research projects. Their role can also be mainly advisory, i.e. they offer opinions related to specific research planning. Such approval or opinion is, in many cases, required by law. In other cases, an ethical opinion or approval may be required by sponsors or by publishers.

You will have to investigate how the ethics review of research is organised in your case and learn to know the demands that apply to your research.

16.3 A risk-based approach

The ethics-related control system is basically aimed at avoiding or mitigating risks. The foremost risks related to research that involves human beings are those that affect the *health, safety and personal integrity* of research participants. Physical injury or negative psychological effects during or after participation, as well as burdens, are common risks. Other risks are instances of breach of personal integrity.

To be acceptable, such risks have to be counterbalanced by the scientific value of the research. The scientific value of a research project will have to be made clear. Often, a power calculation will indicate how to recruit enough research persons to arrive at significant results, but not too many. In an *explorative* study (i.e. if there is not yet enough knowledge to perform a power calculation), it is often relevant to only recruit a limited number of research persons in order to learn how to set up a later experiment with more precision. In a *qualitative* study, the number of participants will have to be settled by other means. If the results are well known already, you should, for ethical reasons, not expose any persons at all to an experiment involving risks.

The primary tasks of the ethics committees are to evaluate specific research projects with regard to the following questions:

- Which are the risks connected to the project?
- Have they been adequately eliminated or mitigated in the research programme?

- Are remaining risks acceptable, i.e. counterbalanced by the scientific value of the study?

Information during recruitment is the fundamental instrument for protecting potential participants. Information should include a description of everything the potential participant needs to know to be able to decide whether to take part or not. *Consent* must be voluntary, explicit and specific to a particular research undertaking. Several international and national documents describe in detail the information that will have to be communicated during the recruitment of research participants. Consent may be redrawn without explanation at any time.

Special concerns have to be made with regard to *vulnerable* people. Everyone who can be assumed to have particular difficulties in safeguarding their rights should be considered vulnerable, and they will have to be treated with particular care when conducting research. Whenever the research results can be achieved without recruiting vulnerable people, this should be done in that way.

To protect the *integrity* of your research participants, you will have to observe the legislation for the protection of personal data. You will have to set up a plan for the data you will collect, for data security and identification restrictions, as well as storage after the project. You will have to inform the participants on the data management and receive their consent.

For potential further use of personal data beyond the project, additional consent will, in many cases, be needed. For instance, if you want to use identifiable material in the future for presentations or published articles, you will need consent from the person in question. Details, exceptions and limitations may apply according to national law.

Many other potential risk sources have been identified for specific research activities and need to be identified and mitigated. In your preparation for your research project, you need to thoroughly analyse the risks that may come up and whether they are necessary or how they can be mitigated or eliminated. In this work, you will have to consult colleagues and ethics committees to identify earlier experiences, practices, and conventions.

In the following, three examples of research ideas are presented with some ethics-related comments.

A researcher wants to study the effects of alcohol consumption on bicycle riding abilities. She plans an experiment in which participants are given gradually increasing doses of alcohol, while their task is to cycle through a route that includes riding straight, making turns and manoeuvring at low speeds.

What is the scientific value of the study?

What is the specific research question? Which variables are going to be studied? Have you or other researchers performed similar experiments? What is known already?

Risk management

Identify the risks and look at ways of performing the study with lower risks. Reduce the risk of accidents and arrange for assistance in case of accidents.

Consider engaging medical expertise in recruitment and in the performance of the experiment. Set up inclusion and exclusion criteria to eliminate risk groups.

Can extra wheels or other means be used to prevent falls? Can helmets and other equipment be used to prevent or mitigate injuries? Can a moving platform be used for more controlled conditions?

Set up a plan to release participants after the experiment with minimal risks, as they are still affected by alcohol.

Information and consent

If you plan to film the research persons during the experiment, you will need to consider the integrity of the participants. Information and consent must include a specific description of what you want to do, including aspects such as the possibility to recognise the participant, analysis, storage, and potential use in presentations of different kinds in the future.

A researcher plans to make behavioural observations by filming interactions at pedestrian crossing. To be able to distinguish facial expressions and informal communication signals, the researcher wants to use high resolution camera. The data collection is done in 'natural settings', i.e. studied participants are not recruited, but are random passers-by.

What is the scientific value of the study?

What are the specific potential scientific results that defend the risks related to high-resolution recording?

The potential scientific value of collecting facial expressions and informal communication signals must be high enough to allow a breach of personal integrity.

Camera surveillance

Camera surveillance is considered ethically sensitive, as it may involve integrity risks. Face recognition is considered particularly sensitive, as identification of a person by means of biometric methods is possible. The recording may also reveal other potentially ethically sensitive matters, such as accidents, illegal behaviour, etc.

Camera surveillance in public places is often subject to approval from the relevant national authority. You will have to check and observe the national legislation and practice.

Risk analysis

The possibility of *indirect identification* of persons will have to be considered, in particular when films or photos are taken. Number plates or face recognition may be used for indirect identification. Can identification be made difficult or impossible at recording or later?

Involvement of humans in research without information and consent is considered particularly sensitive and may not be allowed according to national legislation. Consider revision of the design of the study. Contact with the local ethics committee is required.

A researcher wants to study the effects of traffic safety education in primary schools. He wants to interview children, their parents and teachers before and after the completion of a traffic safety course, as well as to observe children interacting with traffic by following after them with a camera on the way to and from school.

Recruitment of children

Children are considered vulnerable and not yet fully competent to decide about complex or sensitive matters and should be given special care when recruited to take part in research. As a general rule, vulnerable people should only be recruited to research when the results of the research can be of value for themselves or other persons in a similar situation.

In most cases, national legislation based upon international conventions, as well as guiding documents and practice, includes specific rules for the involvement of children in research. Observe national and international practices for setting up easy-to-understand information for children, for involving parents and for deciding the age when the children themselves can decide about their participation.

Observation of unforeseen behaviour

It is difficult to know in advance what will happen during an observation study of this kind. You will need to foresee and analyse the most probable complications and prepare measures to mitigate potential risks, such as accidents or near accidents.

Reference

Good Research Practice (2017), The Swedish Research Council/Vetenskapsrådet https://www.vr.se/english/analysis/reports/our-reports/2017-08-31-good-research-practice.html (accessed 2024-09-13).

Chapter 17

Use of personal data and artificial intelligence–ethical and legal issues

Ulf Görman[1]

17.1 Introduction

As indicated in several earlier chapters in this anthology, traffic safety data are of many different types and studied in a large number of different activities.

Many professional studies make use of already existing data, such as accident data records from police, insurance companies and hospitals, as well as statistical data derived from such databases. Other data are collected for specific purposes, such as self-reports related to accidents by means of questionnaires, interviews and mobile equipment. Still other methods discussed in earlier chapters are site-based traffic conflict observations, as well as behaviour observations, made on the spot or by nomadic devices.

Common for these studies is that all handling of such data includes processing of personal data related to human persons. Many of these studies will handle particularly sensitive data such as information related to injuries or illegal activities. Surveillance systems may include accidents or near accidents and may include or allow face recognition, as well as identification of vehicle registration numbers. This chapter describes the ethical issues related to the processing of personal data in all these cases.

17.2 Proviso

The information in this chapter is by no means intended to be exhaustive. All information on ethical or legal issues in this chapter is aimed to be an initial orientation on the regulations in question, including background, overview and explanations. In order to actually observe the legislation or ethics guidance, you will have to further consult original texts and expert opinions. It has not been possible to cover all provisions and possible circumstances.

European legislation and guiding documents are used here as a model for how ethical and legal issues related to personal data and artificial intelligence (AI) are

[1]Center for Theology and Religious Studies (CTR), Lund University, Sweden

and may be handled. At the time of this publication, the legislation and ethics-related guidance provided by the European Union (EU) is the most detailed and influential regulation in the world. It is directly applicable to all professionals working in an establishment in the EU, regardless of whether the processing takes place in the Union or not. Professionals engaged in international cooperation may need to relate to these regulations, depending on the character of the work. In addition, the efforts made in the European Union are an interesting example for all professionals on how to protect personal integrity.

Some pieces of the text are drawn from or inspired by Görman (2024).

17.3 Historical aspects of human rights and data protection

Ever since the mid-twentieth century, starting in the years immediately after the Second World War, the protection of human persons received attention in a number of national and international declarations, conventions and guidelines. This included the protection of individuals with regard to information about themselves.

The United Nations (UN) had been established in 1945. As early as 1948, the member countries agreed on the Universal Declaration of Human Rights, intended to be a common norm for all peoples and nations, with the aim of promoting respect for rights and freedoms for all people and social organizations. Among else, in Article 12, this declaration states the right to privacy.

In 1950, the newly founded Council of Europe agreed on the European Convention for the Protection of Human Rights and Fundamental Freedoms, where Article 8 made the generally held declaration 'Everyone has the right to respect for his private and family life, his home and his correspondence'.

Over the years, many soft law documents have further pointed out the value and importance of the protection of the individual. Several divergent descriptions such as the value of human beings, human dignity, privacy, integrity, etc., have been used. The different wordings may, to some extent, relate to different value systems and ideologies, but they all point in the same direction.

One of the most recent and detailed, and also most influential, documents is the *Charter of Fundamental Rights of the European Union* (2012/C 326/02). Its Preamble points out the fundamental tenets of the Charter: The Union is founded on the indivisible, universal values of human dignity, freedom, equality and solidarity, based upon the principles of democracy and the rule of law. It places the individual at the heart of its activities by creating an area of freedom, security and justice. The Charter aims at strengthening the protection of fundamental rights by making those rights more visible.

Some of the prime principles relevant to the protection of personal data are the following:

- Human dignity is inviolable. It must be respected and protected (Article 1).
- Everyone has the right to respect his or her physical and mental integrity (Article 3).

- Everyone has the right to respect his or her private and family life, home and communications (Article 7).
- Everyone has the right to the protection of personal data concerning him or her (Article 8).

17.4 Personal integrity – why it matters

What is meant by *personal integrity*? Often described as an aspect of human dignity, this complex, collective term encompasses a person's values, conceptions, opinions and desires, beliefs and mental life. As indicated, personal integrity is protected by declarations of human rights and national laws. The right to personal integrity refers to the right of every human being to have his or her individuality and inner sphere respected and not to be subjected to unwanted intrusion.

Why does personal integrity receive so much attention? This becomes evident if we look at the potential threats to integrity. As soon as you use a database of personal data, you must always consider that for one reason or another, someone may have an interest in the information contained therein:

- to find out who is registered in a database or whom certain interesting information concerns,
- to collect more data on one or more already known or identified individuals, and
- to act on the basis of the information obtained about the individual(s).

The researchers themselves may already have such an interest, anyone else who becomes aware of the contents of the register could also be interested in them. Often this is just a matter of curiosity, but sometimes more serious intentions are at play.

Identification facilitates access to personal data. Both identification and access may violate the right to personal integrity, which is why they are often described as an invasion of privacy or a breach of personal integrity. Common risks related to personal integrity include

- identification
- access to information about a known individual
- undermining personal integrity

Such information may also result in harm to health and safety, for example, if it is used for more far-reaching purposes, e.g.

- discrimination
- slander
- opposition
- exploitation
- fraud
- harassment
- persecution

17.5 The EU General Data Protection Regulation (GDPR)

Within the EU, efforts to achieve greater uniformity and clarity regarding the protection of personal data have been ongoing for a long time. Regulation (EU) 2016/679 on the protection of natural persons with regard to the processing of personal data and the free movement of such data, better known as the GDPR, entered into force on 25 May 2018.

The GDPR is the directly applicable legislation regarding personal data processing in the EU member states that prescribes rules and limits for all processing of personal data. The GDPR defines specific areas where national legislation may add supplementary provisions to or exceptions from certain provisions in the Regulation.

The protection of national persons in relation to the processing of personal data is a fundamental right. The Regulation is intended to contribute to the accomplishment of an area of freedom, security and justice, to economic and social progress and to the wellbeing of natural persons.

17.5.1 What are personal data?

Personal data are any kind of information that directly or indirectly relates to a living natural person. The GDPR defines personal data as

> any information relating to an identified or identifiable natural person ('data subject'); an identifiable natural person is one who can be identified, directly or indirectly, in particular by reference to an identifier such as a name, an identification number, location data, an online identifier or to one or more factors specific to the physical, physiological, genetic, mental, economic, cultural or social identity of that natural person (Article 4(1)).

Name and personal identity number are the most obvious examples of data that make it possible to directly link information to a particular person. In addition, research data may constitute personal data for many other reasons.

Indirect identification means that someone who gains access to the information can find out to whom the information refers, even when the information does not contain direct identifiers. Some oft-mentioned examples of indirectly identifiable data are

- information that by simple means can be linked to a particular person, such as a mobile telephone number, email address, or IP address;
- information that together with one or more additional pieces of information can be uniquely linked to a particular person, such as a residential address or GPS data for a person's domicile; and
- information relating to unusual details.

Indirect identification is also possible if the person who gains access to the material can identify to whom the information refers by using specific pieces of information that each on its own cannot be used to identify a person.

A number of studies have investigated how much information is needed in order to make such an indirect identification. It was estimated that three to five specific pieces of information were sufficient to identify to whom the information referred, even in the case of a large amount of data. For instance, see Farzanehfar *et al.* (2021).

17.5.2 What does processing of personal data involve?

The processing of personal data is

> any operation or set of operations which is performed on personal data or on sets of personal data, whether or not by automated means, such as collection, recording, organisation, structuring, storage, adaptation or alteration, retrieval, consultation, use, disclosure by transmission, dissemination or otherwise making available, alignment or combination, restriction, erasure or destruction (GDPR, Article 4(2)).

In other words, when you get access to information that directly or indirectly can be linked to a living natural person, all processing of this material is considered personal data processing.

17.6 General rules for the management of personal data

The principles of data protection should apply to any information concerning an identified or identifiable natural person. The protection of natural persons should be technologically neutral and should not depend on the techniques used. It should apply to the processing of personal data by automated means, as well as to manual processing if the personal data are contained or are intended to be contained in a filing system.

However, the Regulation does not apply to the processing of personal data by a natural person in the course of a purely personal or household activity and thus with no connection to a professional or commercial activity. Personal or household activities could include correspondence and the holding of addresses or social networking and online activity undertaken within the context of such activities.

The GDPR aims to ensure an equivalent level of protection of natural persons and the free flow of personal data by setting up a number of general principles for processing personal data, which are described in Article 5:

- lawfulness, fairness and transparency
- purpose limitation
- data minimization
- accuracy
- storage limitation

- integrity and confidentiality
- accountability

Most of these general principles are summarized below and are also needed to fully understand the more detailed regulations on lawfulness, fairness and transparency, which are discussed later.

Purpose limitation

Personal data shall be processed for specified purposes. They shall be collected for specified, explicit and legitimate purposes and not further processed in a manner that is incompatible with those purposes. Among else this means, as a main rule, that data may not be kept any longer than necessary for the defined purpose.

However, further processing for, among else, scientific or historical research purposes or statistical purposes shall not be considered to be incompatible with the initial purposes. This means that data collected for research purposes may be stored after the original task has been fulfilled, and they may be re-used for further research in the future.

With regard to the specific rules for research purposes and statistical purposes, see *Safeguards and derogations related to personal data in the public interest or scientific purposes*, later in this text.

Data minimization

The data you collect must be adequate, relevant and limited to what is necessary in relation to the needs of the purpose in question. You are not allowed to collect additional data that are not necessary for this purpose. This limitation applies even though you may think that additional information could have been useful for further scientific or other purposes.

Accuracy

Accuracy is about ensuring that data is carefully and correctly collected and documented. It is also about ensuring that collected data is correct and precise, for instance, that it gives a true idea of the information provided by the data subject.

Accuracy is also about keeping data correctly up to date. This may not be relevant when data are collected in relation to a specific situation.

Storage limitation

Personal data must not be kept in a form that permits the identification of data subjects for longer than is necessary for the purpose. However, they may be stored for longer periods insofar as the personal data will be processed solely for archiving purposes in the public interest, scientific or historical research purposes or statistical purposes.

Integrity and confidentiality

Integrity and confidentiality refer to the *security* of personal data, including protection against unauthorized or unlawful processing, accidental loss, destruction or damage.

Each person or organisation responsible for the processing of personal data must arrange for adequate security measures for the data collected. You need to ask the security experts who are responsible for data security at your workplace to assist you in organising an adequate level of security for your data.

Accountability

Accountability refers to the responsibility for compliance with all items mentioned above. According to GDPR, the natural or legal persons responsible for the processing of personal data must be clearly identified. You must identify the organization and/or person(s) who are responsible for the processing of personal data.

Who is responsible for fulfilment of the regulations in GDPR?

Everyone who is involved in the processing of personal data is obliged to observe the regulations of GDPR. However, you are not alone in this endeavour. GDPR points out a number of persons and organisations with responsibilities in this respect:

- The *Controller* is the natural or legal person, public authority, agency or other body which, alone or jointly with others, determines the purposes and means of the processing of personal data. In most cases, your university or the public authority where you are employed is the controller.
- The controller may have one or several *representatives*, i.e. natural or legal persons who represent the controller. The representative is the contact point for the controller.
- *Data protection officer* (DPO) is a person designated by the controller to monitor compliance with GDPR and to give information and advice with regard to processing of personal data. Public authorities are obliged to designate a DPO.
- *Processor* means a natural or legal person, public authority, agency or other body who processes personal data on behalf of the controller. The processor is the person or unity within or close to the project, who manages the actual processing of personal data.The controller and the processor shall ensure that the DPO is involved, properly and in a timely manner, in all issues that relate to the protection of personal data.

  The controller has the final responsibility that activities involving personal data are correctly performed. To fully understand and apply the GDPR to your activities that involve personal data, you have to consult the controller or the DPO, follow their advice and obtain confirmation for the planned activities.
- The *Supervisory Authority* is an authority responsible at Member State level for monitoring the application of GDPR. The supervisory authority has the task

of supervising and advising the processing of personal data to make sure that it does not lead to an undue breach of personal integrity. Among else, it has the power to carry out inspections and give orders and can impose fines.

Each controller and processor is obliged to cooperate with the supervisory authority. Among else, the controller is obliged to notify *a personal data breach* to the supervisory authority without undue delay. A personal data breach is a security incident, leading to accidental destruction or unauthorized access to personal data, as detailed in Article 4(12).

17.7 Lawfulness

Any processing of personal data must have a valid legal ground as described in GDPR. Article 6(1) contains an exhaustive list of the lawful alternatives. In short, processing must be either based upon consent from the data subject or on one of five specific legal grounds.

Consent is in many cases relevant and required for data processing, with some relevant exceptions discussed later in this text. However, consent is not acceptable as the legal basis of processing personal data, when there is a clear imbalance of power between the controller and the data subject, in particular when the controller is a public authority.

Before you start processing personal data you must carefully investigate whether any of these alternatives may be applicable to your planned activities. You need to make documentation of which alternative you find relevant and applicable to your further processing of personal data.

In many cases, where road traffic data are relevant, the lawfulness of processing personal data can be based on Article 6(1) point (e):

> processing is necessary for the performance of a task carried out in the public interest or in the exercise of official authority vested in the controller.

Research is commonly considered a task that is carried out in the public interest.

17.7.1 Special categories of data

Particular restrictions are applicable for the processing of special categories of data, often also described as *sensitive* personal data. Special categories of data are defined in Article 9(1). These are data revealing

- racial or ethnic origin
- political opinions
- religious or philosophical beliefs
- trade union membership

or involving the processing of

- genetic data
- biometric data for the purpose of uniquely identifying a natural person

- data concerning health
- data relating to a natural person's sex life or sexual orientation

Processing of such data is in principle forbidden but can be processed for a number of specific purposes, listed in Article 9(2). Two such purposes are particularly relevant for the professional work with traffic data:

(g) processing is necessary for reasons of substantial public interest, and

(j) processing is necessary for [. . .] scientific or historical research purposes or statistical purposes in accordance with Article 89(1)

In both cases, there are specific detailed preconditions for this work. In particular, data relevant to the study of traffic safety data may include data concerning health and biometric data. However, be aware that other kinds of restricted data may happen to be included. You may need to set up a policy for the case that unforeseen restricted data will be collected during your work. You will have to consult your controller and/or DPO to set up the data management correctly.

17.7.2 Personal data relating to criminal convictions and offences

The processing of personal data relating to criminal convictions and offences is also restricted (Article 10):

> Processing of personal data relating to criminal convictions and offences, or related security measures based on Article 6(1) shall be carried out only under the control of official authority or when the processing is authorised by Union or Member State law providing for appropriate safeguards for the rights and freedoms of data subjects. Any comprehensive register of criminal convictions shall be kept only under the control of official authority. (Article 10)

Please note that the processing of data related to criminal convictions and offences is more restricted than for special categories of data. National law may have set up additional regulations. Some universities may be official authorities. Consult your DPO and, if needed, the National Supervisory Authority in cases of doubt.

17.8 Fairness and transparency

Adequate *information* to data subjects is one of the most important instruments for fair and transparent processing of personal data, according to GDPR. As a general rule, the data subject shall be informed of the existence of the processing operation and its purposes.

The principle of transparency requires that any information and communication relating to the processing of those personal data be easily accessible and easy to understand, and that clear and plain language be used. It should be transparent to natural persons that personal data concerning them are collected, used, consulted or otherwise processed and to what extent the personal data are or will be processed.

In particular, the specific purposes for which personal data are processed should be explicit and legitimate and determined at the time of the collection of the personal data. The personal data should be adequate, relevant and limited to what is necessary for the purposes for which they are processed.

Article 13 describes the information to be provided where personal data are collected directly from the data subject. You may want to ask them to become involved in some kind of activity, where collection of personal data is a part, or you are only interested in information they may provide, such as opinions or personal details.

In any of these cases, you will have to provide your informants with all applicable information listed in Article 13. You are not allowed to make an ad hoc selection of the information provided.

Article 14 describes the information to be provided where personal data have *not been obtained* from the data subject. The possibility to process data that have not been obtained from the data subject himself of herself is (mainly) relevant when you want to use data that have already been collected and may be taken from some kind of already existing database. This is often labelled *secondary use of data*. As a main rule, you are obliged to inform the data subject also for this kind of data.

Article 14(5) describes the *exceptions from the duty to inform* the data subject. One of these exceptions is related to public interest, scientific or statistical purposes, which are discussed in Article 89. For further information on this, see *Safeguards and derogations related to personal data in the public interest or scientific purposes* later in this text.

Please note that in such cases the controller has a particular responsibility to protect the data subject's rights and freedoms and legitimate interests in other ways, such as making the information publicly available. This can be done in several ways. Traditionally, advertisements in newspapers have been dominating. Today, however, several other methods have become common, such as information on the website of the controller, or information on relevant social media.

17.8.1 The rights of the data subject

The data subject shall have several rights:

- Right of access, i.e. right to obtain from the controller whether personal data are being processed, and if so, access to the personal data and important circumstances (Article 15).
- Right to rectification, i.e. to have wrong information revised or completed (Article 16).
- Right to be forgotten, i.e. to have the information erased, under a few specific circumstances (Article 17).
- Right to restriction, i.e. to have the processing limited, under a few specific circumstances (Article 18).

- The controller shall communicate the rectification, erasure or restriction to each recipient to whom the personal data have been disclosed, unless impossible or disproportionate (Article 19).
- Right to data portability, i.e. the right to have personal data transmitted to another controller under a few specific circumstances (Article 20).
- Right to object to automated decision-making (Article 21).

Member State law may provide derogations from some of these rights for scientific or statistical purposes.

17.9 Consent

Consent from the data subject is a core principle in GDPR. Consent should be given by a clear affirmative act establishing a freely given, specific, informed and unambiguous indication of the data subject's agreement to the processing of personal data relating to him or her.

This could be a written statement, including by electronic means, or an oral statement. It could also include ticking a box when visiting an internet website, choosing technical settings for information society services or another statement or conduct which clearly indicates in this context the acceptance of the proposed processing of his or her personal data. Silence, pre-ticked boxes or inactivity should not constitute consent.

Consent should cover all processing activities carried out for the same purpose or purposes. When the processing has multiple purposes, consent should be given for all of them.

It is often not possible to fully identify the purpose of personal data processing for scientific research purposes at the time of data collection. Therefore, data subjects should be allowed to give their consent to certain areas of scientific research (Recital 33).

Consent is presumed *not* to be freely given if it does not allow separate consent to be given to different personal data processing operations (Recital 43).

As a consequence, it must be clear in the consent form that consent is given to each research activity and each data processing operation separately, for instance, by ticking a separate box for each operation.

Please note that you will need to keep record on the specific consent you have received from each data subject and to be observant of the consent or absence of consent for each processing operation.

17.9.1 Withdrawal of consent

A data subject should have the right to withdraw consent to the processing of personal data concerning him or her at any time, in GDPR also called a 'right to be forgotten'. Consent should not be regarded as freely given if the data subject has no genuine or free choice or is unable to refuse or withdraw consent without detriment. It shall be as easy to withdraw as to give consent.

17.9.2 Information and consent to participation in activities vs. data processing

When persons are recruited to take part in research, they may be involved in specific research activities, which may involve the collection of personal information. It should be noted that these aims are two distinct purposes. The processing of personal data is regulated by the GDPR, while national legislation may include specific conditions for research participation. The requirements for information and consent related to each of these regulations may be different and will both have to be fulfilled.

17.9.3 Safeguards and derogations related to personal data in the public interest or scientific purposes

Archiving purposes in the public interest, scientific and historical research purposes and statistical purposes are privileged activities according to the GDPR.

- *Further processing* for such purposes is considered to be compatible with the initial purpose, i.e. an exception from the purpose limitation (Article 5(1)(b)).
  It is often not possible to fully identify the purpose of personal data processing for scientific research purposes at the time of data collection. Therefore, data subjects should be allowed to give their *consent* to certain areas of scientific research when in keeping with recognised ethical standards for scientific research (Recital 33).
- Personal data may be *stored* for longer periods than necessary for the original purpose, i.e. an exception from the storage limitation (Article 5(1)(e)).
- *Special categories of data* may be processed for such purposes when proportionate to the aim pursued (Article 9(2)(j)).
- *Information* to the data subject on secondary use of personal data for such purposes is not required when such information proves impossible or would involve a disproportionate effort. In such cases, the controller shall take appropriate measures to protect the data subject's rights and freedoms and legitimate interests, including making the information publicly available (Article 14(5)(b)).
- The data subject shall not have the *right to be forgotten* when data processing is necessary for such purposes (Article 17(3)(d)).

However, such data shall be subject to appropriate safeguards. Article 89 mentions in particular:

- Respect for the principle of data minimisation, i.e. that data are *limited* to what is necessary in relation to the needs of the purpose in question;
- That data are *pseudonymised* when appropriate;
- Further processing that does *not permit identification* of data where the purposes can be fulfilled in that manner.

The GDPR mentions in particular *pseudonymisation* as an appropriate safeguard that impedes identification of personal data in connection with research

Article 89(1). Pseudonymisation involves replacing the names of the people concerned with a code number and drawing up a special code list, a so-called code key, and keeping this in a separate location. Pseudonymised data are to be considered identifiable, and the regulations in GDPR are applicable.

Union or Member State law may also provide for derogations from some of the rights of the data subject, when such rights are likely to render impossible or seriously impair the achievement of the specific purposes of the processing and such derogations are necessary for the fulfilment of those purposes. Member States differ substantially in employing such derogations.

Whenever you consider setting up a policy for applying the derogations mentioned above, you need to consult your controller and/or DPO and arrange for a documentation of the decision.

17.10 GDPR on profiling

Some of the regulations of GDPR are directly applicable to the use of AI for processing of personal data.

The data subject should have the right not to be subject to *profiling*, which consists of any form of automated processing of personal data evaluating the personal aspects relating to a natural person.

In particular, this applies to efforts to analyse or predict aspects of the data subject's performance at work, economic situation, health, personal preferences or interests, reliability or behaviour, location or movements, where it produces legal effects concerning them or similarly significantly affects them.

Exceptions to this prohibition can be expressly authorised by Union or Member State law. In any case, such processing should be subject to suitable safeguards, which should include specific information about the data subject.

The controller should use appropriate procedures for the profiling, to ensure, in particular, that factors that result in inaccuracies in personal data are corrected and the risk of errors is minimized, among else discriminatory effects on natural persons on the basis of sensitive personal data. Automated decision-making and profiling based on special categories of personal data should be allowed only under specific conditions.

These provisions are detailed in GDPR Recital 71 and in Articles 4, 13, 14, 15 and 22.

17.11 Anonymous information

Anonymity is not a regulative concept in GDPR, but a derived concept, indicating that when the data subject is no longer directly or indirectly identifiable, it can be described as anonymous.

> The principles of data protection should therefore not apply to anonymous information, namely information which does not relate to an identified or

> identifiable natural person or to personal data rendered anonymous in such a manner that the data subject is not or no longer identifiable. This Regulation does not therefore concern the processing of such anonymous information, including for statistical or research purposes. (Recital 26)

The principles for data protection according to GDPR are not applicable to anonymous information. However, it is important to note that when information is understood as anonymous, this is based upon an assessment of what is reasonable, taking into account available technology, time consumption, expenses, probability, etc. Such circumstances and estimations may change.

As a consequence, when you plan to make secondary use of data for a new purpose, a new estimation may be necessary to evaluate whether the data may be considered anonymous in this new context.

17.12 International projects

Many scholars are involved in international projects, where researchers from multiple countries participate, investigations may be carried out in several countries and data may be transmitted between countries. Whenever you are engaged in such activities, you will have to ensure that the work is carried out in accordance with the regulations in force in each country where any of the activities take place.

For instance, it may be noted that there is a crucial difference between the legislation with regard to the protection of personal data in the EU and the United States. This difference can be described by focusing on the concept of *linked* data. Personal data is considered linked when the data are pseudonymised, and the current holder of the database does not have access to the code for retrieval of the identity of the data subjects. The code is instead kept by someone else.

According to the US Code of Federal Regulations (Title 45), linked data is considered anonymous and can be freely used without protection or permission from the data subject. In contrast, the EU regulation considers such data as personal data, which requires data protection according to the regulations in GDPR. This difference has created many misunderstandings within the scientific community, including ambiguity in the understanding of the concept of anonymity. This difference is also one important reason why the EU does not consider the US as a country with adequate protection of personal data. Accordingly, the exchange of personal data, in particular sensitive personal data, between the EU and the US is subject to restrictions (Bergmann *et al.* 2008).

Data exchange with other countries

Rules for the processing and protection of personal data are different from one country to another. For this reason, the EU monitors the protection of personal data and issues directions for the transfer of personal data between countries. The responsibility for data security rests on the authorities in each country where a portion of the data is handled. However, research participants have the right to know that data are being transferred to another country, and the research participant information shall include information about the transfer of data and where the data are processed.

For projects that take place in part in countries outside the EEA, two conditions have to be considered:

Export of data

Personal data may only be transferred to a country outside the EU/EEA, a so-called *third country*, under certain limited circumstances.

Import of data

It is not uncommon that researchers want to import information containing personal data from a third country. The researcher him- or herself or another person may plan to collect data in connection with field research or research collaborations or have sometimes already collected such data. It also happens that researchers want to reuse information archived in a third country from a previous research project. This may only be done if the research participants to whom the data refer have enjoyed protection corresponding to the EU level of protection for their safety and integrity, in connection with the collection and continued processing of the information.

Whenever you plan to engage in international exchange of personal data, you will need to consult expertise in this respect, such as your DPO.

17.13 Open access to research data

For some years work has been going on to create open access to data that have been collected or generated in connection with research. The purpose of this work is often described in terms of the so-called *FAIR principles*, according to which research data should be Findable, Accessible, Interoperable, and Reusable. This work refers to many types of research data and also includes data derived from research involving humans.

The aim of the Open Data Directive (EU) 2019/1024 is to ensure that, among other things, research data can be reused by being made available in accordance with the principle of *open by default*. The Directive does not affect the legal protection of individuals with regard to the processing of personal data, and the reuse of personal data is only permitted if the principle of purpose limitation is met.

A number of research funding bodies already require that researchers draw up a data management plan (DMP), which, among other things, describes how the FAIR principles are realised and how the protection of personal data is considered.

For researchers involved in work with open access to research data, it is therefore important to investigate to what extent the availability of the information about humans that will be collected or generated within the framework of a project is in line with the principles of the GDPR, with respect to, among else, purpose limitation, information, and consent.

In particular, this applies to sensitive personal data. You will have to consider whether it is at all appropriate and legally possible to make such personal data openly accessible. If making certain data from a research project available proves to be incompatible with the GDPR principles, then these data will have to be excluded from open access. Decisions of this kind need to be documented in the data management plan for the activity.

17.14 What is new with AI?

AI is the theory and development of a group of fast-evolving technologies for using computers to simulate human intelligence and problem-solving capabilities. AI may be able to perform tasks that have so far been considered unique for the human mind and intelligence.

Among else, this includes the analysis of huge amounts of data to

- discover and analyse complex connections
- construct high-level theories
- make predictions
- produce advice or make other suggestions
- make decisions

Such results of machine activities may involve or affect human persons and may influence or replace human decision-making processes.

The evolution of AI has been made possible by major technological advances in computational power and storage capacity, the availability of large amounts of digital data, as well as innovations in scientific methods and engineering, including machine learning, deep learning and artificial neural networks.

It is clear that the prospect of achieving a better understanding of complex problems makes the use of AI attractive also in research and other data analysis based upon on traffic data.

This section describes some major ethics and legal issues that may be relevant if you perform scientific or other analytic or developmental work based upon traffic data and consider the use of AI, irrespective of whether this relates to the development, deployment, or use of AI or AI-produced results. When advice and decisions that affect humans are left to machines, new risks emerge. The ethical concerns are all based upon the understanding of such risks.

17.14.1 The EU AI Act

In 2024, the EU adopted the world's first comprehensive regulation of AI, the EU AI Act. It was approved by the European Parliament on 13 March 2024 and by the EU Council on 21 May 2024. The Act complements the GDPR and the Law Enforcement Directive (Directive (EU) 2016/680). It will become applicable at different times, depending on the character of each regulation.

Initially, in the Preamble, it is explained that AI is a fast-evolving family of technologies that can bring a wide array of economic and societal benefits across the entire spectrum of industries and social activities. By improving prediction, optimising operations and resource allocation, and by personalising service delivery, the use of AI can support socially and environmentally beneficial outcomes and provide key competitive advantages to companies and the European economy.

However, the use of AI with its specific characteristics (e.g. opacity, complexity, dependency on data, and autonomous behaviour) can also adversely affect a number of fundamental rights enshrined in the EU Charter of Fundamental Rights. The AI Act seeks to ensure a high level of protection for those fundamental rights and aims to address various sources of risks through a clearly defined risk-based approach. The AI Act aims to be as technology-neutral and future-proof as possible.

Rules for AI available in the Union market or otherwise affecting people in the Union should be *human-centric*, so that people can trust that the technology is used in a way that is safe and compliant with the law, including the respect for fundamental rights.

The rules set up in the Act aim to cover all stages in the work with an AI system, such as design, development, deployment, placement on the market and use.

The focus of the AI Act is the regulation of certain high-risk AI systems and restrictions on the uses of remote biometric identification systems with regard to the processing of personal data, among else restrictions on the use of AI systems for real-time remote biometric identification in publicly accessible spaces.

Prohibited AI practices

Certain particularly harmful AI practices are prohibited as contravening Union values, while specific restrictions and safeguards are proposed in relation to certain uses of remote biometric identification systems for the purpose of law enforcement.

The prohibitions cover

- Practices that have a significant potential to manipulate persons through *subliminal* techniques beyond their consciousness.
- Systems that *exploit any vulnerabilities* of specific groups such as children or persons with disabilities in order to distort their behaviour in a manner that is likely to cause him or her or another person psychological or physical harm.
- Systems for the *evaluation* or classification of the trustworthiness of natural persons, based upon their social behaviour or personal characteristics, with the social score leading to detrimental, unfavourable, unjustified or disproportionate treatment.
- The use of 'real-time' remote *biometric identification* systems in publicly accessible spaces for the purpose of law enforcement, unless necessary for specific and detailed objectives (Article 5).

AI systems may also facilitate other manipulative or exploitative practices affecting adults. These are not mentioned in the AI Act, as they could be covered by the existing data protection, consumer protection and digital service legislation.

High-risk AI systems

The focus of the AI Act is the regulation of *high-risk* AI systems that pose significant risks to the health and safety of fundamental rights of persons. In line with a risk-based approach, those high-risk AI systems are permitted on the European market subject to compliance with certain mandatory requirements and a conformity assessment set up in advance.

The classification of an AI system as high-risk is based on the intended *purpose* of the AI system, in line with existing product safety legislation. Therefore, the classification as high-risk does not only depend on the function performed by the AI system but also on the specific purpose and modalities for which that system is used.

Two main categories of high-risk AI systems are identified:

1. Those used as *safety components* of certain products.
2. Other stand-alone AI systems with *fundamental rights implications*, explicitly listed in Annex III. This list may be updated in the future.

One of these listed high-risk systems is of particular interest for the work with traffic safety data:

> AI systems intended to be used as safety component in the management and operation of road traffic.

Such systems are placed on par with other AI systems related to the management and operation of the supply of water, gas, heating and electricity.

Requirements for high-risk AI systems

For such high-risk AI systems, a number of specific requirements are considered strictly necessary to mitigate the risks to fundamental rights and safety posed by AI. They are described in great detail under the following topics (Articles 9–15):

- risk management system
- data and data governance
- technical documentation
- record-keeping
- transparency and provision of information to users
- human oversight
- accuracy, robustness and cybersecurity

The technical solutions to achieve compliance with those requirements are not described in the Act but may be provided by standards or other technical specifications or be developed in accordance with general engineering of scientific knowledge at the discretion of the provider of the AI system.

These key requirements are not covered by other existing legal frameworks but reflect a widespread and common approach, as evidenced by many ethical codes and principles, that AI development and use should be guided by certain essential value-oriented principles. They build on several years of analysis and involvement of the High-Level Expert Group on AI (HLEG) and many stakeholders including academics, businesses, social partners, non-governmental organisations, Member States and citizens.

Duties of providers and users of high-risk systems
The AI Act also sets up a number of specific obligations for providers and users of high-risk AI systems.

AI providers will be obliged to provide meaningful information about their systems and the conformity assessment caried out on these systems. Moreover, AI providers will be obliged to inform national competent authorities about serious incidents or malfunctioning that constitute a breach of fundamental rights obligations as soon as they become aware of them, as well as any recalls or withdrawals of AI systems from the market.

The duties of providers and users of high-risk AI systems and other parties, organisation and obligations related to notification, and the harmonisation of standards are all described in Articles 16–29 and cover (among else) the following topics:

- compliance with the requirements for high-risk systems
- quality management system
- technical documentation
- automatically generated logs
- conformity assessment procedure
- registration
- corrective actions in case of lack of conformity
- duty of information to national competent authorities
- cooperation with competent authorities
- certificates and ce-marking of conformity to standards
- transparency

These rules will be enforced through a governance system at Member State level, building on already existing structures, and a cooperation mechanism at Union level with the establishment of a European Artificial Intelligence Board.

The Commission will be in charge of monitoring the effects of the AI Act. It will establish a system for registering stand-alone high-risk AI applications in a public EU-wide database.

Transparency obligations for certain AI systems

For some specific AI systems, only minimum *transparency obligations* are proposed, in particular when chatbots or 'deep fakes' are used (Article 52).

Specific obligations to *inform the persons concerned* apply for

- systems that aim to interact with humans
- emotion recognition systems
- biometric categorisation systems
- systems that generate or manipulated image, audio or video content

These obligations are applicable irrespective of whether these are high-risk systems or not. Exceptions apply for systems authorised to investigate criminal offences.

Other AI systems than high-risk

The AI Act sets up a regulatory framework for high-risk AI systems only, with the possibility for all providers of non-high-risk AI systems to follow a code of conduct. Businesses or public authorities that develop or use any AI applications not classified as high risk only have minimal obligations of information. However, they could choose to join others and together adopt a voluntary code of conduct to follow suitable requirements, and to ensure that their AI systems are trustworthy (Article 69).

The requirements mentioned in this advice are those set up in the AI Act and compulsory for high-risk AI systems.

17.14.2 Ethics guidance for trustworthy AI

A large number of guidance documents related to the use of AI have been published over the years. An overview of and analysis by Jobin *et al.* in 2019, claims that their analysis reveals 'a global convergence emerging around five ethical principles (transparency, justice and fairness, non-maleficence, responsibility and privacy)' (Jobin *et al.* 2019).

So far, the most ambitious and elaborated analysis and guidance on the ethical aspects of AI has been set up by the EU. In 2018, the European Commission appointed a group of experts to provide advice on its AI strategy, the high-level expert group on artificial intelligence (AI HLEG). The results of the work of the AI HLEG have been made public in several documents, above all The Ethics Guidelines for Trustworthy AI, 2019, and The Assessment List for Trustworthy AI (ALTAI) 2020).

The Guidelines are aimed to be a voluntary tool for stakeholders to operationalise their commitment to trustworthy AI by offering a clear, elaborated, and concise description of the key issues of main ethical concerns related to the use of AI, and how they can be mitigated. They explain that AI has the potential to significantly transform society and is seen as a promising means to increase human flourishing, by enhancing individual and societal well-being and the common good, as well as bringing progress and innovation. AI systems need to be human-centric, resting on a commitment to their use in the service of humanity and the common good. However, AI systems also give rise to certain risks that must be handled appropriately and proportionately. This entails seeking to maximise the benefits of AI systems while at the same time preventing and minimising their risks.

> In a context of rapid technological change, we believe it is essential that trust remains the bedrock of societies, communities, economies and sustainable development. We therefore identify Trustworthy AI as our foundational ambition, since human beings and communities will only be able to have confidence in the technology's development and its applications when a clear and comprehensive framework for achieving its trustworthiness is in place (p 4).

17.14.3 Foundations of Trustworthy AI

Fundamental rights

When describing how trustworthiness can be founded, the Guidelines point out the *fundamental human rights* that are described in similar ways in a large number of documents, in the first place, the Universal Declaration of Human Rights adopted by the United Nations General Assembly on 10 December 1948. For the first time in human history, this declaration spells out basic civil, political, economic, social and cultural rights that all human beings should enjoy. It has been followed by a number of international human rights treaties and other instruments, which States are bound to respect. In particular, the Guidelines mention the EU Treaties and the EU Charter, with similar content.

Respect for human dignity

Every human being possesses an intrinsic worth, which entails that people must be treated with respect for them as persons, including protection of physical and mental integrity, personal and cultural identity, and essential needs.

Freedom of the individual

Human beings should remain free to make life decisions for themselves, which entails freedom from intrusion and freedom of expression, right to private life and privacy, mitigation of illegitimate coercion, unjustified surveillance, deception or unfair manipulation.

Respect for democracy, justice and the rule of law

All governmental power must be legally authorised and limited by law, respect equality before the law, the plurality of values and the life choices of individuals.

Equality, non-discrimination and solidarity

Equal respect for the dignity of human beings must be ensured, which entails non-discrimination and respect for potentially vulnerable persons or groups.

Citizens' rights

Citizens benefit from a wide array of rights to public information, public goods and services to society.

Trustworthy AI must safeguard compliance with these fundamental rights. They need to have built-in systems for securing that they are upheld, respected and executed.

AI systems with capacity of independent analysis followed by recommendations or decisions that influence human lives may easily come in conflict with these basic values. In one way or other, such AI systems will have to *learn, integrate, and execute* these values. The system's operations should not generate unfairly biased output, and the outcome should not come in conflict with these fundamental values.

It has many times been noticed that AI systems may copy and execute hidden biases in the data material they are using. Trustworthy AI needs to have effective built-in knowledge to avoid such outcomes.

Wider context

In a wider ethical context, the Guidance points out four ethical principles or ethical imperatives that have to be observed by any AI system that interconnects with humans or information on humans:

Respect for human autonomy

Humans interacting with AI systems must be able to keep full and effective self-determination over themselves, which means that human oversight of the work processes of AI systems must be secured.

Prevention of harm

AI systems must not cause or aggravate harm or otherwise adversely affect human beings. Vulnerable persons should receive greater attention. Particular attention must be paid to asymmetries of power or information.

Fairness

AI systems must be free from unfair bias, discrimination and stigmatisation. Fairness also refers to the ability to seek effective redress against decisions made by AI systems and by the humans operating them.

Explicability

AI processes need to be transparent, to the extent possible to those directly and indirectly affected. Measures for traceability, auditability and transparent communication on system capabilities may be required.

17.14.4 Requirements for trustworthy AI

Based upon these general considerations, the Guidance presents seven detailed requirements of trustworthy AI, briefly summarised below:

- human agency and oversight
- technical robustness and safety
- privacy and data governance
- transparency
- diversity, non-discrimination and fairness
- societal and environmental well-being
- accountability

Human agency and oversight

Human Agency and Autonomy concerns the risks related to AI systems that are aimed at guiding, influencing or supporting humans in decision-making processes.

> Are end-users adequately made aware that they are interacting with an AI system, and that advice etc. is the result of an algorithmic decision? Are procedures in place to avoid that end-users over-rely on the system, or that the system in other ways affect human autonomy? Are adequate precautions taken to minimise risks for addiction or manipulation?

Human Oversight concerns the risks related to systems that are self-learning or in any other way autonomous.

> Are humans capable of intervention in all activities and every decision cycle of the system, including detection and response mechanisms for undesirable adverse effects? Do humans have the possibility to abort an operation when needed, and control measures to halt the self-learning or autonomous character of the AI system?

Technical Robustness and Safety

Resilience to Attack and Security concerns risks related to robustness when facing changes in the operating environment or the presence of other agents that may interact with the system in an adversarial manner.

> Are risks of threats such as technical faults, attacks or misuse identified and mitigated? How exposed is the system for cyber-attacks? How are such attacks mitigated?

General Safety concerns include risks related to weaknesses in design or technical performance and environmental threats.

> Are such risks adequately identified, such as the dependency of the AI system's decisions on its stable and reliable behaviour? Have adequate preparations for fault tolerance been made? Is the possibility of malicious or otherwise inappropriate use analysed and counteracted?

Accuracy is about the capacity of the system to perform its task correctly and with adequate precision.

> Are measures put in place to ensure that the data used to develop the system is of adequate quality and representative for the future tasks of the system? How stable and reliable is the capacity of the AI system to arrive at correct conclusions? What consequences may deficiencies in these aspects have? How have these been mitigated?

Robustness also includes Reliability, Fallback plans and Reproducibility.

> Are processes put in place to monitor if the AI system is meeting the intended goals and to ensure reproducibility? Are verification and validation methods and documentation put in place? Have fall-back plans been defined and tested to address system errors? When a system is using continual learning, have potentially negative consequences been analysed and mitigated?

Privacy and Data Governance

Privacy concerns the potential impact of the AI system on privacy and data protection.

> Has the AI system been analysed with respect to its potential impact on the right to privacy and integrity and data protection? Have adequate mechanisms been established that allow flagging such issues?

Data Governance brings up matters relevant in case the AI system was developed or trained by using personal data, including sensitive personal data, or may be used for such purposes.

> Have all requirements and restrictions related to the use of personal data according to GDPR been observed during development and deployment of the AI system?

Transparency

Traceability refers to the possibility to verify the background or history of the decisions of an AI system.

> Is it possible to trace back which data was used to make a certain decision, and which AI model or rules led to the decisions or recommendations? Have measures been put in place to assess the quality of these outputs? Are logging practices in place to record the decisions and recommendations?

Explainability refers to the ability to explain the technical processes and the reasoning behind the decisions or predictions that the AI system makes.

> Are you able to explain the decisions of the AI system to the users? Are you able to verify that the users understand the decisions and their explanation?

Communication is about whether the AI system's capabilities and limitations has been communicated to the users in an appropriate manner.

> Is it communicated to users that they are interacting with an AI system instead of a human? Are mechanisms established to inform users about the purpose, criteria and limitations of the decisions generated by the AI system?

Diversity, non-discrimination and fairness

Avoidance of Unfair Bias has been proved important, as AI systems may suffer from the inclusion of inadvertent historic bias, incompleteness, and bad governance models. Identifiable and discriminatory bias should be removed in the collection phase where possible.

> Has a strategy been established to avoid creating or reinforcing unfair bias in the AI system, regarding the use of input data as well as for algorithm design? Are processes in place to test and monitor for potential biases? Are educational and awareness initiatives in place to help AI designers and developers in this respect? Is a mechanism in place that allows for the

flagging of issues related to bias, discrimination or poor performance of the AI system?

Accessibility and Universal Design refers to the claims, particularly in business-to-consumer domains, that AI systems should be user-centric and allow all people to use AI products or services, including persons with disabilities. AI systems should address the widest possible range of users, following relevant accessibility standards.

Is it ensured that the AI system corresponds to the variety of preferences and abilities in society? Is the user interface usable by those with special needs or disabilities?

Stakeholder Participation refers to the advisability to consult stakeholders who may be affected by the AI system throughout its life cycle.

Has a mechanism been considered to include the participation of the widest range of possible stakeholders in the AI system's design and development?

Societal and environmental well-being

Environmental Well-being should be encouraged, and the positive and negative impacts of the AI system on the environment should be assessed.

Are there potential negative impacts of the AI system on the environment? Have mechanisms to evaluate the environmental impact of the AI system been established? Are measures to reduce the environmental impact defined?

Impact on Work and Skills is to be foreseen when AI-systems are introduced on the workplace.

Does the AI system impact human work and work arrangements? Are impacted workers and their representatives consulted? Are the impacts of the human work well understood? Are measures taken to counteract de-skilling risks and to provide training opportunities for re- and up-skilling?

Impact on Society at Large or Democracy needs careful consideration, as AI systems may amplify fake news, segregate the electorate, facilitate totalitarian behaviour etc.

May the use of the AI system potentially affect stakeholders or society at large? Have measures been taken to ensure that the AI system does not negatively impact democracy or cause societal harm?

Accountability

Auditability refers to the possibility of making an evaluation of the AI system by internal and external auditors.

> Is it possible to ensure responsibility for the development, deployment and use of the AI system, e.g., to trace the development process, the sources of training data, logging outcomes and impact?

Risk Management refers to the ability to report on actions or decisions that contribute to the AI system's outcome and to respond to the consequences of such an outcome. Relevant interests and values implicated by the AI system and trade-offs should be identified in terms of risks and safety and ethical principles, and adequate redress should be ensured if adverse impact occurs.

> Has risk training been organised? Is any external guidance or third-party auditing processes to oversee ethical concerns and accountability foreseen? Have redress mechanisms been put in place?

How to use these requirements

As indicated in the Guidance, these requirements for trustworthy AI are presented as suggestions for all work in the development, deployment or use of all AI systems with any impact on human persons. They can be used for self-assessment, to investigate that all relevant ethical issues are considered during the development, deployment and further use of an AI system. However, their main importance is as an instrument to explain the trustworthiness of an AI system. Trustworthiness is not about a declaration to establish trust, but it is instead a result of information upon which those affected by an AI system may obtain trust that the system is ethical and robust.

In some areas, the requirements have a stronger role. All research sponsored by the European Commission is subject to an ethics review before granting. In the framework program for 2020–27, Horizon Europe, AI is subject to particular attention in this ethics review process, and specific ethics-related requirements may be included in the Grant Agreement to be observed and fulfilled in the work of the project.

References

Bergmann M., Görman U., and Mathers J. (2008) 'Bioethical considerations for human nutrigenomics'. *Annual Review of Nutrition*. 28:447–467.

Farzanehfar A., Houssiau H., and de Montjoye Y. (2021) 'The risk of re-identification remains high even in country-scale location datasets'. *Patterns*. 2(3):100204.

Görman U. (2024) *Guide to the Ethical Review of Research on Humans, Version 1. 0*, English translation 2024. Uppsala: The Swedish Ethical Review Authority, https://etikprovningsmyndigheten.se/wp-content/uploads/2024/05/Guide-to-the-ethical-review_webb.pdf (accessed 2024-06-30).

Jobin A., Ienca M., and Vayena E. (2019) 'The global landscape of AI ethics guidelines'. *Nature Machine Intelligence*. 1:389–399.

The Assessment List for Trustworthy AI (ALTAI) (2020) https://digital-strategy.ec.europa.eu/en/library/assessment-list-trustworthy-artificial-intelligence-altai-self-assessment (accessed 2024-06-02).
The Ethics Guidelines for Trustworthy AI (2019) https://digital-strategy.ec.europa.eu/en/library/ethics-guidelines-trustworthy-ai (accessed 2024-06-02).

Conclusion

Matúš Šucha[1], Niels Agerholm[2] and Aliaksei Laureshyn[3]

This book comprehensively explores various methodologies and sources used to understand and improve road safety. It delves into the advantages and disadvantages of primary data sources such as police accident records, statistical methods for data analysis, in-depth investigations, healthcare data, self-reporting of behavior, data from automated vehicles, naturalistic driving/cycling/walking studies, site-based traffic conflict observations, and insurance data. The book emphasizes the importance of each data type in providing insights into accident causes, trends, and patterns, thereby guiding evidence-based interventions to enhance road safety. Furthermore, it discusses surrogate safety measures such as traffic conflicts and safety performance indicators (SPIs), highlighting their role in predicting and preventing accidents by proactively identifying risk factors and assessing safety infrastructure. Overall, the book underscores the evolving methodologies and ethical considerations in utilizing diverse data sources to mitigate road accidents and improve traffic safety globally.

In the realm of road safety, understanding the advantages and disadvantages of different accident data types is crucial for developing effective strategies to reduce accidents and improve overall safety.

Police accident records provide comprehensive details about the circumstances of road accidents, such as location, time, vehicle types, and possible causes, making them invaluable for detailed analysis. These records are also an official source of information, ensuring legal documentation and standardization, which enhances their reliability. Moreover, the wide coverage of police reports includes a significant number of incidents, offering extensive datasets for analysis. However, police accident records can suffer from underreporting, particularly for minor accidents or incidents without injuries, leading to potential data gaps. Additionally, the quality of data is dependent on the thoroughness and skill of the reporting officer, and there can be delays in data availability due to the processing and documentation stages.

Statistical methods for accident data analysis enable the identification of patterns and correlations within the data, which can highlight trends and potential risk

[1]Department of Psychology, Polacky University Olomouc, Czech Republic
[2]Danish Road Directorate, Denmark
[3]Department of Technology and Society, Faculty of Engineering, LTH, Lund University, Sweden

factors. These methods support evidence-based decision-making for safety interventions, making them highly valuable for policymakers and researchers. Predictive analysis can also forecast future accidents and assess risk factors, contributing to proactive safety measures. However, the complexity of these methods requires expertise in statistical tools and techniques, posing a barrier to their effective use. The sensitivity of the data can be an issue, as results may be skewed by outliers or inaccuracies, leading to potential misinterpretations. Accurate interpretation is crucial, as misinterpreting statistical outputs can result in incorrect conclusions and ineffective interventions.

In-depth investigations offer detailed insights into the causes and contributing factors of accidents through comprehensive case studies. This approach can identify the most effective interventions to prevent similar accidents in the future, making it highly effective for understanding specific incidents. However, in-depth investigations are resource-intensive, requiring significant investments of time, effort, and expertise. The scope of these studies is typically limited to individual cases, which may not represent broader trends, reducing their generalizability. Additionally, accessing the necessary data and parties involved in the investigations can be challenging, potentially limiting the scope and depth of the analysis. Despite these challenges, in-depth investigations provide critical, detailed information that can lead to targeted and effective road safety interventions.

Healthcare data provides detailed insights into the types and severity of injuries resulting from road accidents, adding depth to accident reports and offering a comprehensive view of the health impacts of these incidents. This data is instrumental in understanding the wider public health consequences of road accidents, helping to shape effective health and safety policies. However, there are significant disadvantages to using healthcare data. Privacy concerns are paramount, as sensitive personal health information must be handled with strict confidentiality. Integrating healthcare data with other types of data can also be complex and challenging. Additionally, there may be coverage gaps, as not all accident victims seek medical care, leading to incomplete data and potential inaccuracies in the overall analysis of road safety trends.

Insurance data provides unique advantages in assessing road safety by offering detailed insights into accident frequency, severity, and contributing factors based on claims and payouts. This data helps identify high-risk behaviors, locations, and demographics, enabling insurance companies and policymakers to implement targeted interventions and risk mitigation strategies. Additionally, insurance data often covers a broad spectrum of incidents, providing a comprehensive overview of road safety trends. However, reliance on insurance data alone may overlook non-reported or under-insured incidents, leading to potential gaps in understanding the full extent of road safety challenges. Moreover, data privacy concerns and the proprietary nature of insurance data can limit accessibility and integration with other sources, hindering collaborative efforts to improve road safety on a broader scale.

Self-reporting of behavior and attitudes in traffic offers valuable insights into personal attitudes and self-reported behaviors of road users, providing direct feedback that can be cost-effectively collected through surveys or questionnaires. This method helps capture subjective experiences and perceptions, which can inform targeted road safety interventions. However, the data's accuracy can be compromised by biases such as social desirability or recall bias, where respondents might not accurately represent their behaviors or attitudes. Additionally, self-reported data may not always reflect actual behavior, and low response rates can affect the representativeness of the data, potentially leading to skewed or incomplete analyses.

Data from automated vehicles presents significant advantages in enhancing road safety by providing high-precision, real-time information on vehicle performance and behavior. This continuous data stream enables informed decision-making for immediate monitoring and analysis of traffic conditions, contributing to proactive safety measures. Moreover, the large volume of data generated by automated vehicles allows for comprehensive analysis, identifying patterns and trends that can inform effective safety interventions. However, managing and processing such large datasets pose challenges in terms of data storage, processing capabilities, and cybersecurity. Privacy concerns also arise regarding the collection, use, and storage of personal data from automated vehicles. Furthermore, the quality and reliability of the data depend heavily on the technology and sensors employed, which can affect its accuracy in certain driving conditions or environments.

Naturalistic driving, cycling, and walking studies offer unparalleled advantages in capturing real-world behavior in natural settings, providing detailed insights into interactions, environmental factors, and traffic dynamics. These studies involve continuous monitoring over extended periods, offering comprehensive data that reflects actual road user behavior. This allows researchers to understand the complexities of traffic interactions and identify specific risk factors that contribute to accidents, leading to more targeted and effective safety interventions. However, these studies are resource-intensive, requiring substantial investments in equipment, data collection, and analysis. Privacy concerns also emerge due to the continuous monitoring of participants, necessitating strict protocols to protect personal data. Additionally, participant bias may influence behavior if individuals are aware they are being observed, potentially altering their natural driving or walking habits.

Site-based traffic conflict observations offer distinct advantages by allowing researchers to proactively identify potential safety issues before they escalate into actual accidents. By focusing on specific locations known for frequent conflicts, these observations provide detailed insights into near-misses and risky behaviors that can inform targeted safety improvements. The non-intrusive nature of this method ensures that traffic flow remains undisturbed during observation, allowing for accurate data collection without disrupting normal road conditions. However, the subjectivity involved in identifying and categorizing traffic conflicts can introduce variability and bias into the data. Training and consistency among observers are crucial to mitigate these issues and ensure reliable results. Additionally, site-based observations may have limited scope in capturing broader trends and systemic issues that contribute to road safety challenges beyond specific locations.

Behavioral observations in traffic provide direct insights into actual behaviors and actions of road users, offering immediate feedback that can inform targeted interventions to improve road safety. By capturing real-time behaviors rather than relying on self-reported data, this method enables researchers to identify unsafe practices and patterns that contribute to accidents. It facilitates the design of specific interventions tailored to address observed behaviors, thereby potentially reducing the occurrence of road incidents. However, behavioral observations can be perceived as intrusive by those being observed, potentially influencing their behavior and affecting the naturalistic validity of the study. Moreover, observations are often limited to small sample sizes and specific locations, which may not fully capture broader behavioral trends or systemic issues affecting road safety on a larger scale.

RSI and EURORAP play crucial roles in enhancing road safety by systematically assessing and improving infrastructure safety standards across road networks. RSI involves comprehensive inspections that identify hazards and deficiencies in road design and maintenance, leading to targeted improvements that mitigate accident risks. EURORAP, on the other hand, evaluates and rates roads based on their safety performance, providing valuable information for policymakers and road authorities to prioritize safety investments. However, these assessments require significant resources and expertise to conduct, posing challenges in terms of funding and implementation. Moreover, the effectiveness of RSI and EURORAP relies on consistent data collection and analysis, which may vary across regions and affect the accuracy and comparability of safety ratings and recommendations.

SPIs offer valuable advantages by providing quantifiable metrics to monitor and evaluate road safety initiatives and outcomes. They allow policymakers and road authorities to track progress, identify areas needing improvement, and benchmark performance across regions and time periods. SPIs provide a comprehensive view of road safety, encompassing infrastructure quality, traffic management, and user behavior, thereby facilitating evidence-based decision-making for safety interventions. However, SPIs require reliable and consistent data sources to ensure accuracy and relevance. Context-specific considerations and varying data availability can pose challenges in standardizing SPIs across different jurisdictions, impacting their comparability and effectiveness in guiding targeted road safety strategies. Careful interpretation and contextual understanding are essential to avoid misrepresenting SPI data and drawing incorrect conclusions that may hinder effective safety improvements.

Looking ahead, the future of road accident data types and surrogate safety measures shows promising developments driven by technological advancements and evolving methodologies. One notable trend is the increasing integration of data from connected and automated vehicles (CAVs). These vehicles generate vast amounts of real-time data on driving behaviors, road conditions, and potential hazards, offering unparalleled insights into accident causation and prevention. As CAV technology matures, there is a growing emphasis on leveraging this data to develop predictive models that can anticipate and mitigate risks before they escalate into accidents. However, challenges such as data privacy concerns and the need for robust data management systems must be addressed to fully harness the potential of CAV data for enhancing road safety.

Another emerging trend is the application of AI and machine learning in analyzing accident data. These technologies enable more sophisticated pattern recognition and predictive analytics, allowing for proactive safety interventions based on real-time and historical data trends. AI-driven algorithms can identify complex correlations in accident data that human analysts may overlook, thereby enhancing the effectiveness of safety measures. However, ethical considerations surrounding the use of AI in handling sensitive personal data and ensuring algorithm transparency and accountability remain critical concerns that need careful navigation in future implementations.

Moreover, there is a growing recognition of the importance of holistic safety indicators and surrogate measures beyond traditional accident data. Surrogate safety measures such as traffic conflicts, near-misses, and SPIs are increasingly adopted to complement accident data analysis. These measures provide proactive insights into potential safety risks and enable preemptive interventions to mitigate hazards before accidents occur. The integration of advanced sensing technologies and data analytics platforms is expected to further enhance the accuracy and reliability of surrogate safety measures, paving the way for more targeted and effective road safety strategies in the future. As these trends continue to evolve, collaboration between stakeholders, including researchers, policymakers, industry leaders, and the public, will be crucial in shaping a safer and more resilient road transport system globally.

Index

Abbreviated Injury Scale (AIS) 6, 19, 67–8
absolute validity 166
accident analysis 158
accident and emergency (A&E) departments 62–3, 82
accident data 5
  accident databases 14
    international databases 15–17
    international experience 17–18
    problem of data harmonization 18–19
  collection 8
    accident reporting process 8–11
    accident reports and in-depth accident investigations 12–14
    self-reported accidents 11–12
  interpret accident data 19
    identification of hazardous locations 22–3
    underreporting problem 23
  road accidents 5
    formal definitions of accident events and accident frequency 5–6
    statistical definition of accident frequency 6–8
accident databases 14
  international databases 15–17
  international experience 17–18
  problem of data harmonization 18–19
accident frequency 20
accident modelling 30
  available knowledge 39
    Highway Safety Manual 39–40
    scientific literature 40–2
  basic definitions 30
  regression technique, model form and goodness-of-fit evaluation 31
    empirical Bayes estimation and role of NB dispersion parameter 36–8
    generalized linear modelling approach 32–4
    goodness-of-fit measures 34–6
    key points on accident modelling 38–9
  regression to mean effect and relationship with exposure 30–1
accident prediction models (APMs) 31
accident rates 20, 30, 287
accident reporting process 8–14
accountability 307, 325–6
accuracy 306
AI-driven algorithms 333
Akaike's Information Criterion (AIC) 35
ambulancc trips 82
annual average daily traffic (AADT) 288–9
anonymous information 313–14
artificial intelligence (AI) 301

ethics guidance for trustworthy AI 320
EU AI Act 316
high-risk AI systems 318–19
other AI systems than high-risk 320
prohibited AI practices 317
transparency obligations for certain AI systems 319
foundations of trustworthy AI 321
fundamental rights 321
wider context 322
requirements for trustworthy AI 322
accountability 325–6
diversity, non-discrimination and fairness 324–5
human agency and oversight 322–3
privacy and data governance 323–4
societal and environmental well-being 325
technical robustness and safety 323
transparency 324
Association for the Advancement of Automotive Medicine (AAAM) 67
auditors 222
auto-insurance industry 138
automated driver assistance systems (ADAS) 210
automated driving 125
AV collected data 130
data received from other vehicles and roadside units 131
motion and vehicle control data 131
own perception sensors data 130
prediction and planning results 131
components of driving task 126
communicating with nearby road users and/or infrastructure 128
execution of planned trajectory 129–30
perceiving surrounding environment 127–8
planning own motion to eliminate risk of collision 129
predicting near-future positions of nearby road users 128–9
data sharing and users 136
AV fleet owners 140
AV manufacturers 140–1
government 137–8
individual AV owners 139
insurance companies 138–9
vehicle owners 139
safety-relevant incidents and their causes 131
AV data for crash analysis 133–4
AV data for traffic conflicts analysis 134–6
causes of conflicts and crashes 132–3
automated driving system (ADS) 136
automated vehicles (AVs) 125
collected data 130
data received from other vehicles and roadside units 131
motion and vehicle control data 131
own perception sensors data 130
prediction and planning results 131
data for crash analysis 133–4
data for traffic conflicts analysis 134–6
fleet owners 140
manufacturers 140–1

autonomous cruise control (ACC) 209
Autopilot system 139
average annual daily traffic (AADT) 30–1, 34
AXA Winterthur crash recorder database 272–3

basic model 239–41
Bayesian approaches 41
behavioural indicators 192
behavioural observations 190
  description of procedures and data 195
    FOT/natural driving studies 204–6
    in-car observation with focus on Vienna driving test 195–9
    on-site observation of behaviour and interaction of road users 201–4
    traffic conflict registrations on site 199–201
  FOT 209
  in-car observations 206–7
  methods/types of observations 192
    observation with technical help or solely by technical equipment 194–5
    obtrusive *vs.* unobtrusive observation 192–3
    stationary observation 193–4
    on traffic simulator 195
  NDS 209–11
  some results derived from on-site observation studies 208–9
  traffic conflict registrations 207–8
  understand traffic behaviour with help of model of Michon 211
Belgian Road Safety Institute 253
Big Data 147
black spot analysis 22
Black-spot Management 232
blood alcohol concentration (BAC) 116
burden of injuries 69–70
business models 147

calibration model 177
California's approach 137
California's program 137
camera surveillance 298
capture–recapture method 73–4
careless responses 103–4
causal model 32
cellular vehicle to everything (C-V2X) 128
child restraint systems (CRS) 243
clarity 228
coding observer 196
combined indexes 173
Common Accident Dataset (CADaS) format 15
communication 132–3
complex causal mechanisms 158
comprehensive data collection 57
computer vision 177
computing 287
conflict severity index 173
connected and automated vehicles (CAVs) 125, 332
consent 311
  information and consent to participation in activities *vs.* data processing 312
  safeguards and derogations related to personal data in public interest or scientific purposes 312–13
  withdrawal of 311
consistency 228
context-based methods 169

controlled data collection techniques 190
cost-effectiveness 102
counterintuitive effects 41
counting serious injuries 68
crash data 189
crashes 3, 134
crash frequency 30
crash modification factor (CMF) 40, 42
crash rate 30
credibility 228–30
cumulative standardized residuals (CSR) analysis 36
cutting-edge AI-driven simulations 41
cyclists 111

data 147, 149–50
  from automated vehicles 331
  GNSS data in 150
  video recording in 150–1
  volume 148
data accessibility 17
data analysis 113–14, 249
data cleaning 113
data coding 113
data collection 203
  methods 246–7
  tools 174
    human observations 174–5
    video analysis and other emerging technologies 176–7
    video recording 175–6
data harmonisation 63
data linkage methods 72
data minimization 306
datasets 2
data sources 246–7
data transformations 18
daytime running lights (DRL) 243
decision-making process 40
Delta-V 173
demographic representation 57
deterministic approach 72
diagnose-treatment combinations (DBC) 62
direct interactions 192
disability-adjusted life years (DALYs) 19, 69
dispersion parameter 36–8
distribution tails 165
Driver Behaviour Questionnaire (DBQ) 109
Dutch Healthcare Authority 64
Dutch Injury Surveillance System (DISS) 65

empirical Bayes (EB) estimation 36–8
empirical Bayes method 24
engine control module (ECM) 130
enumeration factor 288
environmental factors 49, 55
E-Survey of Road users' Attitudes (ESRA) 104, 114–16
ethics-related control 295–6
ethnographic methods 191
ethnography 191
EU AI Act 316
  high-risk AI systems 318–19
  other AI systems than high-risk 320
  prohibited AI practices 317
  transparency obligations for certain AI systems 319
European Commission (EC) 6
European Road Assessment Programme (EURORAP) 3, 231, 332
European Road Safety Observatory (ERSO) 115
European Trendline programme 106

European Union (EU) 66, 302
event 48
explicability 322
exploratory data analysis (EDA) 113–14
exponential model 292
export of data 315
extreme value theory (EVT) 165–6

fairness 322
Fatality Analysis Reporting System (FARS) 14
Federation Internationale Automobile (FIA) 277
field-operational tests (FOT) 204–6, 209, 213–14
Folksam crash recorder database 271–2
forward collision warning (FCW) 209
Front Crash Prevention test 139

gamma function 33
General Data Protection Regulation (GDPR) 10, 76
generalized additive models (GAMs) 41
generalized linear modelling (GLM) approach 32–4, 38
Geographic Information Systems (GIS) 14, 150
German Insurance Association crash database (UDB) 269
German Insurers Accident Research (UDV) 269
Global Burden of Diseases studies (GBD) 70
Global Navigation Satellite System (GNSS) 148, 150
Global Status Report on Road Safety (GSRRS) 115
goodness-of-fit measures 34–6
good research conduct 295
good scientific practice 295
governments 137

Haddon Matrix 56–7
Hawthorne effect 106
healthcare data 59, 62, 330
  hinders for access and use of hospital data 76
  injury classifications 66
    Abbreviated Injury Scale and Maximum AIS 67–8
    Injury Severity Score and New Injury Severity Score 68
    International Classification of Diseases 67
    MAIS3+ as definition of serious injury 71
    other indicators 68–9
  limitations of police accident records 59–62
  linking police and hospital records 71
    data linkage methods 72
    estimation of true number of injuries 72–6
  Netherlands 76
    history 76–7
    linked database 77–83
  potentials of healthcare data for road accident research 63–4
  Sweden 83
    historical note 83
    Strada database 84–8
  types of hospital records, availability 64–6
healthcare records 76
high-risk AI systems 318–19
Highway Safety Manual (HSM) 37, 39–40
hospital data 85–6

Hospital Discharge Register (HDR) 65
Hospital Discharge Reports 63
Hospital Standardised Mortality Ratio (HSMR) 62
human autonomy 322
human factors 48, 54–5
human observations 174–5
human operators 141

IEEE 802.11p 128
import of data 315
in-car observations 206–7
in-depth accident investigations 2, 12–14, 47, 330
 background investigation 48–9
 case analysis 56
  contributing factors 56
  primary cause 56
  use of Haddon Matrix 56–7
 different needs of methods and techniques 48
 interviews of involved persons 52–3
 learning and product development 47–8
 legal evidence 47
 on-scene investigation 49–51
 representativity of collected data 57
  comprehensive data collection 57
  demographic representation 57
  geographic and environmental factors 58
  random sampling 57
  standardized data collection methods 58
 research 48
 retrospective investigation 52
 selection/sampling criteria 52
 Special Crash Investigations 52
 variables in database 54
  crash details 54
  environmental factors 55
  human factors 54–5
  legal and regulatory variables 56
  other variables 56
  post-crash outcomes 55
  vehicle information 54
indicators 237
injury road accident 5
Injury Severity Score (ISS) 14, 19, 68
Injury Surveillance System (LIS) 82
Inspection Team 224
insurance companies 138–9
insurance data 263, 330
 analyzing trends of crashes and injured traffic participants to supplement official statistics 263
  advantage of common long-term injury classification 265
  differences between insurance statistics and official road traffic statistics 263–4
  example of insurance data reports of trends for crashes and injured traffic participants 264–5
 conducting retrospective studies on effectiveness of safety interventions 265–6
 enrichment of insurance data 269
  German Insurance Association crash database 269
  insurance datasets with information from crash recorders 271–4
  People Around the Vehicle crash database 270–1
  Volvo Cars traffic Accident Database 270
 providing consumer information on losses incurred by used cars 266

Folksam 267–9
IIHS, HLDI 266–7
insurance datasets with information from crash recorders 271
AXA Winterthur crash recorder database 272–3
Folksam crash recorder database 271–2
insurance crash data enhancements with sensors and cameras 273–4
Intelligent Speed Assistance (ISA) 151
Intelligent Transportation System (ITS) 128, 151
Intensive Care Unit (ICU) 65
interaction/communication processes 196
International Classification of Diseases (ICD) 14, 67
International Committee on Traffic Conflict Techniques (ICTCT) 159
international comparisons 102
International Road Assessment Programme (iRAP) 232
International Traffic Safety Data and Analysis Group (IRTAD) 13, 16
Internet of Things 147
interpret accident data 19
identification of hazardous locations 22–3
underreporting problem 23

Kalman filter 128–9
Karolinska Sleepiness Scale (KSS) 118
key performance indicators (KPIs) 243, 246

lane deviation warning (LDW) 209
lawfulness 308
personal data relating to criminal convictions and offences 309
special categories of data 308–9
length of hospital stay (LoS) 71
linear regression 107
linked database 77
other medical data 82–3
procedures for linking police and hospital data 77–8
procedures to estimate total number of serious injuries 78–82
using hospital data and severity indicators 78
location of intersection 231
low- and middle-income countries (LMICs) 178

machine learning 110, 177
manipulation 105
Maximum Abbreviated Injury Scale (MAIS) 6, 19, 67–8, 80
meta-analysis 292
Michon’s model 211, 214
Model Minimum Uniform Crash Criteria Guideline (MMUCC) 17–18
motion and vehicle control data 131
motion prediction methods 169

National Intensive Care Evaluation (NICE) 66, 83
natural driving studies (NDS) 204–6, 209, 214
naturalistic data 190
naturalistic driving study (NDS) 135, 147, 331
data for 149–50
GNSS data in 150
video recording in 150–1
examples of NDS based on video recording 152
examples of projects based on GNSS data 151–2
methods 148–9

pitfalls regarding data analyses 153–4
type of analyses 152–3
Naturalistic Teenage Driving Study (NTDS) 152
NB dispersion parameter 33, 36–8
Netherlands Institute for Health Services Research (NIVEL) 65
Network Safety Management 232
New Car Assessment Programme (NCAP) 232
New Injury Severity Score (NISS) 19, 68

objectivity 201, 204, 206
observation of unforeseen behaviour 299
occupant injuries 267, 270
Occupational Health and Safety (OHS) 278
Olomouc 202–3
On Board Unit (OBU) 151
on-scene investigation 49–51
operational design domain 138
operational thresholds 125
organisational traffic safety data 278–9
Organization for Economic Cooperation and Development (OECD) 15
original equipment manufacturer (OEM) 47–8
outcome-based indicators 172–3
outcome indicators 278
over dispersion 8

particular attention 97
path planning 133
Pay As You Speed (PAYS) project 151
Pearson χ^2 statistics 34
pedestrian accidents 87–8
pedestrians 111
PENDANT 75
People Around the Vehicle crash database (PAV) 270–1
perception algorithm 130, 132
performance indicators 278–9
performance tracking 234–5
period-prevalence surveys 111
personal data 3, 301
anonymous information 313–14
consent 311
information and consent to participation in activities *vs.* data processing 312
safeguards and derogations related to personal data in public interest or scientific purposes 312–13
withdrawal of 311
EU General Data Protection Regulation 304–5
fairness and transparency 309–11
GDPR on profiling 313
general rules for management of 305
accountability 307
accuracy 306
data minimization 306
integrity and confidentiality 307
purpose limitation 306
storage limitation 306
historical aspects of human rights and data protection 302–3
international projects 314
data exchange with other countries 314–15
export of data 315
import of data 315
lawfulness 308
personal data relating to criminal convictions and offences 309

special categories of data 308–9
open access to research data 315–16
personal integrity 303
proviso 301–2
pilot study 203
pitfalls regarding data analyses 153–4
Poisson distribution 7–8, 32, 37
Poisson model estimations 33
police accident records 329
police records 63
police reporting 74
policymakers 134
post-crash outcomes 55
post-encroachment time (PET) 170, 200
post-event 48
Potential Collision Speed (PCS) 173
pre-crash matrix (PCM) 13
predictive method 39
pre-event 48
principal component analysis (PCA) 114
Probability of a Collision Course (PCC) 170
property damage only (PDO accidents) 5, 60
public support for policy measures 116–17
public transportation systems 149

quantifiability 241
Quenault's driving-test 196
questionnaire surveys 100
additional advantages of using online panels for 101–2
advantages of 100
design flexibility 101
eliminating interviewer effects 101
enriching data collected 101
facilitating longitudinal studies 101
flexibility and broad application 100
identifying careless respondents 101
measuring unobservable constructs 101
reducing costs 100
reducing social desirability bias 101
limitations of 102–4

random sampling 57
Reaction Level Scale 14
recall bias 103, 112
recruitment of children 299
reduced representativity 102–3
registered data 205–6
regression modelling 165
regression-to-the-mean (RTM) 20, 23, 29–30
relative validity 166
reliability 166, 201, 204, 206
representativeness 242
representativity 102
research ethics
ethics-related control 295–6
good research conduct 295
risk-based approach 296–9
retrospective investigation 52
risk analysis 298–9
risk-based approach 296–9
risk management 298, 326
risk mapping 233–4
risk of permanent medical impairment (RPMI) 69–70
road accidents 5
formal definitions of accident events and accident frequency 5–6
statistical definition of accident frequency 6–8

road alignment 228
Road Assessment Programme (RAP) 221, 231
  performance tracking 234–5
  risk mapping 233–4
  Star Rating 232–3
road fatality 6
road function 228
road network safety analysis 22–3, 289
Road Safety Audit 221
Road Safety Inspections (RSIs) 3, 221, 332
  benefits and costs of 222–3
  carried out 223
  completion meeting 226
  defined 222
  good RSI practice guidelines 227
  organisational issues 224–5
  partners involved in 224
  perform RSI 224
  preparatory work 225
  report 226
  site visit 225–6
  typical safety deficiencies identified by 228
    clarity 228
    consistency 228
    credibility 228–30
    location of intersection 231
    road alignment 228
    road function 228
    roadside features 230–1
    traffic signals 231
    visibility 230
road safety management 238–9
road safety performance indicators 24
roadside features 230–1
roadside surveys 107
road traffic accidents 23
robust privacy framework 137
route 197

Safe System mindset shifts 178
Safe System principle 161
SafetyAnalyst 38
SafetyCube project 70
safety managers 134
SafetyNet project 75, 243, 245
safety performance functions (SPFs) 31, 40–1
safety performance indicators (SPIs) 237, 329, 332
  challenges and future needs 256–8
  examples of SPI evaluations 253
  four main dimensions of 244
  good practice in using 251
    as part of national road safety policies and plans 251–2
    at sub-national level 252–3
  international recommendations 249–51
  methodological rules for using 239
    basic model 239–41
    criteria and conditions for using 241–2
    principles for data collection and analysis for estimating 245–9
    subdivisions 242–5
  as predictors of road safety outcomes 253–6
  in road safety management 238–9
sampling approach 245, 247–9
scaled deviance (SD) 34
selection bias 20
self-declared approaches 97
self-reported accidents 11–12
self-reported data

advantages and limitations of using questionnaire surveys in road safety research 100
additional advantages of using online panels for questionnaire surveys 101–2
advantages of using questionnaire surveys 100–1
limitations of questionnaire surveys 102–4
specific limitations of using online panels 104–5
suitability of self-reported data collection for road safety research 105
ESRA initiative 114–16
good practice in self-reported data collection 110
data analysis 113–14
ethical considerations when questioning people in surveys 114
survey design 110–13
measuring drowsy driving 118
measuring road user behaviour through self-reporting 105
example comparing self-reported and observed behaviour 107–9
future directions for self-reported data on behaviour in traffic 109–10
international acceptance of data on self-reported behaviour 109
methods for measuring road user behaviour 105–7
overview of main data collection methods for 98–100
public support for policy measures 116–17
in research 97
sources and references 97
Traffic Safety Culture Index 117
self-reporting 97
of behavior and attitudes in traffic 330
semantic techniques 200
sensor data 130
sensors 132
single-vehicle crashes 192
site-based traffic conflict observations 331
sleepiness 118
social desirability bias 103, 112–13
Society of Automotive Engineers (SAE) 126
space mean speed 292
spatio-temporal long-short-term memory (ST-LSTM) 129
Special Crash Investigations (SCIs) 52
standardized data collection methods 58
Star Rating 232–3
stationary observation 193–4
statistical analysis 114
statistical methods for accident data analysis 329–30
Strada database 84
accessing Strada 86–8
data matching in 86
hospital data 85–6
police data 84–5
privacy in 86
Strategic Highway Research Program 2 (SHRP 2) 152
surrogate measures of safety (SMoS) 157
concept of severity 161–2
data collection tools 174
human observations 174–5
video analysis and other emerging technologies 176–7
video recording 175–6

hierarchy of events 159–61
historical note 158–9
indicators, measures and techniques 166
combined indexes 173
measures related to evasive actions 171–2
outcome-based indicators 172–3
PET family 170–1
TCTs 173–4
TTC family 167–70
limitations of accident data 157–8
reliability and validity 166
traffic conflicts and accidents 164–6
traffic conflicts and exposure 162–4
Surrogate Safety Assessment Model 179
survey design 110
compensation of recall bias 112
compensation of social desirability bias 112–13
designing survey questions 110–11
sampling approach 111–12
two types of questionnaire surveys for measuring behaviour 111
Sustain 75
Swedish method 265
Swedish Traffic Accident Data Acquisition (STRADA) system 14, 83
Swedish Transport Agency (*Transportstyrelsen*) 83, 86
Swerving 172

technical equipment 205–6
Tesla's servers 139
time headway warning (THW) 209
time-to-collision (TTC) techniques 167–70, 200
Time-to-Zebra indicator 169
total vehicle mileage
definition and theory 288
estimation of 288–91
speed 291–2
traffic accidents 1
traffic conflict registrations 199, 207–8
methodology 200
quality criteria 201
types of 200
traffic conflicts 126, 132, 329
and accidents 164–6
and exposure 162–4
traffic conflict techniques (TCTs) 159, 173–4
traffic conflict theory 160
traffic counts 287
traffic engineers 138
traffic safety 1, 157, 189, 280–1
Traffic Safety Culture Index (TSCI) 117
traffic safety footprint 277–8
challenges and opportunities related to organisations and traffic safety 279–81
future and potential 281
organisational traffic safety data 278–9
traffic safety problem 285–6
traffic signals 231
traffic simulator 195
Trans-European Road Network (TEN-T) roads 221
transparency 324
traumatic brain injury (TBI) 67
travel surveys 287
trip-based prevalence surveys 111

unconscious behaviour bias 103–4
United Nations Economic Commission for Europe (UNECE) 15
unknown injuries methods 89

validity 166, 201, 204, 206
vehicle factors 49
vehicle information 54
vehicle inspections 287
vehicle owners 139
vehicle-to-infrastructure (V2I) communication 128
vehicle-to-vehicle (V2V) communication 128, 179
video data 151
video recording 150–1, 175–6
video surveillance 148
Vienna driving test 196
violations 196
virtual loops 176
visibility 230
vital patient parameters 62
Volvo Cars traffic Accident Database (VCTAD) 270

wearable sensors 148
World Health Organisation (WHO) 15, 67, 115, 277

Years Lived with Disability (YLD) 69
Years of Life Lost (YLL) 69–70, 80

zebra crossing 203

www.ingramcontent.com/pod-product-compliance
Lightning Source LLC
LaVergne TN
LVHW011841200926
848484LV00002B/382
* 9 7 8 1 8 3 9 5 3 0 4 5 6 *